Atlas of Entomopathogenic Fungi

Cordyceps tuberculata on moth (Sphingidae), Indonesia.

Hirsutella versicolor on leafhopper (Cicadellidae), Kenya.

Hirsutella jonesisii on green planthopper (Nephotettix), Indonesia.

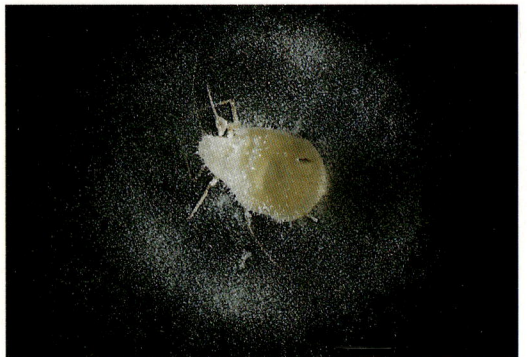

Erynia neoaphidis on *Aphis* (Aphididae), France.

Conidiobolus obscurus on *Acyrthosiphon pisum* (Aphididae), France.

Entomophthora muscae on fly (Diptera), Denmark.

Sporodiniella umbellata on cicada nymph (Cicadidae), Papua New Guinea.

Erynia gammae on lepidoptera larva (Noctuidae), Switzerland.

Erynia radicans on *Dicyphus pallidus* (Miridae), Switzerland.

Erynia culicis on Coddis fly (Diptera), UK.

Erynia conica on *Simulium* sp. (Diptera), UK.

Nomuraea rileyi on rice looper (Lepidoptera), Indonesia.

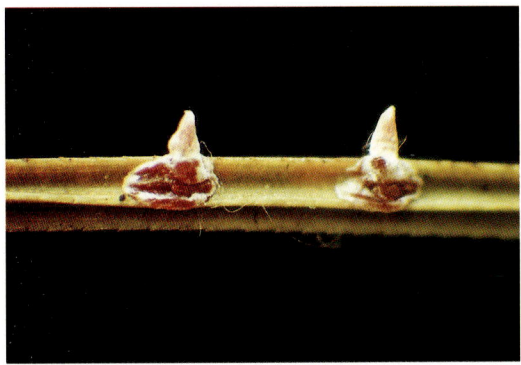
Fusarium coccophilum on pine-needle scales (*Phenacaspis*), Honduras.

Cordyceps variabilis on Coleoptera larva (Elateridae), Canada.

Cordyceps unilateralis on *Camponotus* ant, Costa Rica.

Entomophaga aulicae on lepidoptera larva, Canada.

Cordyceps sp. on adult beetle (Coleoptera), Brazil.

Cordyceps militaris on *Lepidoptera purpa*, Netherlands.

Robert A. Samson   Harry C. Evans   Jean-Paul Latgé

Atlas of Entomopathogenic Fungi

Springer-Verlag
Berlin Heidelberg New York London Paris Tokyo

Wetenschappelijke uitgeverij Bunge
Utrecht

Cordyceps dipterigena on robberfly (Asilidae), Indonesia.

Metarhizium anisopliae var. *majus* on third instar larvae of *Oryctes rhinoceros*, Philippines.

Hymenostilbe sp. on Gryllidae, Ecuador.

Cordyceps locustiphila on Locustidae, Ecuador.

Cordyceps sp. on Psychidae larva, Brazil.

Erynia delphacis on green leafhopper (*Nephotettix virescens*), Indonesia.

Cordyceps sp. on cockchafer larva (Scarabidae), Brazil.

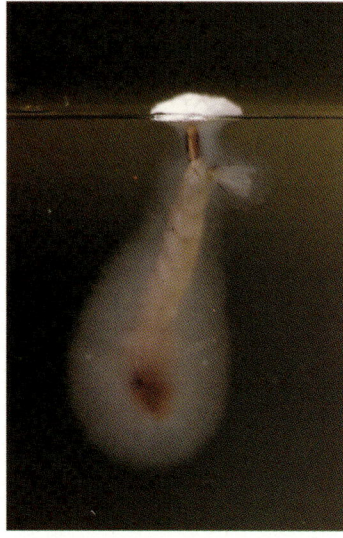
Tolypocladium cylindrosporum on mosquito larva (*Aedes aegyptii*), USA.

Gibellula species on spider (Salticidae), Brazil.

Polycephalomyces ramosus on fly, the Netherlands.

Cordyceps curculionum on weevil adult (Curculionidae), Brazil.

Aschersonia cubensis on scale insect (Lecaniidae), Brazil.

Aschersonia aleyrodis on citrus whiteflies (Aleyrodiidae) USA.

Gibellula pulchra on spider (Salticidae), Ecuador.

Robert A. Samson
Centraalbureau voor Schimmelcultures,
Baarn, The Netherlands

Harry C. Evans
CAB International Institute of Biological Control,
Silwood Park, Ascot, United Kingdom

Jean-Paul Latgé
Institut Pasteur, Paris, France

Sole distribution rights outside the Netherlands granted to
Springer-Verlag Berlin Heidelberg New York London Paris Tokyo

ISBN 3-540-18831-2
Springer-Verlag Berlin Heidelberg New York
ISBN 0-387-18831-2
Springer-Verlag New York Berlin Heidelberg

ISBN 90 6348 4135 / CIP
© Wetenschappelijke uitgeverij Bunge, 1988
Printed in The Netherlands

Preface

A knowledge of insect diseases (whether or not it has to do with control) is of fundamental and far-reaching importance in the study of insect ecology.
E. A. Steinhaus (1949).

As biological control is becoming more acceptable as a practical science and the dangers of the long-term use of chemical pesticides are fully appreciated, there has been a resurgence of interest in employing fungal pathogens to combat insect pests. New production and application techniques combined with a greater understanding of both fungal and insect ecology have meant that biological insecticides can now compete on a more equal footing with traditional chemical pesticides. Consequently, agrochemical companies are beginning to take account of entomopathogens as both viable and economic propositions and to invest more resources in their research and development, particularly to improve the formulations of well-documented pathogens, possibly by genetic manipulation, and to test more obscure or novel ones. It is for these reasons that an atlas or guide to the entomopathogenic fungi is considered necessary, not only to highlight and consolidate the many contributions of past mycologists but also to aid and hopefully to stimulate present and future workers in the field, many of whom will come from disciplines other than mycology. These fungi have either been included within general reviews of insect pathogens, and hence have received somewhat superficial treatment taxonomically, or conversely, they have been described in mycological monographs essentially over-specialised and unavailable to non-mycologists. This book, therefore, is an attempt to combine the information relating to the taxonomy, ecology and physiology of this important group of organisms in a format readily assimilable by agriculturists, biotechnologists, chemists and entomologists, but which at the same time can serve as an identification guide useful to both practical field scientists and specialist mycologists or pathologists in the laboratory.

The only comparable 'popular' book in English solely devoted to this subject was written by Cooke in 1892 under the evocative title *Vegetable wasps and plant worms*, and is essentially a mycological treatise but skilfully presented to appeal to a wider audience. We hope that our book can go some way towards replacing this classic publication and thereby attain at least some of Cooke's earlier aspirations: 'In its present form I hope that the book will be welcome alike to the entomologist and the mycologist, and assist them in their respective studies'.

We restrict the term 'entomopathogenic fungi' to those genera or species which are proven pathogens of insects or for which circumstantial evidence exists concerning their pathogenicity. The looser term 'entomogenous fungi' is avoided here because this also relates to fungi growing on or colonising insect substrates, either facultatively or obligately, but which are strictly non-pathogenic. Hence the commensal parasites of insects such as the Laboulbeniales are excluded by definition from the true entomopathogenic fungi and consequently will not be dealt with here. However, it is felt justified in including the obligate pathogens of spiders (araneopathogenic fungi) within any treatment of the entomopathogenic fungi since many of them are related to those on insects and they exploit a similar ecological niche and hence face the same problems of host penetration, colonisation and preservation in addition to sporulation on and dissemination from the host. Moreover, many of the metabolites necessary to achieve these ends are probably common to both groups of fungi. Similarly, fungi also attack arthropods other than insects and spiders; some groups, for example, are common pathogens of marine invertebrates. These mycopathogens will not be discussed here with the exception of the crayfish pathogen *Aphanomyces*, because of its relevance to recent progress in invertebrate immunology (see Chapter 4). The readers interested in these pathogens should consult Lightner (1981).

The fungal genera *Beauveria* and *Metarhizium* are well known to the majority of entomologists who would have come across these names in the literature or assigned them to the fungi occurring on their specimens, depending on whether the 'mould' exhibited a white or a green colouration. We wish to demonstrate to those who may be involved with any aspect of arthropod population dynamics, be it the ecologist, the practical agriculturist or the pesticide sales executive, that the range and variety of entomopathogenic fungi encompasses considerably wider dimensions than this relatively narrow framework upon which most research has been concentrated and that there exist many more genera which have received little or no attention but which conceivably may contain species exploitable by man as sources of biocontrol agents or of useful metabolites. At the very least we shall dispel some of the myths surrounding entomopathogenic fungi, increase the levels of knowledge, and thereby redirect or divert them into the mainstream of mycology: '... the study of entomogenous fungi being a well known if somewhat esoteric, mycological by-way' (Ainsworth, 1981).

Acknowledgements
We are very grateful to the following individuals who provided data and material for illustrations: Stanizlaw Balazy, Mildred Blackwell, Paul Brey, Jorgen Eilenberg, Brian Federici, Dave Malloch, Dan Molloy, Jacqie Pendland, David Perry, Chris Prior, Clay McCoy, Steve Moss, Bert Orr, J. Pillai, Michiel Rombach, J. P. Skou, Kenneth Söderhäll, George Soares, Tony Sweeney, Christine Tarrant, Alain Vey, John Webster and Neil Wilding. A special word of thanks should be expressed to Siegfried Keller for his generous loan of the beautiful colour slides of the Entomophthorales. We wish to thank Richard Hall, Drion Boucias and Kenneth Söderhäll for reviewing the manuscripts of some chapters; Mrs. Ans Spaapen-de Veer for invaluable help with the typing, and Francis Snippe-Claus and Ellen Mul for assisting with the drawings and the photography respectively. Finally we are indebted to our families for their support and patience while working on the preparation of the book.

SPRING 1988   THE AUTHORS

Contents

Preface ix

1 Introduction 1

2 Taxonomy of entomopathogenic fungi 5

3 Illustrations of common fungal pathogens 17

4 Fungal pathogenesis 128

5 Natural control: ecology and biology 140

6 Biotechnology (*J. P. Latgé and R. Moletta*) 152

7 Biological control: past, present and future 165

Bibliography 173

Index 185

Chapter 1

Introduction

General

One definition of an atlas is a collection of illustrative plates in a volume. As it should be, therefore, this is the main thrust of the present publication and is presented in Chapter 3. However, since this Atlas is meant to be consulted by specialists and non-specialists alike, we feel that the latter would have an equally strong interest in the function of entomopathogenic fungi as well as in their form. Consequently, additional chapters have been included within the volume which, hopefully, synthesise the information relating to all aspects of entomopathogenic fungi – in the modern sense, reviewing the state of the art – and which also can orientate the more committed reader to the relevant, specialised literature.

In Chapter 2, the taxonomy of the major genera is outlined with simple keys for identification of the important entomopathogenic genera. The taxonomy of entomopathogenic fungi can be said to be in a state of flux. The current taxonomical status of the difficult groups is discussed. In his review of the fungal parasites of vertebrates, Ainsworth (1968) stated that '… most fungal infections of man and higher animals, whether endogenous or exogenous, are opportunistic'. In contrast, the fungi described here are, in the main, highly specialised obligate or facultative parasites which have co-evolved with their hosts over a considerable period of time. Indeed, two fossil records of entomopathogenic fungi from amber in the Oligocene-Miocene boundary (appr. 25 million years old) have been reported recently (Poinar & Thomas, 1982; 1984). The fungi represent, therefore, an old selection pressure but the extreme diversity and abundance of arthropods, particularly insects, demonstrate that they have competed successfully with such microbial parasites. This has been achieved through the development of a series of barriers (physical and chemical) to counteract microbial infections. These resistance mechanisms and the ways in which entomopathogenic fungi overcome them are discussed in Chapter 4. However, it should be conceded that our present knowledge of fungal pathogenesis in relation to arthropod hosts is rudimentary compared with that of microbial pathogens of vertebrates; for the greater majority of the fungal genera described here, it is totally lacking.

The study of entomopathogenic fungi in nature has also been neglected (Ferron, 1981) but in Chapter 5, we attempt to collate what little information is available on the ecology and biology of the host-pathogen relationships in both primary (i.e. unexploited) and agricultural ecosystems. This consists mainly of field observations rather than experimental data.

In Chapter 6, we review the biotechnology available for the industrial production of those mycopathogens likely to be used in biological control programmes.

In the final Chapter, we analyse the past and present experimental data pertaining to man's efforts to utilise entomopathogenic fungi for the control of arthropod pests of agricultural and medical importance, and cast an eye to the future. It should be emphasized that we recognise two distinct types of control: natural control, which occurs without man's conscious intervention; biological control, involving the manipulation of organisms by man. These definitions correspond closely to those outlined by Steinhaus (1956) but contrast with the broad concept of biological control as accepted by van den Bosch (1971), which includes naturally-occurring control in addition to classical control. The most obscure interpretation of biological control, in this case applied to plant pathogens, is that proposed by Cook & Baker (1983): '… the reduction of the amount of inoculum or disease-producing activity of a pathogen accomplished by or through one or more organisms other than man'. One could be forgiven for assuming that man plays no part in the events. By designating two easily comprehensible, and in essence readily separable, types of control, we aim to clarify the situation for the general reader.

Historical

We are indebted to the scholarly reviews of Steinhaus (1956; 1975) for much of the early history of entomopathogenic fungi. He concludes that insect diseases were probably first observed by sericulturists in the Orient, and he quotes early Japanese accounts (about 900 AD) of muscardine silkworms (i.e. infected with *Beauveria bassiana*) being used for the treatment of palsy or paralysis. Apparently, such diseased larvae were also valued as antiseptics, constituting a cure for sore throat, toothache, wounds and abscesses. Possibly the earliest account of the genus *Cordyceps* also dates back to the ninth century: '... certain trees on the island Sombrero in the East have large worms attached to them underground ...' (in Gray, 1858). According to Steinhaus, the first illustration of an entomopathogenic fungus was published in 1726 and probably depicts the 'Chinese plant worm' *(Cordyceps sinensis* on a lepidopteran larva). Gray (1858) and later Cooke (1892) painted an interesting historical picture of this species which was held in high esteem by the Emperor's physicians in China for several centuries, who knew it under the name 'summer herb – winter worm'. The preparation of the drug, the effects of which are compared to those of ginseng, was suitably exotic, being stuffed into a duck and then slowly roasted. The duck meat into which the fungal products were supposedly absorbed was then eaten. Similarly, in his description of *Cordyceps robertsii* on Lepidoptera larvae from New Zealand, Gray noted that the natives make use of fungal-infected insects; eating the fruitbodies and extracting a pigment for tattooing purposes. He describes rubbing the extracts into the wounds and it is feasible that this was actually functioning as an antiseptic. Gray's book contains a wealth of such information, both of mycological and entomological value, as well as superbly executed illustrations of entomopathogenic fungi.

As Steinhaus correctly points out, the prominent fruitbodies of species of *Cordyceps* probably first excited the interest of early field naturalists, however, it is clear from the common names that the relationship between insect and fungus was confused. Gray & Cooke both quote early eighteenth century descriptions of *Cordyceps* from the West Indies: 'By the latter end of July the tree arrives at its full growth, and resembles a coral branch... and bears several little pods, which, dropping off, become worms, and from them flies, like the English caterpillar.' Indeed, a contemporary French collector took this as evidence of transmutation: 'He sees in these plant-animals a proof of the passage and mutation of animal species into the vegetable, and reciprocally from the vegetable to the animal' (Cooke, 1892). Torrubia, an ecclesiastical naturalist, recorded the presence of dead wasps in Cuba in 1749, and his description of the diseased insects ('trees'), probably *Polistes* infected with *Cordyceps sphecocephala*, is given in full by Steinhaus (1956) who also includes the original illustration. Although much has been published since, this fanciful but compelling drawing is still considered worthy of a place in this book (see frontispiece).

Cooke (1892) in referring to *Cordyceps* on weevils from South America stated thus: 'Foreign collectors or collectors abroad, have generally a keen eye to the monetary value of such varieties and are seldom satisfied with their weight in gold to part with them.' Petch (1937) commented cynically on this valuation '... it may be noted that their weight in gold would not be of great value, even at current rates.' Nevertheless, there is no doubt that examples of *Cordyceps*-infected insects were much prized by early Victorians, and professional naturalists, such as the entomologist, H. W. Bates, and the botanist, R. Spruce, derived part of their incomes from such specimens gathered during their travels in South America. Some specimens in the Kew Herbarium still bear the name of the broker who handled the monetary transactions.

Steinhaus (1956) cites eighteenth century European accounts of diseased insects, predominantly flies, in which the causal agents would appear to be members of the Entomophthorales. However, the belief that microorganisms could induce disease in animals had not been propounded and the theory of spontaneous generation was the order of the day. Ironically, it was eventually through a classic investigation of the white muscardine disease of silkworms that the germ theory of disease was established. The pioneer who laid the foundations for the study of infectious diseases was A. Bassi, working in Italy. By careful experimentation, he satisfied, and in fact preempted, Koch's postulates, presenting his results in 1834 (Major, 1944; Ainsworth, 1956). It was undoubtedly this work which later influenced the Rev. Berkeley in helping to solve the mystery of the potato blight in Ireland. Another eminent scientist of the period whose contributions had a profound effect not only on insect pathology but also on the whole spectrum of microbial diseases was E. Metchnikoff. His work in the early 1870's in Russia on the green muscardine *(Metarhizium anisopliae)* prepared him for his subsequent outstanding discoveries relating to cellular immunity. Although it was by now generally accepted that fungi could invade, kill and then colonise arthropod hosts, the mechanism involved was open to speculation. For example, Gray (1858) in his description of *Cordyceps robertsii* stated that: '... the parasite becomes connected with the caterpillar by means of the seed being taken in with the food and thus passing into the interior of the insect ...'. Cooke (1892) similarly followed the popular belief that fungal spores are ingested during feeding and that germination and infection only occur within

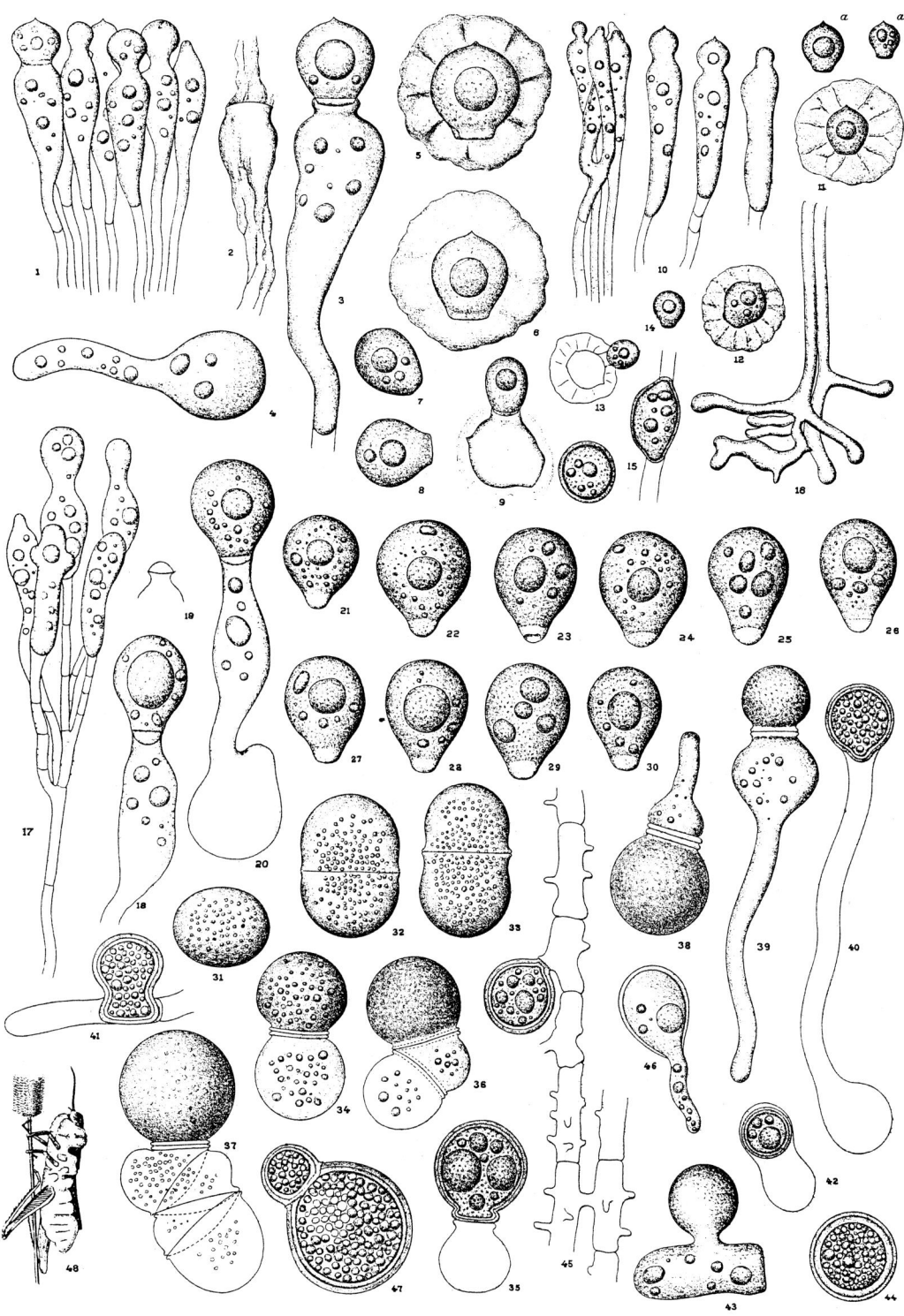

FIGURE 1-1. One of the very detailed drawings of *Entomophthora* by R. Thaxter (1888).

the host: 'The idea that the spore finds, or bores, its way into the interior after it becomes attached to the external surface is one which cannot be entertained.' It is now known, of course, that most entomopathogenic fungi invade via the cuticle and that infection from ingested spores is of rare occurrence (Roberts & Humber, 1984). The latter part of the nineteenth century saw much activity directed towards the taxonomy of entomopathogenic fungi with significant contributions from the great mycologists of the era: M. J. Berkeley, F. Cohn, G. Fresenius, A. Giard, P. C. Hennings, L. Quelet, P. A. Saccardo, C. L. Spegazzini, L. R. Tulasne and P. Vuillemin, but probably the most significant was the monumental work of R. Thaxter on the Entomophthorales (Thaxter, 1888). There then followed a period of intensive research on the use of entomopathogenic fungi in controlling agricultural pests (1895-1925), and this applied aspect will be expanded upon in Chapter 4. However, such interest gradually waned and there then followed a 'lull' period (1925-1960), despite the considerable efforts of E. A. Steinhaus. Nevertheless, during this time significant advances were made in the taxonomy of entomopathogenic fungi, particularly by T. Petch in the Old World and E. B. Mains in the New World. Another significant landmark, was the monographic treatment of the genus *Cordyceps* by Kobayasi (1941). With the reawakening of interest in biological control, and the more detailed study of arthropod populations and their habitats, the numbers of newly described genera and species are rapidly increasing, whilst taxonomic and nomenclatural problems in established genera are also being resolved. The study of entomopathogenic fungi can rightly be said to be in a dynamic state. Whether or not the historical pattern of ups and downs will be repeated, will depend largely on man's ability to harness their unique insecticidal properties.

Chapter 2

Taxonomy of entomopathogenic fungi

Cooke (1892) listed four groups of fungi parasitic upon insects: *Cordyceps* and allied '*Isaria*' conidial states (anamorphs); Entomophthorales; Laboulbeniales; and opportunistic fungi, such as species of *Cladosporium* and *Penicillium*. As previously mentioned, we exclude the Laboulbeniales, since they are not implicated in host mortality, although it is possible that they may indirectly affect insect populations by reducing reproductive capacity. We refer the interested reader to the classic publications by Thaxter, spanning the years 1896-1931. As in plant pathology, the opportunistic fungi are more difficult to define: an essentially saprophytic or weakly pathogenic fungus may at times become associated as the causal agent of epizootics in predisposed insects. The results of subsequent laboratory tests to establish pathogenicity should be treated with caution as should the entomopathogenic status of many of these fungi. We include in this chapter, therefore, only those fungal genera which contain proven or purported entomopathogens and which consistently show adaptations to such a habit (table 2-1). We have made an exception for the Trichomycetes, because there are reports (Sweeney, 1981; Moss & Descals, 1986) indicating that some species can be lethal. Because of current interest in this fungal group, the reader is referred to descriptions of these fungi provided by Lichtwardt (1986) and illustrations of examples of the major groups are included in Chapter 3. A notable exclusion is the basidiomycete genus *Septobasidium* Pat., comprehensively reviewed by Couch (1938). Although members of this large genus actively parasitise scale insects, causing infertility, and probably premature senescence in a proportion of the population, evidence indicates that they live in a state of mutual symbiosis with the insect at the expense of the host plant. Also excluded are the ectoparasitic fungi and mycoparasitic species such as *Filobasidiella depauperata* and *Melanospora parasitica*.

The taxonomy of the entomopathogenic fungi has received increased interest since the 1970's. Before this renaissance, our knowledge

Table 2-1
List of genera containing entomopathogenic species

Chytridiomycota – Chytridiales
Coelomycidium
Myiophagus

Chytridiomycota – Blastocladiales
Coelomomyces

Oomycota – Lagenidiales
Lagenidium

Oomycota – Saprolegniales
Leptolegnia
Couchia

Zygomycota – Entomophthorales
Conidiobolus
Entomophaga
Entomopththora
Erynia
Massospora
Meristacrum
Neozygites

Zygomycota – Mucorales
Sporodiniella

Ascomycota
Ascosphaera
Atricordyceps
'*Calonectria*'
Cordycepioideus
Cordyceps
Hypocrella
Myriangium
Nectria
Podonectria
Torrubiella

Deuteromycota
Acremonium
Akanthomyces
Aschersonia
Aspergillus
Beauveria
Culicinomyces
Engyodontium
Funicularis
Fusarium
Gibellula
Hirsutella
Hymenostilbe
Metarhizium
Nomuraea
Paecilomyces
Paraisaria
Pleurodesmospora
Polycephalomyces
Pseudogibellula
Sorosporella
Sporothrix
'*Stilbella*'
Tetracrium
Tetranacrium
Tilachlidium
Tolypocladium
Verticillium

Mycelia sterilia
'*Aegerita*'

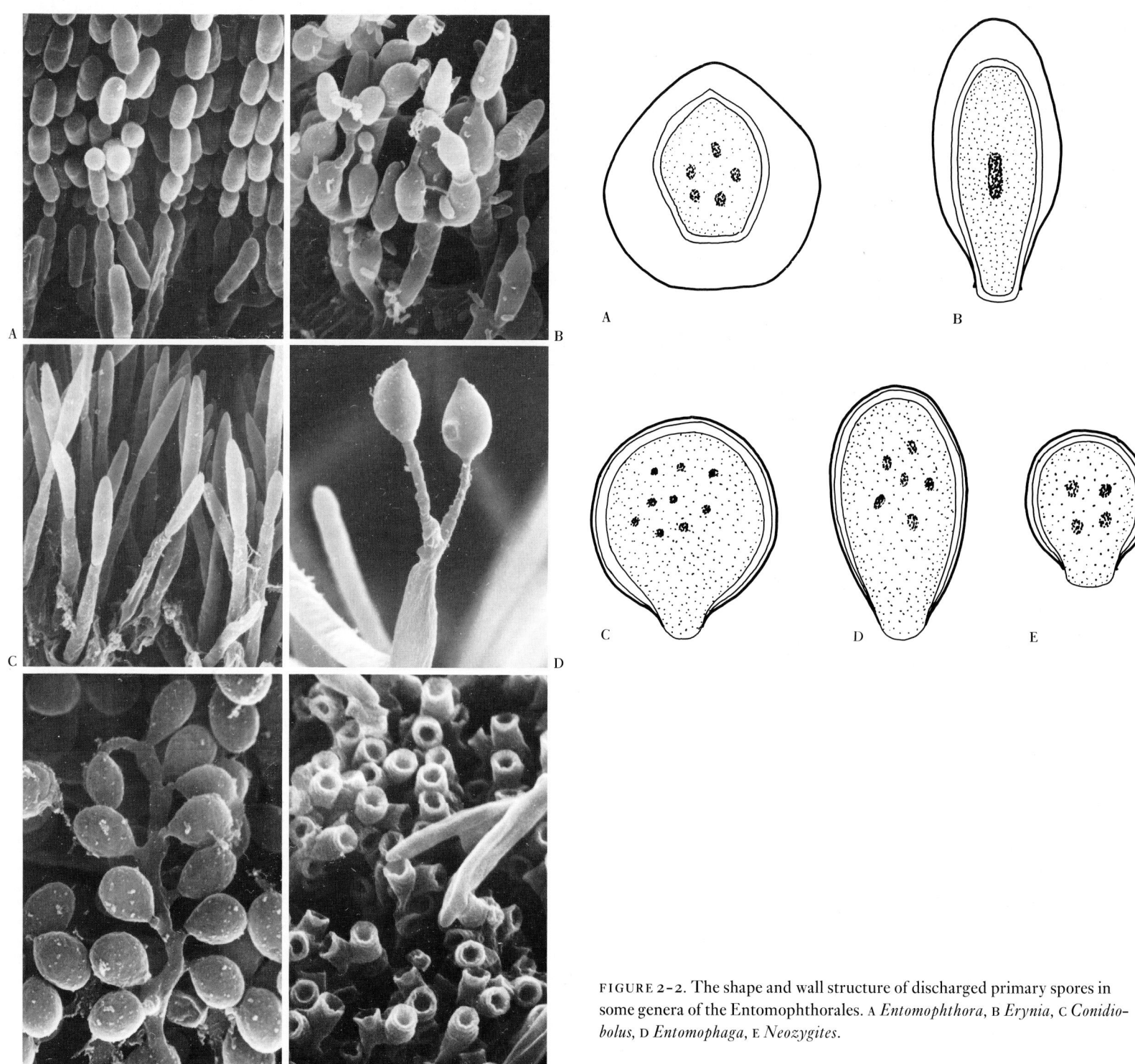

FIGURE 2-2. The shape and wall structure of discharged primary spores in some genera of the Entomophthorales. A *Entomophthora*, B *Erynia*, C *Conidiobolus*, D *Entomophaga*, E *Neozygites*.

left:
FIGURE 2-1. Examples of phialidic (A–D) and sympodial (E,F) conidiogenesis in entomopathogenic Deuteromycetes. A *Metarhizium anisopliae* (× 2000), B *Culicinomyces clavisporus* (× 2800), C *Aschersonia aleyrodis* (× 3500), D *Hirsutella* species (× 3500), E *Beauveria bassiana* (× 7000), F *Hymenostilbe formicarum* (× 5000).

of the fungi parasitising arthropods was relatively poor and few scientists published in this field. Tom Petch (1870-1948) can be considered as the pioneer on systematics of the entomopathogenic fungi; publishing numerous papers over a period of more than 25 years in which he described many new taxa, a significant proportion of which still remains valid.

The majority of the fungi parasitic on arthropods can be readily recognized, because the taxonomic criteria for separating the species are mostly based on morphological features. Direct identification on the host is therefore often possible by preparing a simple microscopic mount. In some cases arthropod cadavers must be placed in humid chambers to allow further development and subsequent sporulation of the fungus. Isolation of the fungus on artificial media can be carried out, and may be helpful for determination (see Samson, 1982).

The mode of production of asexual or sexual propagules is a key basis for classifying the fungi. In the genera of Entomophthorales the shape of the primary spores is essential for generic recognition, whilst without details of the conidiogenesis of the deuteromycetous entomopathogens a correct identification is impossible. For a detailed account of conidiogenesis, the reader is referred to Cole & Samson (1979). Examples of the most important types of conidium formation are given in fig. 2-1. The taxonomy of the Entomophthorales has been recently redefined based on wall ontogeny during spore formation and discharge on a substratum together with the shape and nuclear characteristics of the spores (Remaudiere & Keller, 1980; Humber, 1981; Latgé et al., 1988 b) (see fig. 2-2).

Many fungi are characterized by their pleomorphic life cycles, producing more than one independent form or spore stage. Particularly in the Ascomycetes and associated deuteromycete genera, more than one spore form may occur. In many genera, such as *Gibellula* and *Hirsutella*, several anamorphs may exist (synanamorphs). Although separate naming of each synanamorph is not necessary according to the Botanical Code of Nomenclature, it is practical to have names available, particularly when synanamorphs can occur separately on different hosts or at different stages of development of the fungus (as for example the *Granulomanus* anamorph of *Torrubiella* spp.). Besides these vegetative propagules, other structures such as rhizoids, resting spores, hyphal bodies or sclerotia may be important for characterization of the fungus. Sometimes protoplasts (Entomophthorales) and blastospores (a yeast-like phase in deuteromycetous species, formed in the haemocoele or in submerged culture) are produced and may help in the classification of a species or a genus.

Generic descriptions

Short descriptions of the important entomopathogenic genera are included here. Some poorly known or taxonomically unclear genera (e.g. *Aegerita, Peziotrichum, Calonectria*) are omitted, while the characteristics of some taxa (*Fusarium, Nectria*) have been slightly adjusted towards their entomopathogenic habit. After each genus, the most relevant and/or recent references to a detailed treatment and description of species are given.

CHYTRIDIOMYCOTA - CHYTRIDIALES

Coelomycidium Debaisieux
Thalli in fat body cells or circulating in haemolymph, globose, thin-walled, multinucleate, cleaving into numerous uniflagellate zoospores; resistant sporangia thick-walled, with many prominent reserve droplets, multinucleate, germinating with release of numerous uniflagellate zoospores. On blackflies.
Ref. Debaisieux, P., La Cellulle 30: 249-277, 1920
Weiser, J. & Z. Zizka, Ceska Mykol. 28: 159-162 and 227-232, 1974.

Myiophagus Thaxter ex Sparrow
Vegetative thallus endozoid, coenocytic, dissociating to produce free sporangia at maturity; sporangia mostly globose with slightly thickened wall, forming 1-5 exit papillae; zoospores uniflagellate; resting sporangia mostly globose, reticulate, breaking open upon germination to release zoospores.
Ref. Sparrow, F. K., Mycologia 31: 439-444. 1939
Karling, J. S., Am. J. Bot. 35: 246-254, 1948.

CHYTRIDIOMYCOTA - BLASTOCLADIALES

Coelomomyces Kaillin
Mycelium changing into thick-walled and often ornamented resistant sporangia, yellow to yellow-brown, releasing numerous zoospores upon germination: zoospores uniflagellate, infecting copepods, ostracods, or arthropods other than insects: biflagellate zygotes encysting in dipteran host, forming haustoria and penetrating exoskeleton. In aquatic dipteran larvae.
Ref. Bland, C. E. & J. N. Couch, Can. J. Bot. 51: 1325-1330, 1973
Bland, C. E. et al. In: Microbial control of pests and plant diseases (H. D. Burges), Academic Press, pp. 129-162, 1981.

OOMYCOTA - LAGENIDIALES

Lagenidium Schenk
Mycelium parasitic in haemocoele of mosquito larvae, coarse and thick, coenocytic at first, later septate and forming segments becoming zoosporangia or sex organs; partially differentiated contents of zoosporangia extruded through thin evacuation tube to thin-walled vesicle formed outside host body; zoospores laterally biflagellate and reniform; zygotes (oospores) thick-walled with large subcentric oil droplet, forming several biflagellate zoospores.
Ref. Sparrow, F.K. Aquatic Phycomycetes, Ann Arbor, Univ. Michigan Press. 1960
Couch, J.N. & S.V. Romney, Mycologia 65: 250-252. 1973
Brey, P.T. & G. Remaudière, Bull. Soc. Vector Ecol. 10: 90-97, 1985.

OOMYCOTA - SAPROLEGNIALES

Leptolegnia de Bary
Zoosporangia terminal, not swollen but long, with basal septum and terminal papilla; zoospores biflagellate, formed in single file; oogonia occasionally numerous, on short lateral branches, surface often spinulose, containing a single oospore with one or more accentric oil droplets.
Ref. Sparrow, F.K., Aquatic Phycomycetes, Ann Arbor, Univ. Michigan Press, 1960
Seymour, R.L. Mycologia 76: 670-674, 1984.

ZYGOMYCOTA - ENTOMOPHTHORALES

Conidiobolus Brefeld
Synonyms: *Boudierella* Costantin; *Delacroixia* Saccardo
Sporangiophores mostly unbranched; primary spores globose to pyriform with broadly rounded apex and prominent papilla, multinucleate, nuclei with a prominent central nucleolus and no apparent heterochromatin forcibly discharged by eversion of papilla, wall layers not separated after discharge; secondary spores similar to primary and forcibly discharged, or forming numerous globose forcibly discharged microspores or cylindrical and passively dispersed from capillary secondary sporangiophore; resting spores (zygospores) often present; sometimes production of terminal azygospores, chlamydospores or villose spores with numerous short hair-like appendages.
Ref. Tyrrell, D. & D.M. MacLeod, J. Invert. Path. 20: 11-13. 1972
King, D.S., Can. J. Bot. 54: 45-65, 1285-1296, 1976 and ibid. 55: 718-729, 1976.
MacLeod, D.M. & E. Muller-Kogler. Mycologia 65: 823-893, 1973.

Entomophaga Batko
Synonym: *Eryniopsis* Humber
Sporophores simple; primary spores usually pyriform or ovoid, with a prominent conical papilla, multinucleate, nuclei with big patches of heterochromatin and no apparent central nucleolus, stained with aceto-orcein, forcibly discharged, wall layers not separated after discharge; secondary spores like primary but smaller (except for *E. caroliniana*); vegetative stages without walls.
Ref: Batko, A., Bull. Polon. Acad. Sci. Ser. Sci. Biol. 12: 323-336. 1964
Humber, R.A, Mycotaxon 13: 191-240, 1981
MacLeod, D.M. & E. Muller-Kogler. Mycologia 65: 823-893, 1973.

Entomophthora Fresenius
Synonym: *Empusa* Cohn
Sporangiophores simple, generally club-shaped; primary spores with prominent apical point and broad, flat basal papilla, with 2-12 (to ca. 40) nuclei, forcibly discharged wall layers completely separated upon impact, spores attached to substrate by intramural mucus; surrounded by and attached to substrate by halo of discharged cytoplasm; secondary spores not prominently apiculate, with smaller, round papillae, discharged by papillar eversion; resting spores arising by lateral budding from parental hypha; vegetative stages without wall, amoeboid or short and rod-like protoplasts.
Ref. MacLeod, D.M. et al., Mycologia 68: 1-29. 1976
Samson, R.A. et al., Can. J. Bot. 57: 1317-1323, 1979
Waterhouse, G.M. & B.L. Brady, Bull. Br. mycol. Soc. 16: 113-143, 1982.

Erynia Nowakowski
Synonyms: *Zoophthora* Batko, *Strongwellsea* Batko & Weiser
Sporangiophores mostly branched; primary spores ovoid to elongate, uninucleate, with prominent heterochromatin, forcibly discharged by papillar eversion, basal papilla rounded to conical with outer wall layer partially separated after discharge; secondary spores similar to primary or more globose, forcibly discharged by papillar eversion or with elongate capillispores at top of capillary secondary sporangiophore; vegetative stages: hyphal bodies or protoplasts; rhizoids usually present to secure host to substrate.
Ref. MacLeod, D.M. & E. Muller-Kogler, Mycologia 65: 823-893, 1973
Remaudière, G. & G.L. Hennebert, Mycotaxon 11: 269-321, 1980
Remaudière, G. & S. Keller, Mycotaxon 11: 323-338, 1980
Humber, R.A. & I. Ben-Ze'ev, Mycotaxon 13: 506-516, 1981
Ben-Ze'ev, I. & R. Kenneth, Mycotaxon 14: 456-475, 1982
Ben-Ze'ev, I., Mycotaxon 25: 1-10, 1986; ibid. 27: 263-269, 1986.

Massospora Peck
Spores formed on simple sporangiophores in irregular cavities in ab-

domen of gregarious cicadas, 1-6 nucleate, passively dispersed upon exposure by disarticulation of abdominal exoskeleton of living insect; resting spores thick-walled, often distinctly reticulate; vegetative cells amoeboid protoplasts, becoming hyphal bodies filling terminal abdominal segments.
Ref. Soper, R.S., Mycotaxon 1: 13-40, 1974; ibid 13:50-58, 1981.

Meristarcum Drechsler
Synonym: *Tabanomyces* Couch et al.
Sporangiophore simple, mostly four-celled, formed by an upright hypha and bearing several spores; primary spores globose, papillate, uninucleate, with a prominent central nucleolus, forcibly discharged by papillar eversion, wall layers not separated; secondary spores like primary; resting spores (zygospores) globose to ovoid, smooth-walled and hyaline.
Ref. Couch, J.N. et al., Proc. Natl. Acad. Sci. USA 76: 2299-2302, 1979
Humber, R., Mycotaxon 13: 191-240, 1981

Neozygites Witlaczil
Synonym: *Triplosporium* (Thaxter) Batko
Sporangiophores not branched; primary spores mostly pear-shaped, rounded with truncate papilla, often with four nuclei with obvious heterochromatin, but not staining with aceto-orcein, forcibly discharged by papillar eversion, wall layers not separated; secondary spores almost exclusively on long curved capillary germ tubes; if present resting spores with black epispore.
Ref. MacLeod, D.M. & E. Muller-Kogler, Mycologia 65: 823-893, 1973
Remaudiere, G. & G.L. Hennebert, Mycotaxon 11: 269-321, 1980
Remaudière, G. & S. Keller, Mycotaxon 11: 323-338, 1980
Ben-Ze'ev, I. & R. Kenneth, Mycotaxon 14: 456-475, 1982.
Note. Ben-Ze'ev, I.S. & R. Kenneth (Mycotaxon, 1987) proposed a new genus *Thaxterosporium* for *N. turbinatum*.

ZYGOMYCOTA – MUCORALES

Sporodiniella Boedijn
Sporangiophores erect, yellow-brown, large, up to 1 cm high, bearing an apical whorl of branches which rebranch to produce whorls of 4-6 sporangioferous branches; each terminal branch a sterile spine bearing a single sporangium; sporangia globose, 25-50 μm diameter, brown; sporangiospores (sub)globose, hyaline, with finely roughened wall; zygospores globose, warted, black.
Ref. Evans, H.C. & R.A. Samson, Can J. Bot. 55: 2981-2983, 1977.

ASCOMYCOTA

Ascosphaera Spiltoir & Olive
Synonym: *Pericystis* Betts
Ascomata, cyst-like, globose with thin wall, often darkly pigmented, containing numerous asci and numerous hyaline smooth-walled, one-celled ascospores remaining in a distinct 'spore ball'. Anamorph: not known.
Ref. Spiltoir, C.R. & L.S. Olive, Mycologia 47: 238-244, 1955
Skou, J.P., Friesia 10:1-24. 1972; ibid 11: 62-74. 1975
Gochenauer, T.A. & S.J. Hughes, Can. Entomol. 108: 985-988, 1976
Skou, J.P., Mycotaxon 15: 467-499, 1982.

Atricordyceps Samuels
Stromata pulvinate, sessile, dark green, completely covered by ovoidal ascomata. Asci cylindrical to narrowly fusiform, apically with massive non-amyloid plug, 8-spored, paraphysis filiform. Ascospores acerose, 12-14-(20) septate, disarticulating into halves at the median septum, while still in ascus, hyaline, smooth-walled partspores. Anamorph: *Harposporium*.
Ref. Samuels, G.J., New Zealand J. Bot. 21: 171-176, 1983.

Cordycepioideus Stifler
Clavae erect, simple or branched, arising from mycelium filling host body; ascomata immersed in stroma, apex sterile; asci clavate, deliquescing at maturity, 2- or 8-spored; ascospores broadly ellipsoid-fusoid, with 7-13 transverse septa at maturity, often adhering slightly pigmented. Anamorph: probably *Paecilomyces*-like.
Ref. Stifler, C.B., Mycologia 33: 82-86, 1941
Blackwell, M. & R.L. Gilbertson, Mycologia 73: 358-362, 1981; ibid 76: 763-765, 1984.

Cordyceps Fries
Synonyms: *Ophiocordyceps* Petch; *Torrubia* Leveille
Host body filled with hyphal bodies or mycelial mass giving rise to one or more stromatic clavae; stromata erect and columnar, with clavate or globose fertile apical portion and sterile stipe or with ascomata scattered on surface of clava; ascomata flask-shaped, immersed in stroma; asci long cylindrical, with prominent apical thickening penetrated by a fine pore, 8-spored; ascospores filiform, multiseptate, often disarticulating into one-celled partspores. Anamorphs: *Akanthomyces, Hirsutella, Hymenostilbe, Nomuraea, Paecilomyces, Paraisaria, Pseudogibellula, Sporothrix, 'Stilbella'* and *Verticillium*.
Ref. Kobayasi, Y., Sci. Rep. Tokyo Bunrika Daigaku, Sect. B. 5: 53-260, 1941; and Trans. mycol. Soc. Japan 23: 329-364, 1982

Mains, E. B., Bull. Torrey Bot. Club 81: 492-500, 1954 and Mycologia 50: 169-222, 1958

Kobayasi, Y. & D. Shimizu, Kew Bull. 31: 557-566, 1977 and Iconography of vegetable wasps and plant worms, Hoikusha Publ. Osaka, Japan, 1983

Samson, R. A. et al., Proc. Kon. Ned. Acad. Wetenschap. Ser. C. 85: 589-605, 1983

Evans, H. C. & R. A. Samson, Trans. Br. mycol. Soc. 79: 431-454, 1982; ibid 82: 127-150, 1984.

Hypocrella Saccardo

Stroma superficial, plate-like or cushion-shaped; ascomata globose to pyriform, immersed in stroma; asci cylindrical to filiform with distinct apical thickening; ascospores filiform, hyaline, multiseptate, often disarticulating into cylindrical partspores. On scale insects and whiteflies. Anamorph: *Aschersonia*.

Ref. Petch, T., Annls Roy. Bot. Gard. Peradenyia 7: 167-178, 1921

Mains, E. B., Mycopath. Mycol. Appl. 11: 311-326, 1959.

Myriangium Montagne & Berkeley

Stroma black with lighter interior, simple and cushion-shaped or lobed; asci scattered, often occurring singly, globose, 8-spored, with two wall-layers separating at dehiscence; ascospores ellipsoidal to oblong, multiseptate with transverse and longitudinal septa, hyaline to subhyaline. Anamorph: not known.

Ref. Petch, T., Trans. Br. mycol. Soc. 10: 45, 1924

Miller, J. H., Mycologia 30: 158, 1938, ibid 32: 587, 1940

Von Arx, J. A., Persoonia 2: 421-475, 1963.

Nectria (Fries) Fries

Synonyms: *Corallomyces* Berkeley & Curtis; *Sphaerostilbe* Tulasne & Tulasne

Ascomata mostly brightly coloured, single or clustered on scale insect or superficial on or partially immersed in stroma or loose subiculum covering host; asci cylindrical to clavate, thin-walled, 8-spored; ascospores ellipsoid to fusoid, 2 to 3-septate, hyaline. Anamorphs: *Acremonium, Fusarium, Stilbella, Verticillium*.

Ref. Petch, T., Trans. Br. mycol. Soc. 7: 89-169, 1921

Booth, C., The genus *Fusarium*, Commonwealth Mycol. Inst., 1971; 235 pp.

Rossman, A., Mycol. Pap. 150: 1-164, 1983.

Podonectria Petch

Ascomata mostly lightly coloured, superficial, hairs often present on ascoma wall; asci long, clavate to cylindrical, with distinct two-layered wall (bitunicate), 8-spored; ascospores long, clavate to cylindric, multiseptate. Anamorph: *Tetracrium, Tetranacrium, (?) Peziotrichum*.

Ref. Petch, T., Trans. Br. mycol. Soc. 12: 44-52, 1927

Rossman, A., Mycologia 69: 355-391, 1977 and Mycotaxon 7: 163-182, 1978; compare also Kobayasi, Y. & D. Shimizu, Bull. Nat. Sci. Mus. Ser. B. 3: 93-97, 1977

Torrubiella Boudier

Stroma absent or poorly developed as light- to brightly coloured subiculum covering host; ascomata brightly coloured, often elongate, conical to pyriform, wall often covered with conidiogenous structures of the anamorph; asci long cylindrical, 8-spored, with prominent apical thickening penetrated by a fine canal; ascospores filiform, multiseptate, disarticulating at maturity into one-celled partspores. Anamorphs: *Akanthomyces, Gibellula, Granulomanus, Paecilomyces, Pseudogibellula, Tilachlidium* and *Verticillium*.

Ref. Petch T., Trans. Br. mycol. Soc. 9: 108-128, 1923

Kobayasi, Y., Trans. mycol. Soc. Japan 23: 329-364, 1982

Kobayasi, Y. & D. Shimizu, Iconography of vegetable wasps and plant worms, Hoikusha Publ., Osaka, Japan, 1983.

DEUTEROMYCOTA

Acremonium Link

Synonym: *Cephalosporium* Corda

Conidiophores simple, mostly consisting of solitary awl-shaped phialides, producing conidia in slimy heads or rarely in short, dry chains. Conidia one-celled (rarely two-celled), variously shaped, hyaline. Vegetative mycelium hyaline, or brightly coloured, occasionally aggregating into cord-like strands. Teleomorphs: *Torrubiella, Cordyceps, Nectria*.

Ref. Petch, T., Trans. Br. mycol. Soc. 10: 152-182, 1925

Gams, W., *Cephalosporium*-artige Schimmelpilze, Gustav Fischer Verlag, 1971

Balazy, S., Bull. Soc. Amis Sci. Let. Poznan, Ser. D. 14: 101-137, 1973.

Akanthomyces Lebert

Synonym: *Insecticola* Mains

Conidiophores synnematous; synnemata often compact, cylindrical or clavate, covered with a hymenium-like layer of phialides. Phialides with inflated or cylindrical basal part and short neck, conidia one-celled, smooth, hyaline, in short fragile chains.

Teleomorphs: *Cordyceps, Torrubiella*.

Ref. Mains, E. B. Mycologia 42: 566-589, 1950

Samson, R. A. & H. C. Evans. Acta bot. Neerl. 23: 28-35, 1974.

Aschersonia Montagne
Stroma cushion- or cup-shaped, mostly brightly coloured with localized spots showing conidial masses, containing pycnidia with conidiophores and paraphysis; conidiophores slender, branched, consisting of thin-walled, mostly awl-shaped phialides; conidia hyaline, one-celled, mostly fusiform, smooth-walled. Teleomorph: *Hypocrella*.
Ref. Petch, T., Ann. Roy. Bot. Gard. Peradeniya 7: 167-278, 1921
Mains, E. B., Lloydia 22: 215-221, 1959, and J. Insect. Path. 1: 43-47, 1959.

Aspergillus Michelli
Synonym: *Sterigmatocystis* Cramer
Conidiophores erect, unbranched, terminating in a swollen ovoid to clavate apex (vesicle); phialides flask-shaped, directly on vesicle (uniseriate) or on metulae (biseriate); conidia one-celled, in long dry chains, smooth or roughened, hyaline to darkly pigmented. Teleomorphs: several but none associated with insects.
Ref. Raper, K. B. & D. I. Fennell, The genus *Aspergillus*. Williams & Wilkins, Baltimore, 1965
Note: Most species are saprophytes. *A. parasiticus* is regularly recorded from insects.

Beauveria Vuillemin
Conidiophores consisting of whorls and dense clusters of conidiogenous cells, hyaline, smooth-walled; conidiogenous cells sympodial, short and globose or flask-shaped with an apical denticulate rachis giving a distinctly zig-zag appearance; conidia one-celled, hyaline, thin-walled, globose to ellipsoidal. Teleomorph:? *Cordyceps*.
Ref. De Hoog, G. S., Stud. Mycol. 1: 1-41, 1972
Samson, R. A. & H. C. Evans, J. Invert. Path. 39: 93-97, 1982.
Note. Von Arx (Mycotaxon 25: 153-158, 1986) transferred all species of *Tolypocladium* to *Beauveria*. This taxonomic view is, however, debatable and not followed here.

Clathroconium Samson & Evans
Vegetative mycelium hyaline, sparse; conidiophores not distinct, consisting of blastic conidiogenous cells on vegetative mycelium, terminal or intercalary, producing one or more conidia on short broad denticles; conidia large, and consisting of an open ball-shaped network of filamentous cells, yellow-brown, rough-walled. On spiders. Teleomorph: not known.
Ref. Samson, R. A. & H. C. Evans, Can. J. Bot. 60: 1577-1580, 1982.

Culicinomyces Couch et al.
Conidiophores erect arising from host body, verticillate, consisting of whorls of phialides; phialides flask-shaped with more or less elongated neck, producing slimy heads; conidia one-celled, hyaline, smooth-walled, clavate to cylindrical. On larvae of mosquitoes and related dipterans. Teleomorph: not known.
Ref. Couch, J. N. et al., Mycologia 66: 374-379, 1974
Goettel, M. S. et al., Mycologia 76: 614-625, 1984.

Engyodontium de Hoog
Conidiophores hyaline, growing along substrate or somewhat erect, with branches arranged in subverticillata pattern; conidiogenous cells awl-shaped and phialidic or polyblastic, often terminating in a geniculate rachis and forming conidia on short denticles; conidia smooth-walled, hyaline, one-celled. Teleomorph: not known.
Ref. De Hoog, G. S., Persoonia 10: 33-81, 1978
Gams, W. et al., Persoonia 12: 135-147, 1984.

Funicularis Baker & Zaim
Conidiophores usually short, often branched, mostly grouped on funicles or occasionally solitary on hyphae. Conidia thallic, irregularly cylindrical to allantoid, smooth-walled, one-celled, sympodially arranged on conidiophores. On mosquito larvae. Teleomorph: unknown.
Ref. Zaim, M. et al., J. Invert. Path. 34: 199-202, 1979

Fusarium Link
Synonyms: *Atractium* Link, *Discofusarium* Petch; *Lechnidium* Giard; *Microcera* Desmazieres; *Pseudomicrocera* Petch
Conidiophores solitary or aggregated in sporodochia or synnemata-like structures, consisting of whorls of phialidic conidiogenous cells; phialides flask-shaped to slender, often with more openings (polyphialides) and producing two types of conidia *1* banana-shaped macroconidia, with prominent basal footcels and with one to several transverse septa, usually released in slimy heads or spore masses, *2* microconidia one-celled, small, ovoid to cylindric, produced in heads or in chains. Teleomorph: *Nectria*.
Ref. Booth, C., The genus *Fusarium*. Commonwealth Mycol. Inst., 1971; 235 pp.
Nelson, P. E. et al., *Fusarium* species, an illustrated manual for identification. Pennsylvania St. Univ. Press, 1983; 193 pp.
Burgess, L. W. & C. M. Liddell, Laboratory Manual for *Fusarium* research, Univ. Sydney, 1982; 162 pp.
Gerlach, W. & H. Nirenberg, Mitt. Biol. Bundesanst. Ld. u. Forstw. Berlin-Dahlem 209: 1-406, 1983.

Gibellula Cavara
Conidiophores occasionally mononematous but mostly distinctly synnematous, septate, rough-walled, with a small terminal vesicle on

which phialides are borne on metulae; synnemata often loosely arranged and not compact; phialides, mostly cylindrical or clavate with short necks, smooth-or rough-walled, producing conidia in chains; conidia one-celled, smooth-walled. *Granulomanus*-synanamorph may occur at base of synnemata or on hyphae covering host body. On spiders. Teleomorph: *Torrubiella*.

Ref. Petch, T., Annls mycol. 30: 386-393, 1932
Mains, E. B., Mycologia 42: 306-321, 1950
Samson, R. A. & H. C. Evans. Acta bot. Neerl. 22: 522-528, 1973
Kobayasi, Y. & D. Shimizu, Kew Bull. 31: 557-566, 1977.

Granulomanus de Hoog & Samson
Conidiophores simple, consisting of whorls of polyblastic conidiogenous cells, single or in whorls, flask-shaped to clavate, with one or more terminal projections. Conidia one-celled, long and rod-like or gently curved, with blunt ends, hyaline, smooth-walled; associated with *Gibellula* anamorph. On spiders. Teleomorph: *Torrubiella*.

Ref. Samson, R. A. & H. C. Evans, Proc. K. Ned. Acad. Wetenschap. Ser. C. 80: 128-134, 1977
De Hoog, G. S., Persoonia 10: 33-81, 1978

Hirsutella Patouillard
Synonyms: *Desmidiospora* Thaxter; *Synnematium* Speare; *Trichosterigma* Petch; *Troglobiomyces* Pacioni
Conidiophores mononematous or synnematous; synnemata erect, cylindrical or slightly tapered, varying from short and verrucose to long and hair-like, sometimes branched; conidiogenous cells phialidic, solitary or crowded in hymenium-like layer along synnema or borne directly on hyphae emerging from host body, mostly with inflated basal portion with one or more slender, thin necks; conidia one- or two-celled, hyaline, variously shaped, often like the segment of a citrus fruit, formed singly to a few often with distinct mucus covering; chlamydospores sometimes present, large, flattened, disc-like, thick-walled; sclerotia produced by some species. Teleomorph: *Cordyceps, Torrubiella* and '*Calonectria*'.

Ref. Speare, A. T., Mycologia 12: 62-76, 1920
Mains, E. B., Mycologia 43: 691-718, 1951
Samson, R. A. et al., Mycologia 72: 359-377, 1980
Minter, D. W. & B. L. Brady, Trans. Br. mycol. Soc. 74: 271-282, 1980
Pacioni, G., Trans. Br. mycol. Soc. 74: 239-245, 1980
Evans, H. C. & R. A. Samson, Trans. Br. mycol. Soc. 79: 431-454, 1982; ibid. 82: 127-150, 1984
Evans, H. C. & R. A. Samson, Can. J. Bot. 60: 2325-2333, 1982
Minter, D. W. et al., Trans. Br. mycol. Soc. 81: 455-471, 1983
Note: The genus *Hirsutella* contains many species, of which several are still undescribed. The taxonomy of *Hirsutella* is complex and still not fully understood. Besides the macroscopic synnematous species, many mononematous taxa parasitising minute arthropods exist.

Hymenostilbe Petch
Synnemata simple, rarely branched, white to dark-brown. Conidiophores in dense hymenium-like layers along synnema or on stromata of the teleomorph. Conidiogenous cells polyblastic, cylindrical to clavate, apically crowded with distinct denticles. Conidia solitary often apiculate, one-celled, smooth-or rough-walled, hyaline. Teleomorph: *Cordyceps*.

Ref. Samson, R. A. & H. C. Evans, Proc. K. Ned. Acad. Wetenschap. Ser. C. 78: 73-80, 1975;
Mains, E. B. Mycologia 42: 566-589, 1950.

Metarhizium Sorokin
Conidiophores in compact to nearly stromatic patches, mostly mononematous but also synnematous when occurring on arthropods buried in soil or other substrates; conidiogenous cells phialides in whorls, often arranged in a candle-like fashion, clavate to cylindrical; conidia one-celled, smooth-walled, hyaline to slightly coloured, forming long chains often aggregated into prismatic columns. Teleomorph: not known.

Ref. Tulloch, M., Trans. Br. mycol. Soc. 66: 407-411, 1976
Hai-li, G. et al., Acta Mycol. Sinica 5: 177-184, 1986
Rombach, M. C. et al., Mycotaxon 27: 87-93, 1986

Nomuraea Maublanc
Conidiophores mononematous or synnematous, verticillate, bearing dense whorls of short branches with clusters of phialidic conidiogenous cells; phialides short, more or less flask-shaped to cylindrical without a distinct neck; conidia one-celled, smooth-walled, green to purple, in dry divergent chains. Teleomorph: *Cordyceps* (for *N. atypicola*).

Ref. Samson, R. A., Stud. Mycol., Baarn 6: 1-119, 1974
Kish, L. et al., J. Invert. Path. 24: 154-158, 1974

Paecilomyces Bainier
Synonyms: *Spicaria* Harting, '*Isaria*' (in part.)
Conidiophores erect, mononematous, but species on insects often synnematous, verticillate, bearing whorls of divergent branches and phialides; phialides flask-shaped or with swollen basal part, abruptly tapering into a distinct neck; conidia one-celled, hyaline to slightly pigmented, produced in dry divergent chains, smooth-walled or occasionally spiny. Teleomorph: *Torrubiella, Cordyceps*.

Ref. Samson, R. A., Stud. Mycol., Baarn 6: 1-119, 1974
Bissett, J., Fungi Canadense, no. 153-159, 1979.

Paraisaria Samson & Brady
Conidiophores hyaline with both verticillately branched and terminal whorls of branches and conidiogenous cells, aggregated into loose feather-like synnemata. Conidiogenous cells phialides, flask-shaped, frequently sympodially proliferating to give rise to 1 to 4 phialide necks. Conidia one-celled, hyaline, smooth-walled, produced in slimy heads. On lepidopterous larvae. Teleomorph: *Cordyceps*.
Ref. Samson, R.A. & B.L. Brady, Trans. Br. mycol. Soc. 81: 285-290, 1983.

Pleurodesmospora Samson et al.
Conidiophores erect, simple, consisting of fertile hyphae with terminal and intercalary conidiogenous cells; conidiogenous cells phialidic, reduced, resembling short, peg-like denticles, often more than one per conidiogenous cell, producing short chains; conidia one-celled, smooth-walled, hyaline. Teleomorph: not known.
Ref. Samson, R.A. et al., Persoonia 11: 65-69, 1980

Polycephalomyces Kobayasi
Synnemata often branched, compact, white or yellow-brown, bearing yellow slimy conidial masses apically. Conidiogenous cells phialides on verticillate whorls or in chains. Phialides more or less awl-shaped, often intercalary phialides also present. Conidia slimy, one-celled, small (less than 5 μm long), yellowish. Teleomorph: not known.
Ref. Kobayasi, Y., Sci Rep. Tokyo Bunrika Daigaku, Sect. B, 5: 53-260, 1941
Samson, R.A. et al., Proc. Kon. Ned Acad. Wetenschap C, 84: 289-301, 1981
Seifert, K.S., Stud. Mycol., Baarn 27: 1-235, 1985

Pseudogibellula Samson & Evans
Conidiophores synnematous or mononematous, arising from mycelial covering of host, consisting of a rough-walled, septate stipe terminating in an apical whorl of vesicles, conidiogenous cells and branches; synnemata more or less cylindrical, with loosely arranged conidiophores; conidiogenous cells polyblastic, sympodial, with inconspicuous scars, flask-shaped to cylindrical; conidia one-celled, smooth-walled, hyaline, apiculate. Teleomorph: *Torrubiella*.
Ref. Samson, R.A. & H.C. Evans, Acta bot. Neerl. 22: 522-528, 1973

Sorosporella Sorokin
Synonym:? *Syngliocladium* Petch
Chlamydospores present in host haemocoele, when mature globose, often brightly coloured, cohering in solid masses or becoming separated and powdery at maturity, germinating to produce simple synnemata bearing phialides, forming one-celled, hyaline conidia in heads *(Syngliocladium)*.
Ref. Speare, A.T., J. Agric. Res. 8: 189-194, 1917 and ibid. 18: 399-439, 1920
Petch, T., Trans. Br. mycol. Soc. 25: 250-265, 1942

Sporothrix Hektoen & Perkins ex Nicot & Mariot
Conidiophores mononematous, but species on insects often distinctly synnematous, consisting of whorls of elongated conidiogenous cells; conidiogenous cells, sympodial, forming single conidia on distinct denticles, often conidiogenous cells becoming septate; conidia one-celled, apiculate, hyaline. Teleomorph:? *Cordyceps*.
Ref. De Hoog, G.S., Stud. Mycol., Baarn 7: 1-84, 1974

Stilbella Lindau
Synonym: *Stilbum* Tode, *Tilachlidiopsis* Keissler pro parte
Synnemata compact, sometimes regularly branched, with distinct terminal head and compact sterile stipe, brightly coloured or black; conidiophores concentrated in apical head, forming a dense hymenium-like layer of conidiogenous cells; conidiogenous cells phialidic, clavate to cylindrical, producing conidia in slime or dry, but not in distinct chains; conidia one-celled, smooth-walled, hyaline. Teleomorph: *Cordyceps*.
Ref. Mains, E.B., Mycologia 40: 402-416, 1948 and Bull. Torrey Bot. Club. 78: 122-133, 1951
Samson, R.A. et al., Proc. Kon. Ned Acad. Wetenschap C, 84: 289-301, 1981
Note. The genus *Stilbella* has been redefined by Seifert (Stud. Mycol. Baarn 27: 1-235, 1985). The entomogenous species, accommodated in *Stilbella* and *Tilachlidiopsis* and described here, should be transferred to a new genus.

Tetracrium Hennings
Conidiophores densely crowded and arranged in cushion-shaped, brightly coloured sporodochia, consisting of blastic conidiogenous cells producing large multi-armed conidia; conidia consisting of a small globose cells from which 2-7 but usually 3-4 tapering multiseptate arms arise. On scale insects. Teleomorph: *Podonectria*.
Ref. Petch, T., Trans. Br. mycol. Soc. 7: 18-40, 1921
Rossman, A., Mycotaxon 7: 163-182, 1978; compare also Evans, H.C. & R.A. Samson, Can. J. Bot. 64: 2098-2103, 1986.

Tetranacrium Hudson & Sutton
Stromata with flask-like pycnidia, containing blastic conidiogenous cells producing large multi-armed conidia; conidia consisting of a small globose cells from which several tapering multiseptate arms arise. On scale insects. Teleomorph: *Podonectria*.
Ref. Sutton, B., Mycol. Pap. 132: 1-43, 1973
Rossman, A., Mycotaxon 7: 163-182, 1978

Tilachlidium Preuss

Synnemata cylindrical, tapering toward the apices, consisting of loosely arranged verticillate conidiophores; conidiogenous cells phialidic, awl-shaped, completely covering the synnema, often occurring in whorls on short stalks along the synnema, producing conidia in slimy heads; conidia one-celled, smooth-walled, hyaline. Teleomorph: *Torrubiella*.

Ref. Mains, E. B., Mycologia 43: 691-718, 1951
Gams, W. *Cephalosporium*-artige Schimmelpilze, Gustav Fischer Verlag, 1971
Evans, H. C. & R. A. Samson, Can. J. Bot. 60: 2325-2333, 1982.

Tolypocladium Gams

Conidiophores erect, mononematous, verticillate to irregularly branched, bearing whorls of branches and phialidic conidiogenous cells; phialides with globose to flask-like basal portion, narrowing abruptly to distinct cylindrical neck which usually bends away from axis of conidiogenous cell; conidia one-celled, hyaline, smooth-walled, formed in slimy heads. Teleomorph: not known.

Ref. Gams, W., Persoonia 6: 185-191, 1971
Bissett, J., Can. J. Bot. 61: 1311-1329, 1983
Samson, R. A. & G. G. Soares, J. Invert. Path. 43: 133-139, 1984
Note: See under *Beauveria*.

Verticillium Nees

Synonym: *Acrostalagmus* Corda

Conidiophores erect or not distinctly differentiated from vegetative hyphae, mostly verticillate with loose whorls of phialidic conidiogenous cells; phialides mostly awl-shaped, sometimes slightly inflated at the base; conidia, one-celled, hyaline, smooth-walled, formed in slimy heads or occasionally in chains. Teleomorph: *Cordyceps, Torrubiella*.

Ref. Gams, W., *Cephalosporium*-artige Schimmelpilze, Gustav Fischer Verlag, 1971
Balazy, S., Bull. Soc. Amis Sci. Let. Poznan, Ser. D 14: 101-137, 1973

Keys to the most important entomopathogenic fungal genera

Entomophthorales

1 Spores typically bell-shaped, complete separation of the wall layers upon impact on a substratum, multinucleate with nuclei with prominent heterochromatin ... *Entomophthora*

1a Spores not bell-shaped, but globose, ovoid, pear-shaped or elongate, wall layers not or only partially separated ... 2

2a Spores ovoid to elongate, partial separation of the wall layers upon impact on a substratum (inner and outer layer not separated on a corolla surrounding the papillum), uninucleate with large patches of heterochromatin ... *Erynia*

2b Spores globose to pear-shaped, walls not separated upon impact on the substratum, multinucleate ... 3

3a Spores globose with conical papillae, containing small nuclei, not stained with aceto-orcein with a prominent central nucleolus and no apparent heterochromatin ... *Conidiobolus*

3b Spores not globose, containing large nuclei ... 5

4a Spores pyriform to ovoid, with conical papillae, containing nuclei with large patches of heterochromatin and no apparent central nucleolus, staining with aceto-orcein ... *Entomophaga*

4b. Spores pear-shaped, rounded, with truncate papillae, mostly containing four nuclei not staining with aceto-orcein; secondary spores mainly capillispores with curved capillisporophores ... *Neozygites*

Ascomycota

1a Ascospores one-celled, remaining in distinct spore balls; mostly associated with bees ... *Ascosphaera*

1b Ascospores septate, not remaining in distinct spore balls ... 2

2a Ascospores multiseptate with both transverse and longitudinal septa; asci globose ... *Myriangium*

2b Ascospores multiseptate with only transverse septa, often disarticulating into partspores; asci cylindrical to clavate ... 3

3a Ascospores ellipsoidal to fusiform, 2-3 septate ... *Nectria*

3b Ascospores long, clavate to cylindrical, multiseptate ... 4

4a Asci clavate to cylindrical, with distinct two-layered wall ... *Podonectria*

4b Asci long, filiform, without two-layered wall, but with distinct apical thickening ... 5

5a Stroma prominent, erect, mostly clavate with distinct fertile apical portion and fertile stipe ... *Cordyceps*

5b Stroma absent or not distinct and erect, but superficial ... 6

6a Ascospores fusiform, up to 30 μm long, 6-9 septate, not disarticulating into partspores ... *'Calonectria'*

6b Ascospores filiform, mostly approximately as long as the ascus, often disarticulating into partspores ... 7

7a Ascomata not in stromata, but superficially embedded in loosely arranged mycelium, on various arthropods ... *Torrubiella*

7b Ascomata in superficial stromata, mostly plate-like or cushion-shaped, on scale insects or whiteflies ... *Hypocrella*

DEUTEROMYCOTA

1a Conidia produced by phialides in chains or in slimy heads ... 2

1b Conidia produced by sympodial conidiogenous cells, often showing distinct scars or denticles ... 19

2a Conidia in dry, often long chains ... 3

2b Conidia not in chains but in slimy heads ... 9

3a Phialides simple, consisting of denticle-like outgrowths of the fertile hyphae ... *Pleurodesmospora*

3b Phialides conspicuous, mostly awl- or flask-shaped and on distinct conidiophores ... 5

4a Conidiophores consisting of an unbranched stipe terminating in a vesicle bearing conidiogenous cells and/or metulae ... 4

4b Conidiophores without a vesicle ... 6

5a Conidiophores mostly united in distinct synnemata, on spiders ... *Gibellula*

5b Conidiophores arising singly from host body, mostly not on spiders ... *Aspergillus*

6a Phialides arranged along synnemata as in a hymenium ... *Akanthomyces*

6b Phialides differently arranged ... 7

7a Conidiophores closely packed in sporodochia or single; conidia in columns ... *Metarhizium*

7b Conidiophores loosely arranged in synnemata or single, mostly verticillate, conidia in long divergent chains ... 8

8a Phialides very short necked, conidiophores bearing dense whorls of branches and phialides ... *Nomuraea*

8b Phialides with distinct necks; conidiophores with irregularly or verticillately branched elements ... *Paecilomyces*

9a Conidiophores arranged in pycnidia on scale insects or whiteflies, conidia usually fusiform ... *Aschersonia*

9b Conidiophores not arranged in pycnidia ... 10

10a Conidia one-or more septate, usually curved and banana-shaped with distinct footcell ... *Fusarium*

10b Conidia non-septate, or not curved and banana-shaped ... 11

11a Synnemata present with distinct fertile mucoid head and stipe ... 12

11b Synnemata absent or without distinct head ... 14

12a Phialides cylindrical, in a dense layer only covering apical part of synnema, conidia mostly longer than 5 μm ... *'Stilbella'*

12b Phialides of different shape, not only covering the apical part of the synnema, conidia mostly less than 5 μm ... 13

13a Phialides solitary or crowded in hymenium-like layer along synnema, with a swollen basal part abruptly tapering to one or more long thin necks; conidia single or few and typically covered with a slime sheath ... *Hirsutella*

13b Phialides mostly in whorls, often in chains, more or less awl-shaped; synnemata brightly coloured, bearing slimy conidial masses apically; conidia not distinctly covered wiith slime ... *Polycephalomyces*

14a Phialides typically awl-shaped ...   15

14b Phialides not awl-shaped, but flask-shaped, or with inflated base and distinct neck ...   17

15a Phialides forming one or only a few conidia usually covered with distinct mucilaginous layer ...   *Hirsutella*

15b Phialides forming many conidia in heads ...   16

16a Phialides solitary, not on verticillate conidiophores ...   *Acremonium*

16b Phialides in whorls, borne on verticillately branched conidiophores ...   *Verticillium* (compare also *Engyodontium*)

17a Phialides with an inflated base and a thin neck curved away from the main axis ...   *Tolypocladium*

17b Phialides flask-shaped with one or more straight necks ...   18

18a Phialides with one neck, conidia clavate; on mosquitoes and related Diptera ...   *Culicinomyces*

18b Phialides flask-shaped with more than one neck, conidia of various shapes ...   *Paraisaria*

19a Conidiogenous cells often cylindrical with crowded scars or denticles, in dense hymenium-like layer along distinct synnemata ...   *Hymenostilbe*

19b Conidiogenous cells not in dense hymenium ...   20

20a Conidiophores with stipe terminating in a vesicle bearing metulae and conidiogenous cells...   *Pseudogibellula*

20b Conidiophores without stipe and swollen vesicles ...   21

21a Conidiophores forming sporodochia and producing conidia with several multiseptate arms ...   *Tetracrium*

21b Conidiophores not in sporodochia, conidia one-celled ...   22

22a Conidia long, rod-shaped, mostly borne on warted conidiogenous cells, associated with *Torrubiella* and *Gibellula* on spiders ...   *Granulomanus*

22b Conidia of different shape, conidiogenous cells smooth-walled ...   23

23a Conidiogenous cells elongate and slender with inconspicuous, terminal or lateral scars ...   *Sporothrix*

23 Conidiogenous cells flask-shaped, with a swollen basal part terminating in a zig-zag rachis ...   *Beauveria*

Chapter 3

Illustrations of common fungal pathogens

In the following chapter the most common and important entomopathogenic fungi are illustrated. At present, several hundred fungal taxa are known which are able to parasitise arthropods and this number is still steadily increasing. It was, therefore, a difficult task to select the taxa to be illustrated in this Atlas, but we hope that the illustrations of the important representatives are provided. The illustrative plates are meant to aid recognition at the generic level. For specific identification, the reader should consult the relevant monographs and papers, which contain taxonomic descriptions and keys. The relevant papers are listed under each genus in Chapter 2. For keys and taxonomic descriptions of the Trichomycetes the reader is referred to Lichtwardt (1986).

List of taxa

1 *Coelomomyces dodgei* Couch
2 *Coelomomyces* spp.
3 *Coelomycidium simulii* Debaisieux
4 *Lagenidium giganteum* Couch
5 *Aphanomyces astaci* Schikora
6 *Legeriomyces* sp. (Trichomycetes, Harpellales)
7 *Asellaria aselli* Scheer ex Moss & Lichtwardt/ *Orchesellaria mauguioi* Manier ex Manier (Trichomycetes, Asellariales)
8 *Amoebidium parasiticum* Cienck (Trichomycetes, Amoebidiales)
9 *Enterobryus* sp./ *Taeniella carcini* Leger & Dubosq (Trichomycetes, Eccrinales)
10 *Conidiobolus apiculatus* (Thaxter) Remaudiere & Keller
11 *Conidiobolus coronatus* (Constatin) Batko
12 *Conidiobolus major* (Thaxter) Remaudiere & Keller
13 *Conidiobolus obscurus* (Hall & Dunn) Remaudiere & Keller
14 *Entomophaga aulicae* (Reich.) Humber
15 *Entomophaga caroliniana* (Thaxter) Samson, Evans & Latgé, comb. nov. (Basionym: Empusa caroliniana Thaxter - Mem. Boston Soc. Nat. Hist. 4: 167, 1888).
16 *Entomophaga grylli* (Fres.) Batko
17 *Entomophaga tenthredinis* (Fresenius) Samson, Evans & Latgé, comb. nov. (Basionym: Entomophthora tenthredinis Fres. - Abhandl. Senckenb. Naturf. Ges. Frankfurt AM. 2: 205, 1858)
18 *Entomophthora muscae* (Cohn) Fres.
19 *Entomophthora muscae* (Cohn) Fres.
20 *Entomophthora planchoniana* Cornu
21 *Erynia aquatica* (J.F. And. & Anagnostakis) Humber/ *Entomophthora culicis* (Braun) Fres.
22 *Erynia castrans* (Batko & Weiser) Remaudiere & Keller/ *Massospora cicadina* Peck
23 *Erynia conica* (Nowakoski) Remaudiere & Hennebert
24 *Erynia dipterigena* (Thaxter) Remaudiere & Hennebert/ *E. blunckii* (Lakon ex Zimmerman) Remaudiere & Hennebert
25 *Erynia elateridiphaga* (Turian) Humber
26 *Erynia gammae* (Weiser) Samson, Evans & Latge, comb. nov. (Basionym: Tarichium gammae Weiser - Ceska Mykol. 19: 204, 1965)
27 *Erynia neoaphidis* Remaudiere & Hennebert
28 *Erynia radicans* (Brefeld) Humber et al.
29 *Erynia rhizospora* (Thaxter) Remaudiere & Hennebert/ *E. plecopteri* Descals & Webster
30 *Erynia virescens* (Thaxter) Remaudiere & Hennebert
31 *Neozygites adjarica* (Tsintasadze & Vartapetov) Remaudiere & Keller/ *N. fumosa* (Speare) Remaudiere & Keller
32 *Neozygites fresenii* (Nowakowski) Remaudiere & Keller
33 *Sporodiniella umbellata* Boedijn
34 *Ascosphaera aggregata* Skou
35 *Ascosphaera apis* (Maassen ex Clausen) Olive & Spiltoir

36 *Calonectria pruinosa* Petch
37 *Cordycepioideus bisporus* Stifler/ *C. octosporus* Blackwell & Gilb.
38 *Cordyceps australis* (Speg.) Sacc.
39 *Cordyceps calocerioides* Berk. & Curt.
40 *Cordyceps gunnii* (Berk.) Berk.
41 *Cordyceps lloydii* Fawcett
42 *Cordyceps martialis* Speg.
43 *Cordyceps militaris* (L.:Fr) Link
44 *Cordyceps nutans* Pat.
45 *Cordyceps polyartha* Moeller
46 *Cordyceps sobolifera* (Hill.) Berk. & Br.
47 *Cordyceps tuberculata* (Leb.) Maire
48 *Cordyceps unilateralis* (Tull.) Sacc.
49 *Cordyceps unilateralis* (Tull.) Sacc.
50 *Hypocrella amomi* Rac.
51 *Nectria flammea* (Tul.) Dingley
52 *Myriangium duriaei* Mont. & Berk.
53 *Podonectria coccicola* (Ellis & Everh.) Petch
54 *Torrubiella arachnophila* (Johnston) Mains
55 *Torrubiella carnata* Moureau
56 *Torrubiella rubra* Pat. & Lagerh.
57 *Akanthomyces aculeatus* Lebert
58 *Akanthomyces gracilis* Samson & Evans
59 *Akanthomyces pistillariiformis* (Pat.) Samson & Evans
60 *Aschersonia aleyrodis* Webber
61 *Aschersonia cubensis* Berk. & Curt.
62 *Aschersonia turbinata* Berk.
63 *Aspergillus parasiticus* Speare
64 *Beauveria bassiana* (Bals.) Vuill.
65 *Beauveria bassiana* (Bals.) Vuill./ *B. brongniartii* (Sacc.) Petch
66 *Beauveria amorpha* (Hohn.) Samson & Evans/ *B. velata* Samson & Evans
67 *Culicinomyces clavisporus* Couch et al.
68 *Engyodontium aranearum* (Cavara) Gams et al.
69 *Fusarium coccophilum* (Desm.) Wollenw. & Reinking
70 *Gibellula alata* Petch
71 *Gibellula leiopus* (Vuill.) Mains
72 *Gibellula pulchra* (Sacc.) Cavara
73 *Granulomanus* state
74 *Hirsutella citriformis* Speare
75 *Hirsutella entomophila* Pat.
76 *Hirsutella jonesii* (Speare) Evans & Samson
77 *Hirsutella sausserei* (Cooke) Speare
78 *Hirsutella thompsonii* Fischer
79 *Hirsutella versicolor* Petch
80 *Hymenostilbe dipterigena* Petch
81 *Hymenostilbe formicarum* Petch
82 *Hymenostilbe muscaria* Petch
83 *Hymenostilbe* species
84 *Metarhizium album* Petch
85 *Metarhizium anisopliae* (Metschn.) Sorokin var. anisopliae
86 *Metarhizium anisopliae* (Metschn.) Sorokin var. anisopliae
87 *Metarhizium anisopliae* (Metschn.) Sorokin var. majus Tulloch
88 *Metarhizium flavoviride* Gams & Roszypal
89 *Nomuraea atypicola* (Yasuda) Samson
90 *Nomuraea rileyi* (Farlow) Samson
91 *Paecilomyces amoeneroseus* (P. Henn.) Samson
92 *Paecilomyces cicadae* (Miquel) Samson
93 *Paecilomyces farinosus* (Holm) Brown & Smith
94 *Paecilomyces lilacinus* (Thom) Samson
95 *Paecilomyces tenuipes* (Peck) Samson
96 *Paraisaria dubia* (Delacr.) Samson & Brady
97 *Pleurodesmospora coccorum* (Petch) Samson et al.
98 *Polycephalomyces ramosus* (Peck) Mains
99 *Pseudogibellula formicarum* (Mains) Samson & Evans
100 *Sporothrix insectorum* de Hoog & Evans
101 *Sporothrix isarioides* (Petch) de Hoog & Evans
102 *Stilbella buquetii* (Mont. & C. Robin) Samson & Evans var. buquetii
103 *Stilbella buquetii* (Mont. & C. Robin) Samson & Evans var. formicarum
104 *Tetracrium coccicolum* Hohn.
105 *Tilachlidiopsis nigra* Yakusiji & Kumazawa
106 *Tilachlidium liberianum* (Mains) Mains
107 *Tolypocladium cylindrosporum* Gams
108 *Verticillium lecanii* (Zimm.) Viegas
109 *Verticillium lecanii* (Zimm.) Viegas

PLATE 1. *Coelomomyces dodgei*
A Hyphae in the copepod, *Acantocyclops vernalis*, B sporangia in *Anopheles quadrimaculatus* (×10), C sporangia (×800), D zygotes (ca. 7 μm in diam.), encysted in *Anopheles quadrimaculatus* (×800).

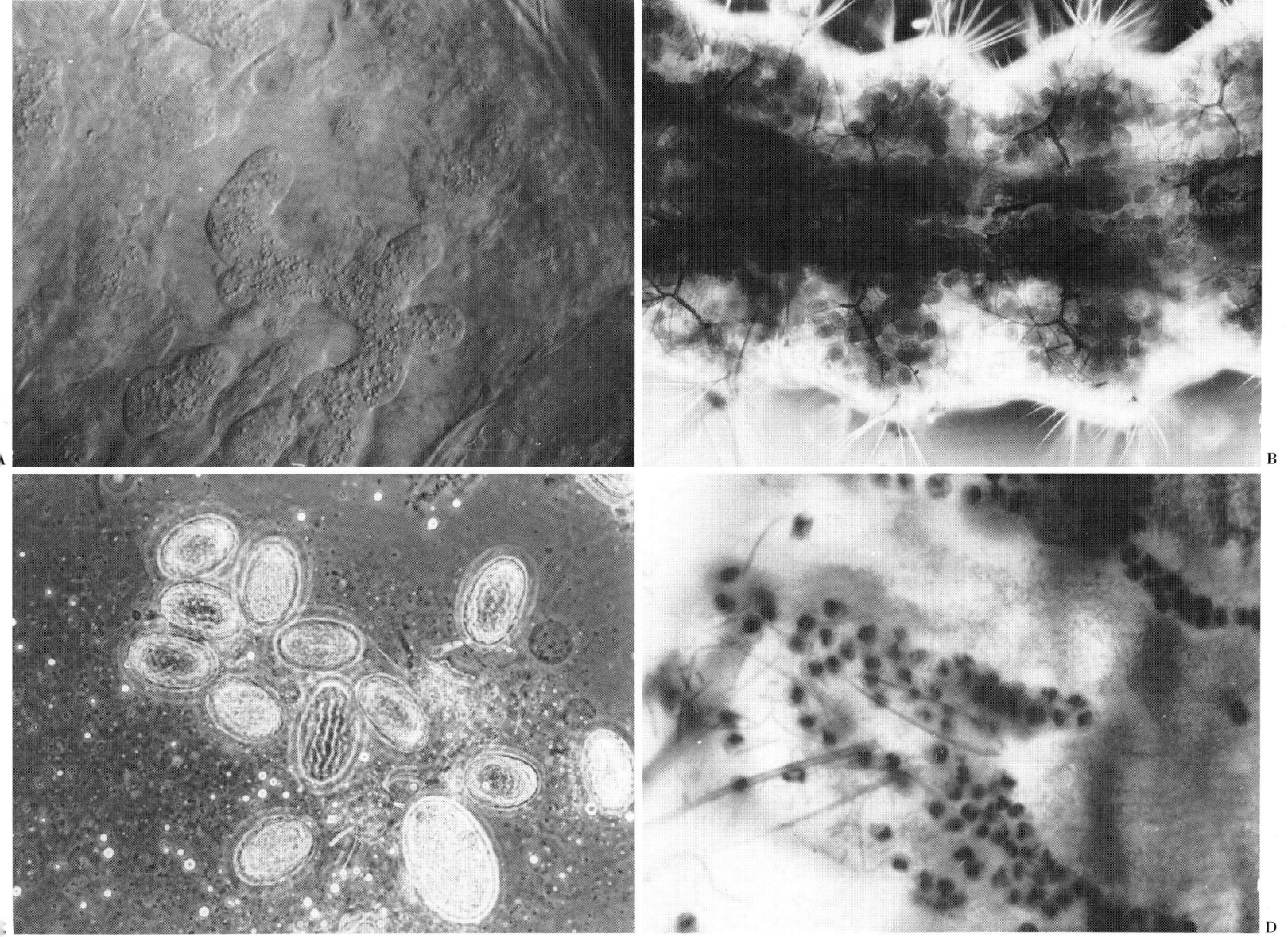

PLATE 2. *Coelomomyces* spp.
A–B Sporangia of *C. opifexi* in mosquito host, C sporangia of both *C. punctatus* (small) and *C. psorophorae* (large) (×800), D–E zoospores of *C. punctatus* either uni- or biflagellate (last type rarely produced) (×1000).

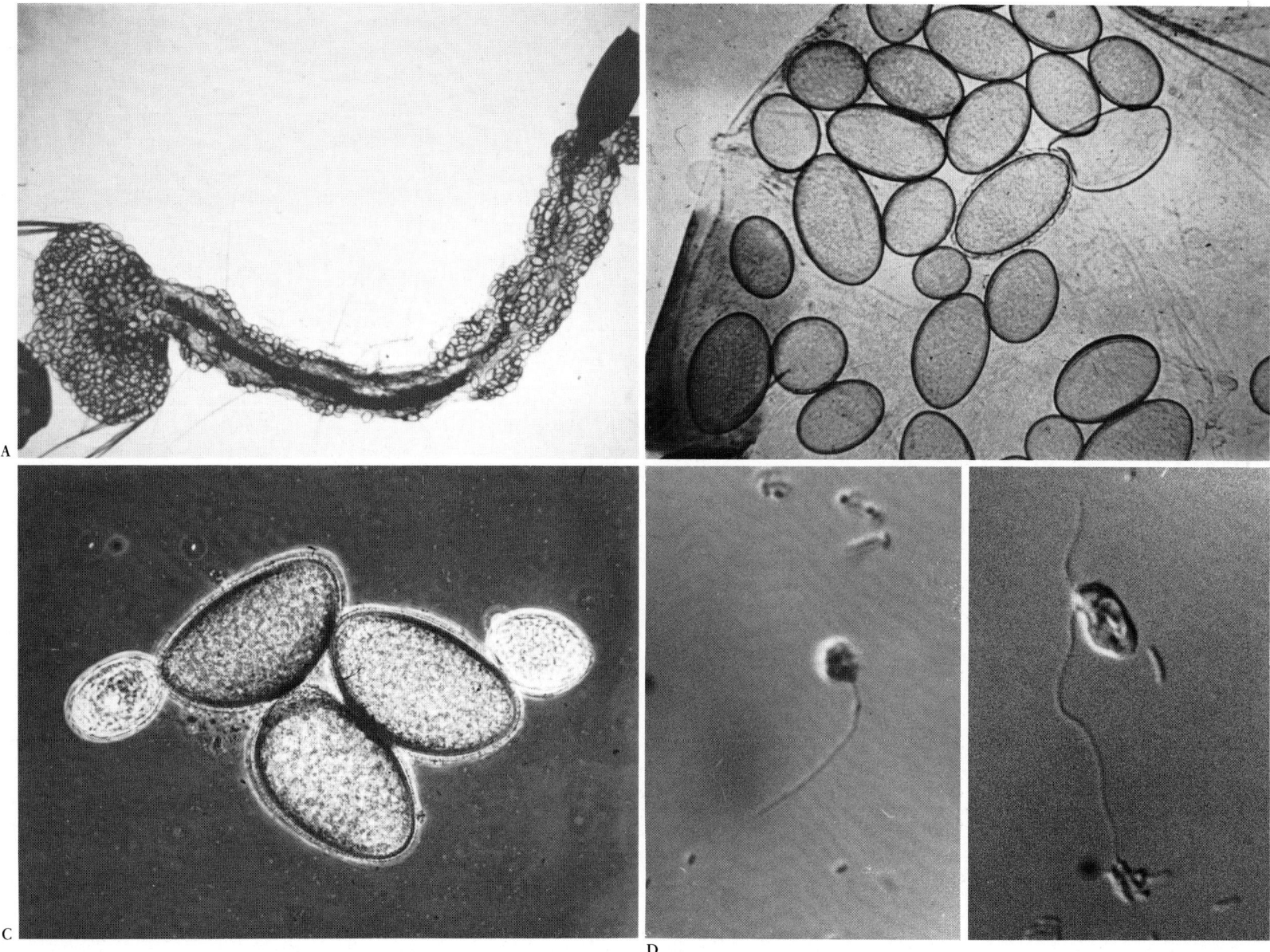

PLATE 3. *Coelomycidium simulii*
A Prosimilium larva with latent infection, B sporangium (×150),
D–E zoospores (×600 and ×1000 resp.).

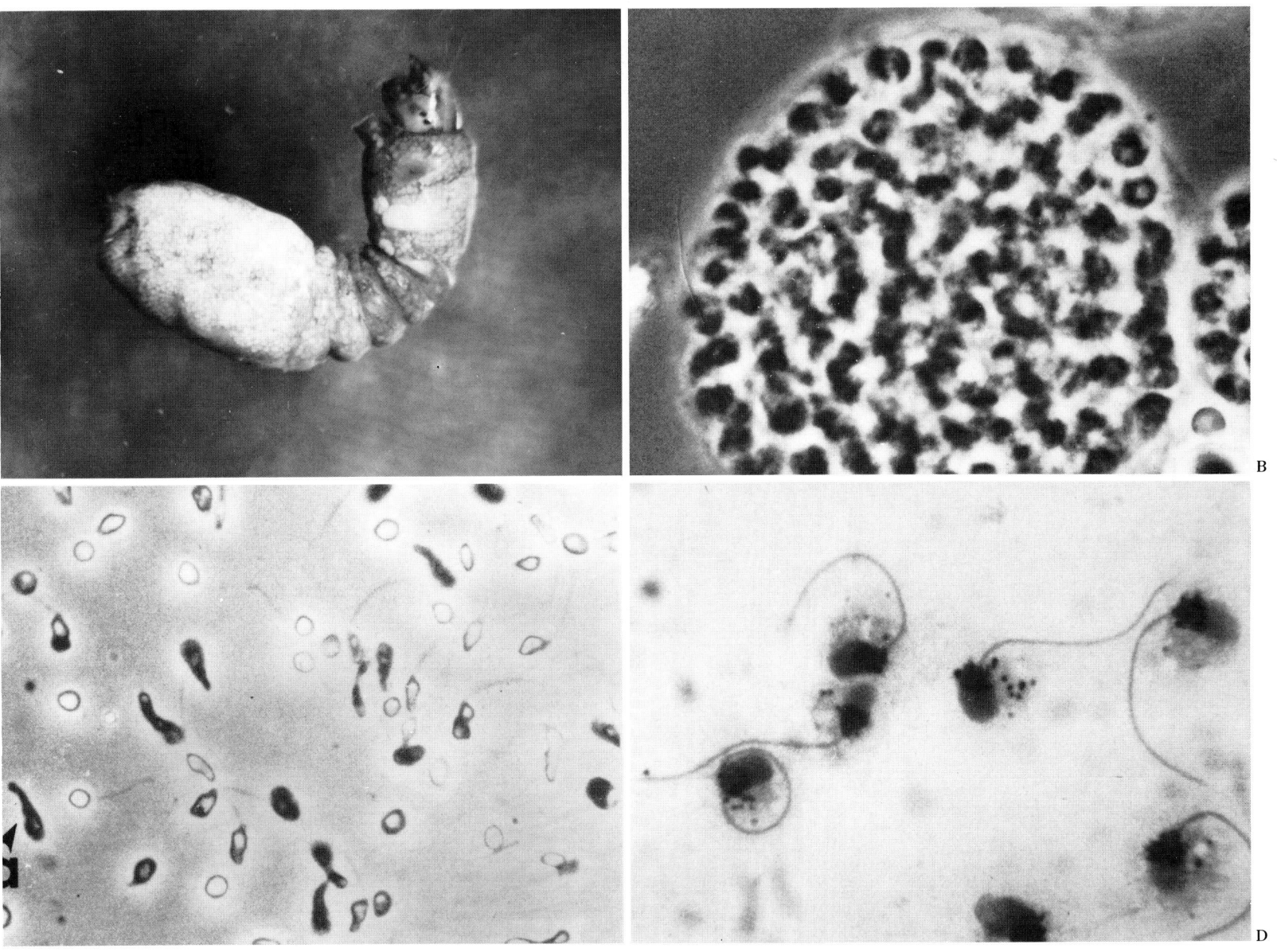

PLATE 4. *Lagenidium giganteum*
A Hyphae invading an anal gill of a larva of *Aedes aegyptii* (×350), B sporangia with exit tube and zoospore formation (×350), C biflagellate zoospore (×2500), D encysted zoospore on cuticle (×500), E oospore formation in oogonia (×1000)

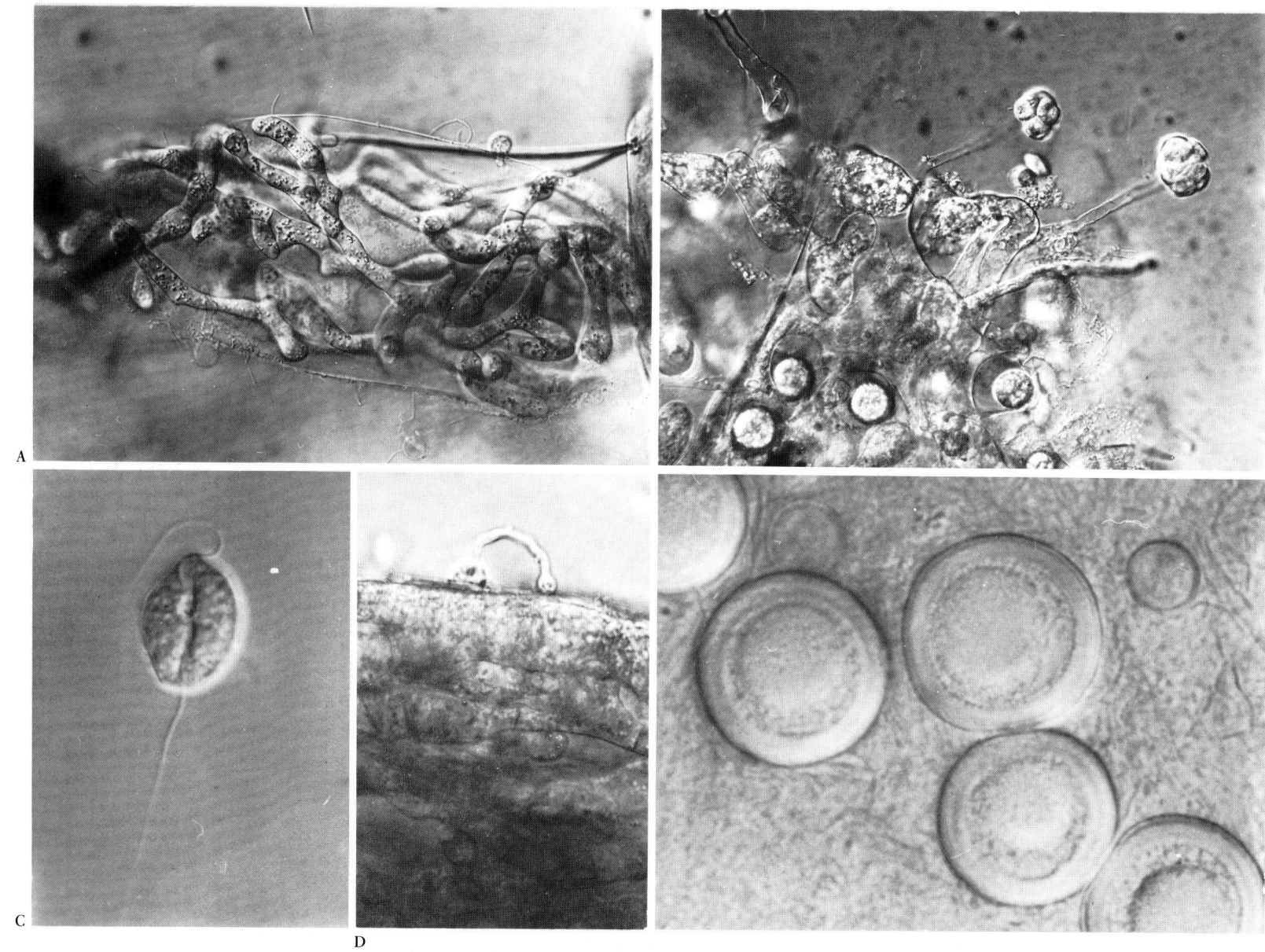

PLATE 5. *Aphanomyces astaci*
A Sporeballs in sporulating hyphae, B zoospore formation and release (× 400), C-E zoospore development (× 800), F melanized hyphae in the cuticle of a crayfish, *Pacifastacus leniusculus* (× 400)

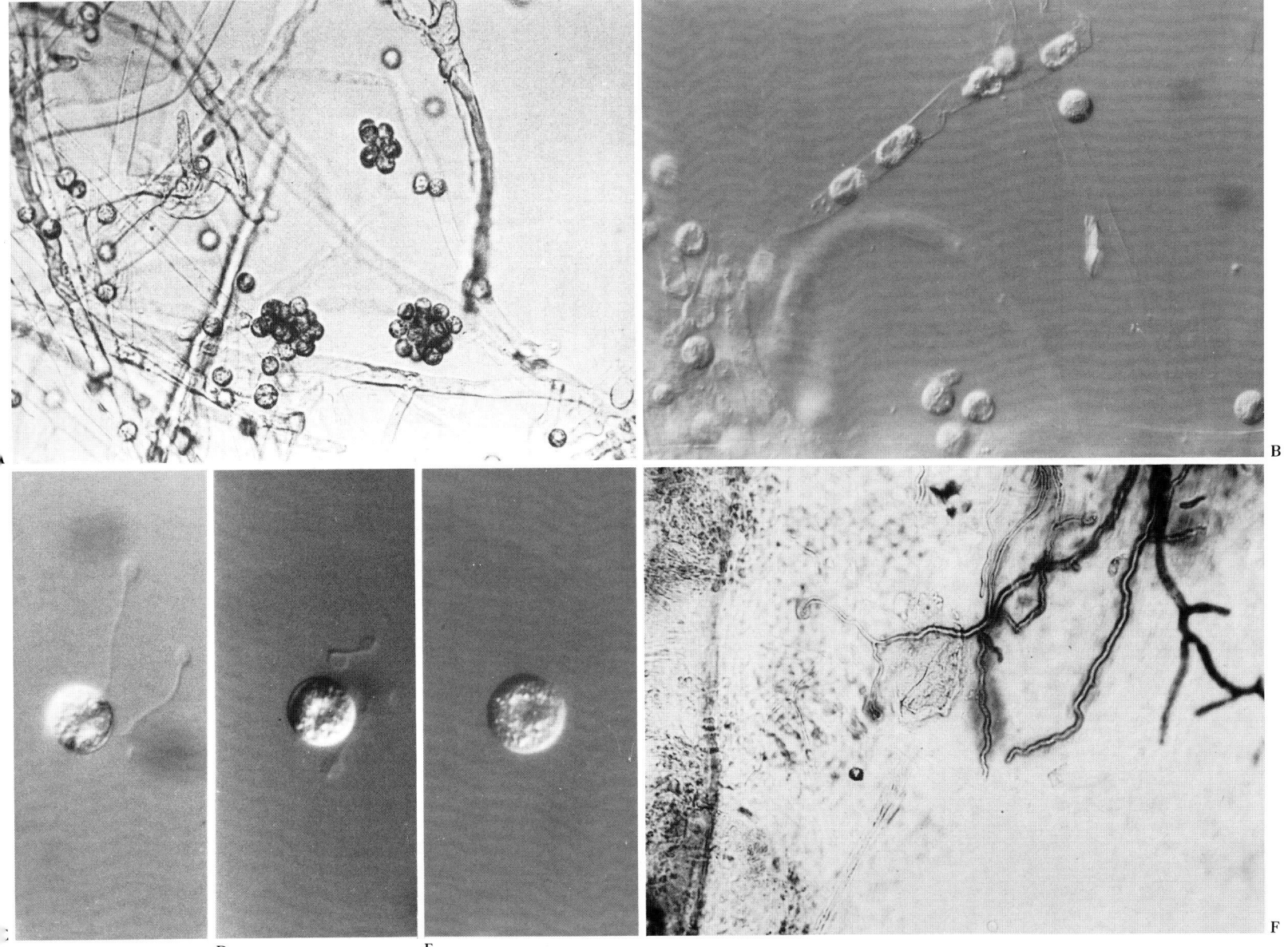

PLATE 6. *Legeriomyces* sp. (Trichomycetes, Harpellales)
A Branched thallus attached to hindgut of a mayfly nymph, B trichospores, C generative cell subtending a trichospore, D released trichospore with two appendages, E zygospore (all × 1100, except A: ×225).

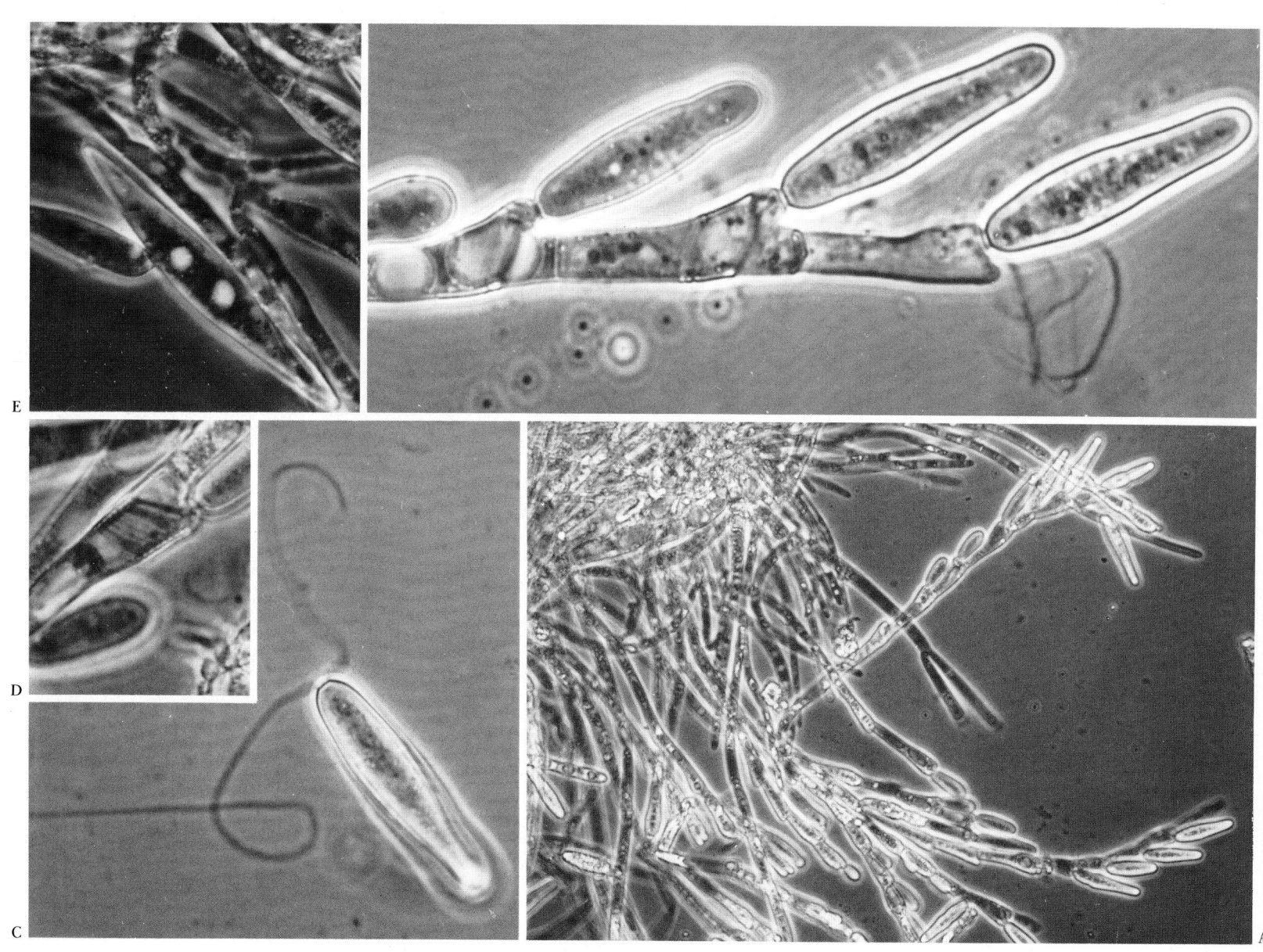

PLATE 7. *Asellaria aselli / Orchesellaria mauguioi* (Trichomycetes, Asellariales)

A–C *Asellaria aselli*. A Thallus attached to hindgut of *Asellus aquaticus* (×330), B bifurcated holdfast cell (×1400), C arthrospore reproduction from lateral branch of thallus (×2300). D–F *Orchesellaria mauguioi*, D arthrospore production from lateral thallus attached to *Isotomorus palustris* (×525), E germinated arthrospore (×1550), F holdfast region of a mature thallus (×550).

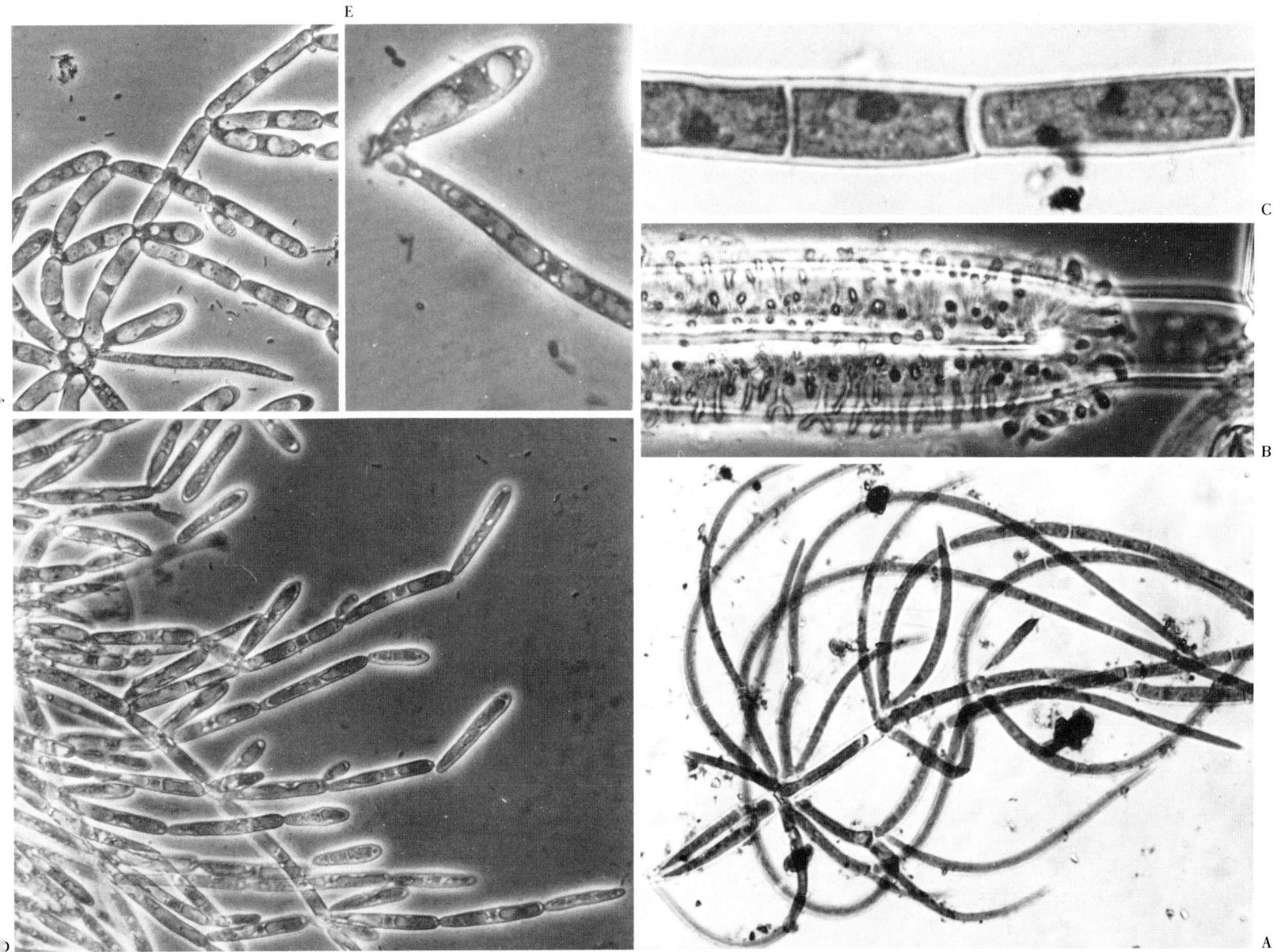

PLATE 8. *Amoebidium parasiticum* (Trichomycetes, Amoebidiales) A Unbranched vegetative thalli attached to cuticle of fresh water insect (× 10 and × 700 resp.), B secreted holdfast of thallus base (× 1400), F thallus cleaved into rigid sporangiospores (× 700), E released sporangiospore (× 700), F attached sporangiospore (× 700), G amoeboid sporangiospore (× 1500), H encysted amoebae (cystospores) (× 400).

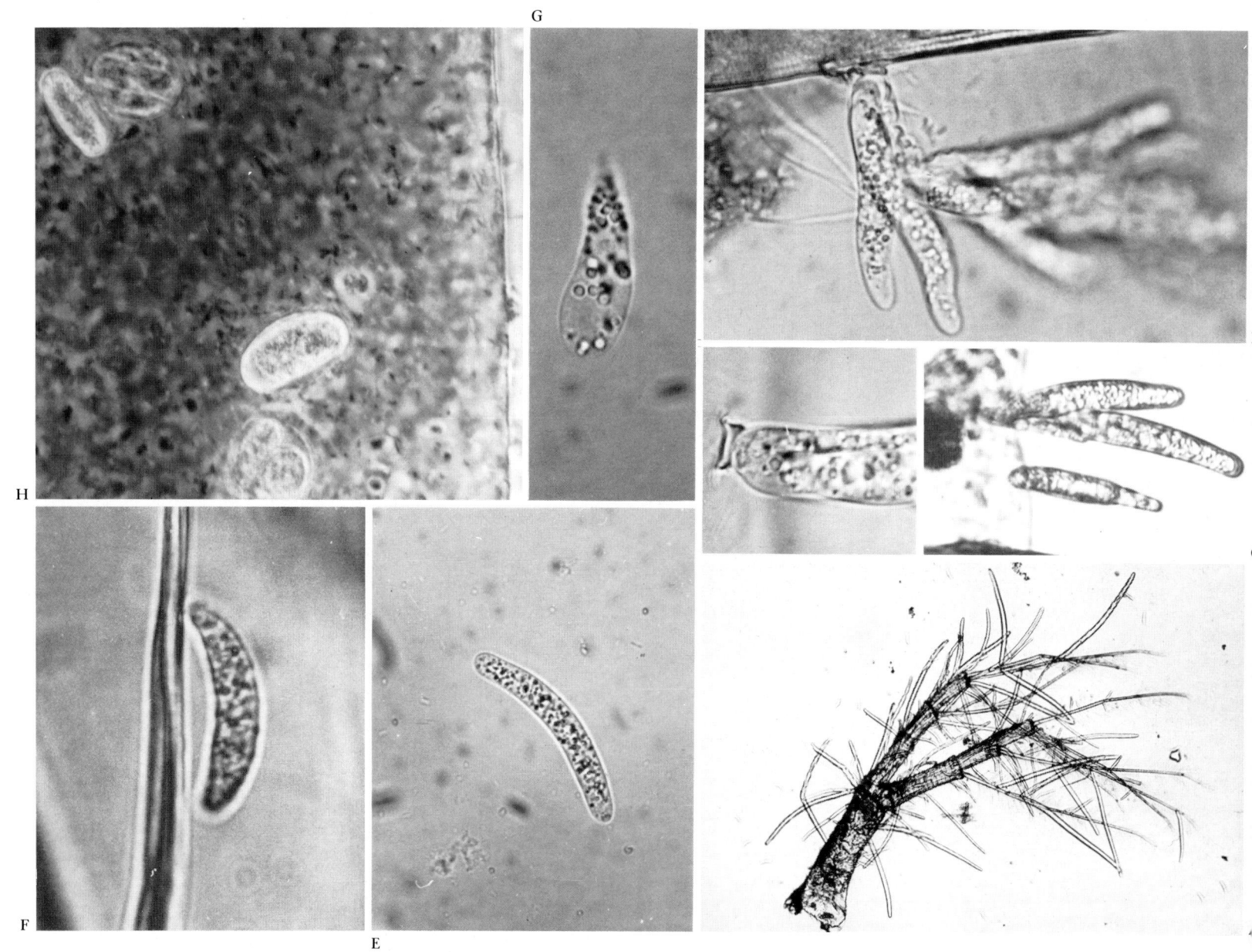

PLATE 9. *Enterobryus* sp. / *Taeniella carcini* (Trichomycetes, Eccrinales)

A-E *Enterobryus* sp., A-B branched thalli attached to cuticle lining the hindgut of *Thyropygus* sp. (×23 and ×54 resp.), C sporulating thallus with uninucleate dispersive spores (×320), D germinated spore with developing holdfast (×320), E secreted holdfast of mature thallus on hindgut cuticle (×650), F-G *Taeniella carcini*, F released spore with two appendages (×1500), G sporulating region of thallus from the hindgut of a marine decapod (×400).

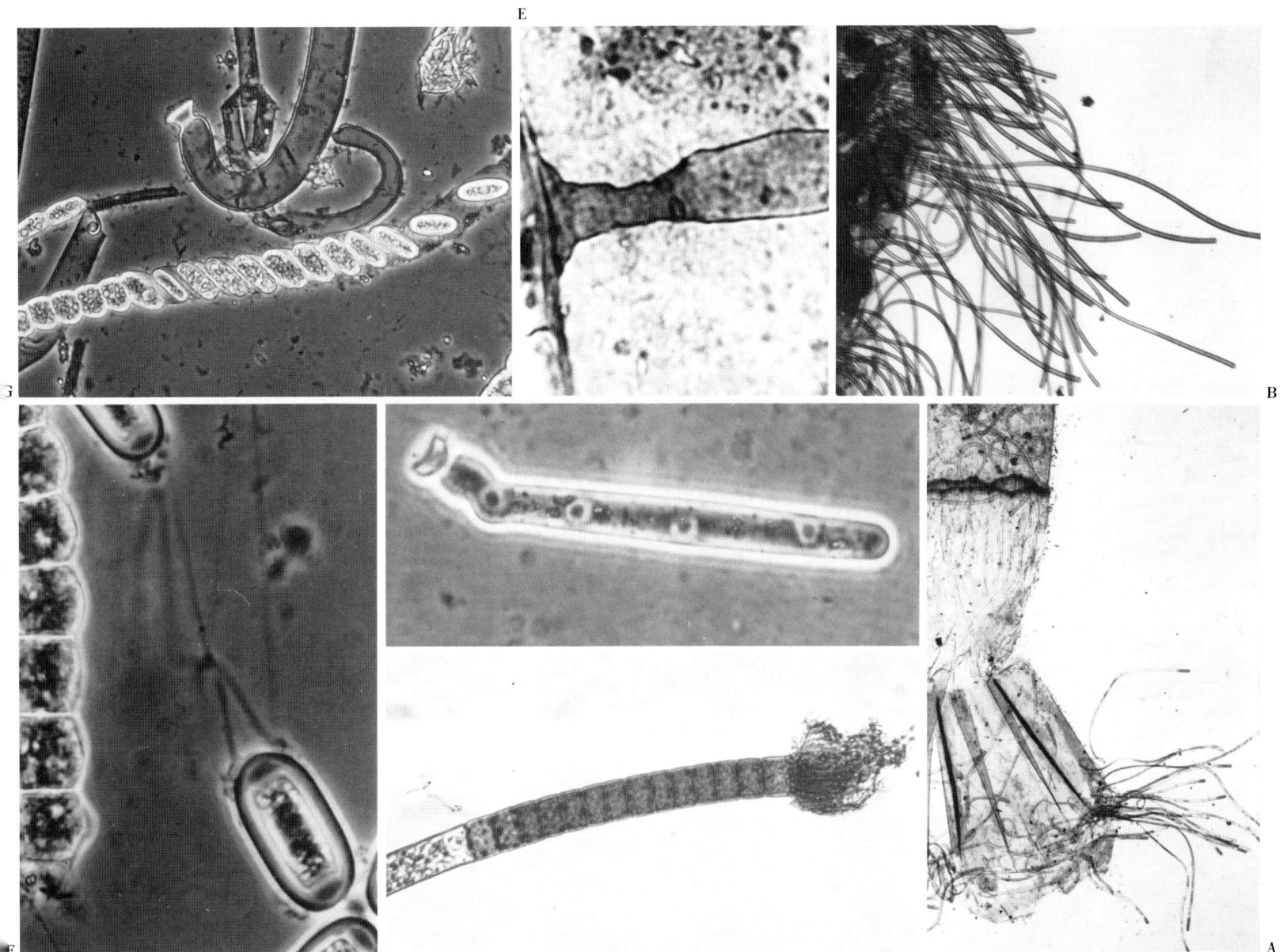

PLATE 10. *Conidiobolus apiculatus*
A On beetle (*Propylaea*, Coccinellida), B rhizoids on a *Sciaridae* sp., C primary spore (×1800), D secondary spore (×1800), E secondary spore formation (×500), F multigerm tubes (×500).

PLATE II. *Conidiobolus coronatus*
A On beetle larvae, B primary spore (×1100), C secondary spore (×1100), D villose spores (×750), E microconidia (×400).

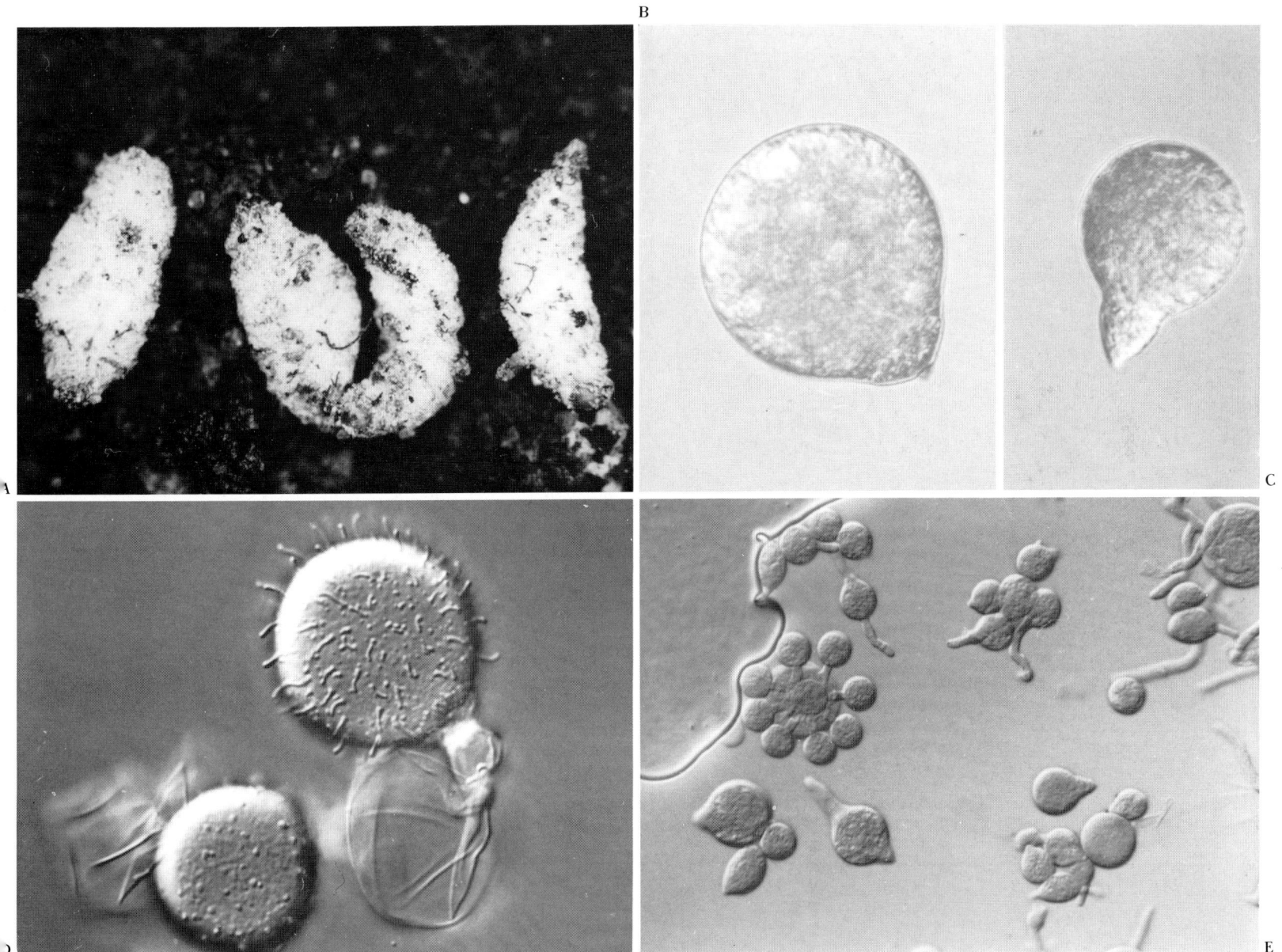

PLATE 12. *Conidiobolus major*
A On dipteran insect, B on host showing rhizoids, C primary spore (× 700), D secondary spore formation (× 700), E resting spores (× 500).

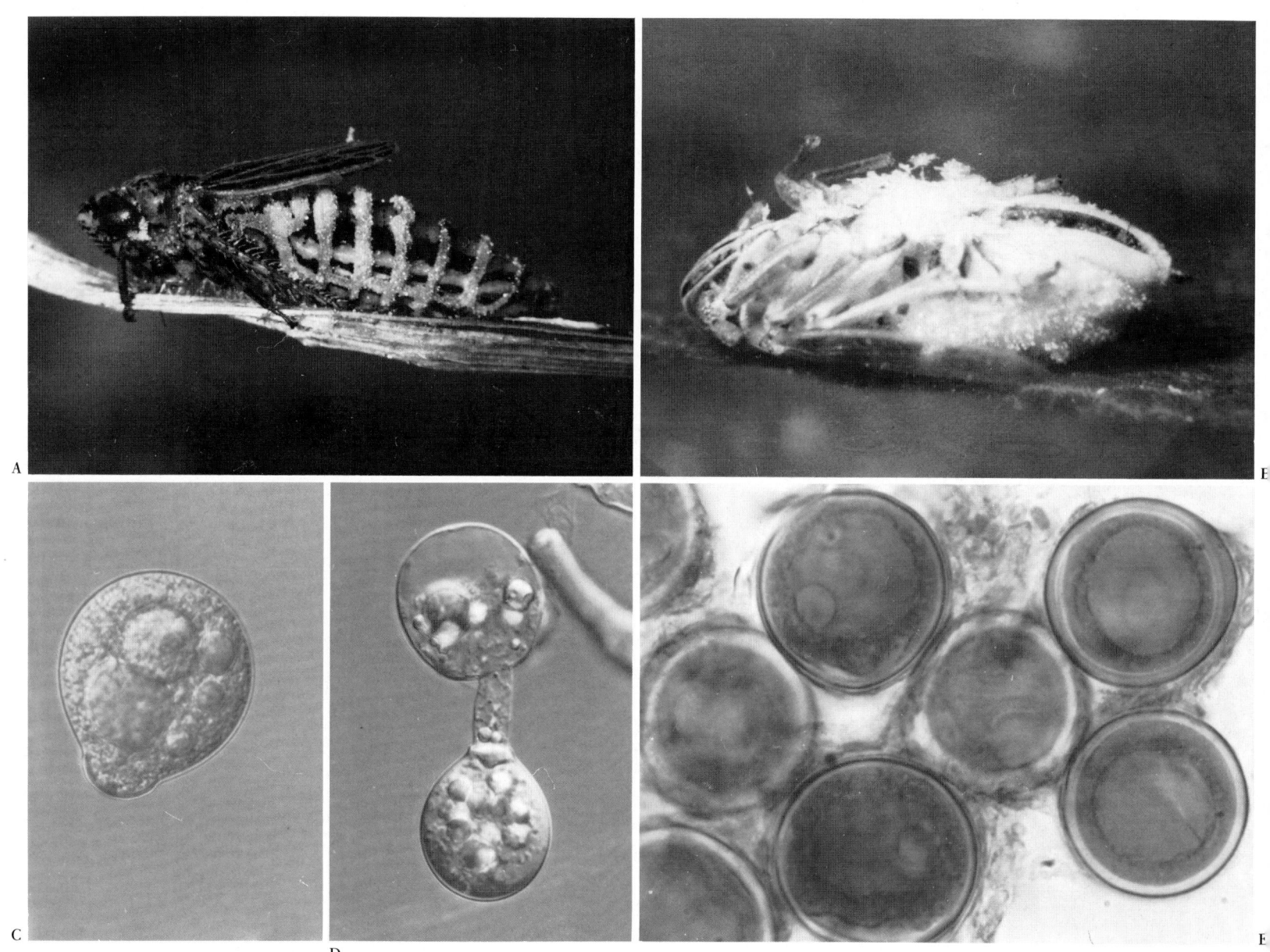

PLATE 13. *Conidiobolus obscurus*
A On aphid, B primary spore formation on the host, C primary spore (× 1000), D secondary spore (× 1000), E multigerm tubes (× 600), F resting spore (× 700).

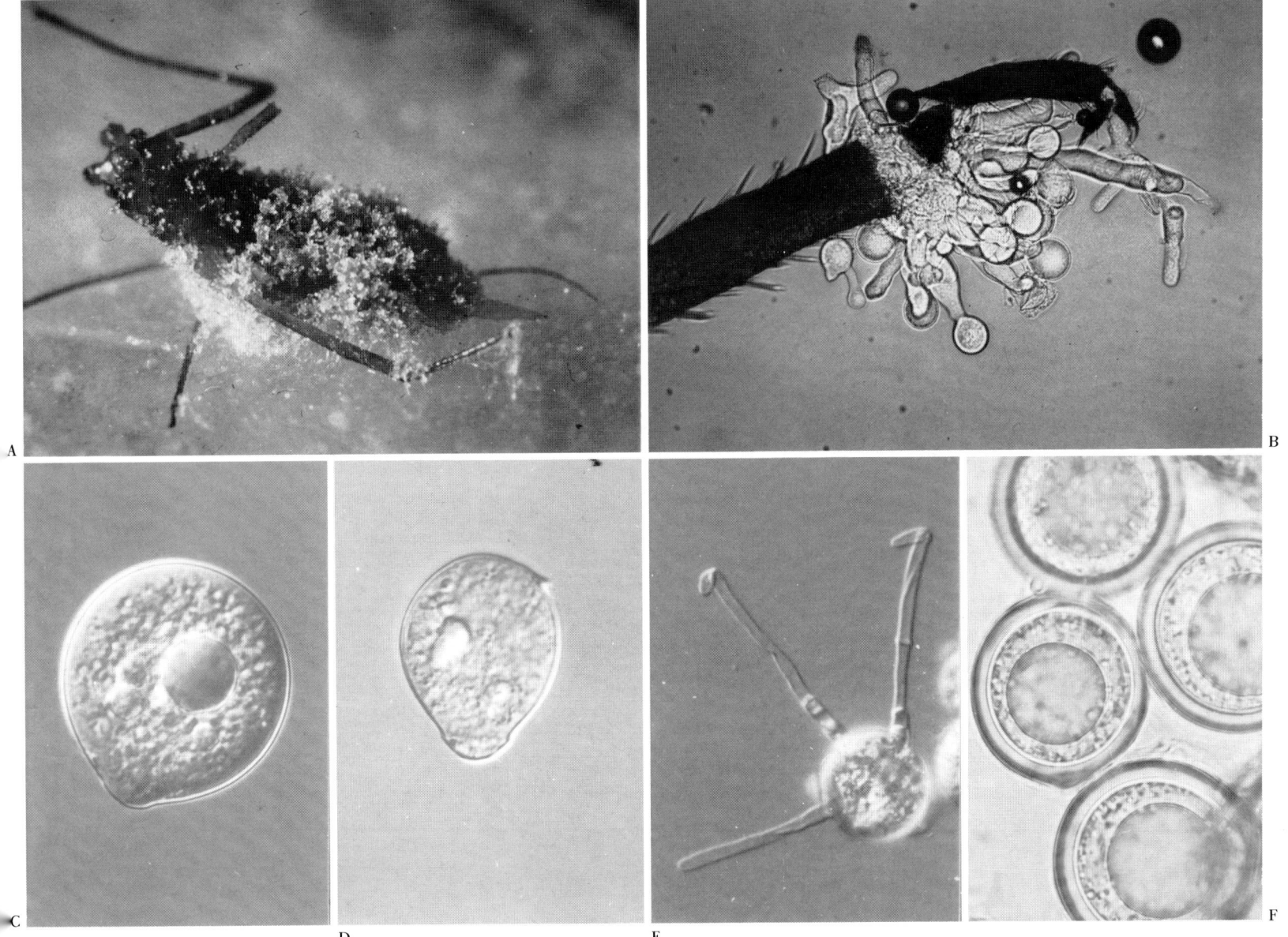

PLATE 14. *Entomophaga aulicae*
A On lepidopteran larva, B secondary spore (×900), C primary spore (×900), D-F protoplast stages from in vitro cultures (×1000), G resting spores (×500)

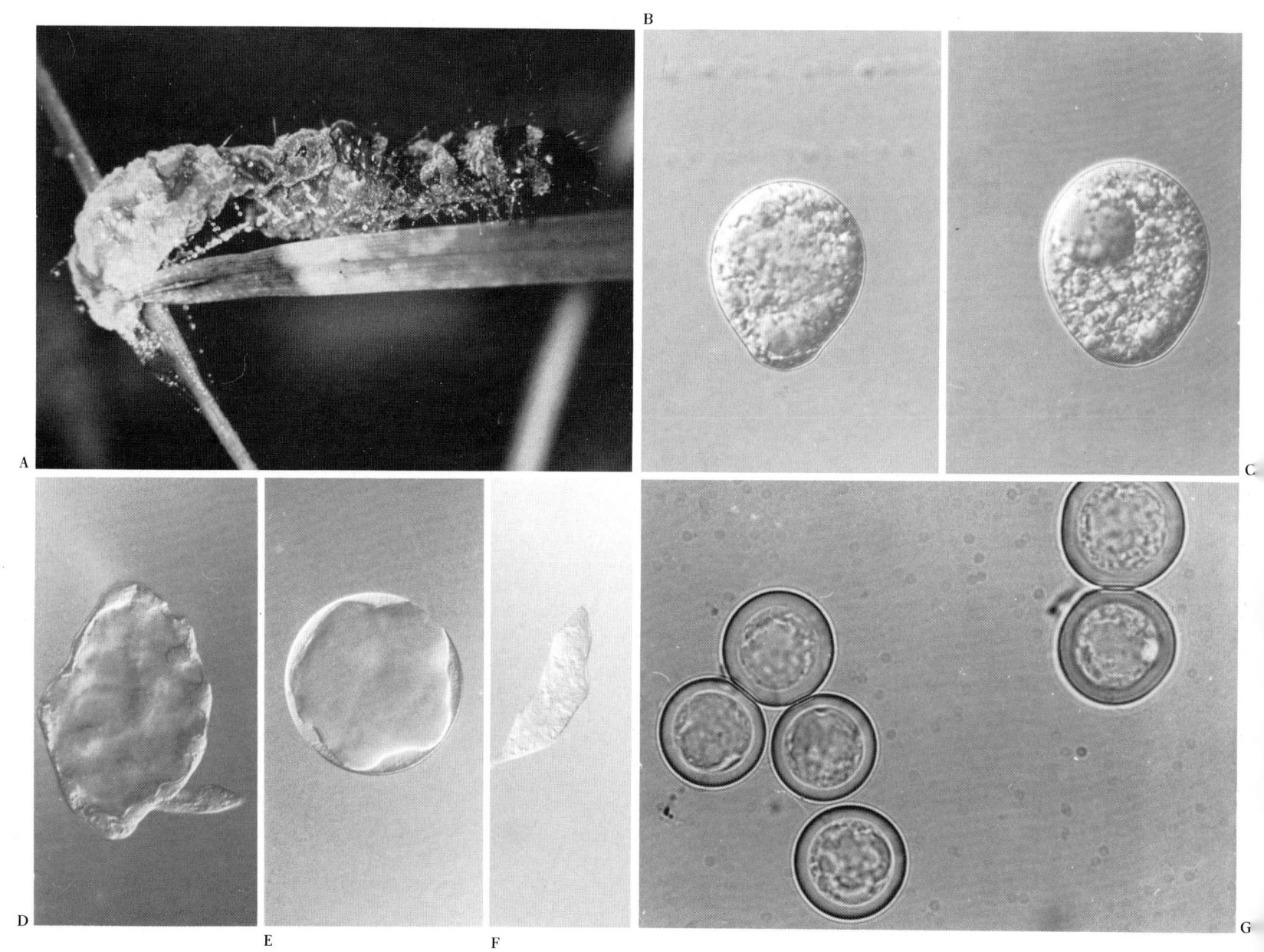

PLATE 15. *Entomophaga caroliniana*
A On *Tipula paludosa*, B primary spores, C pyriform and fusiform secondary spores, D resting spores (all ×500).

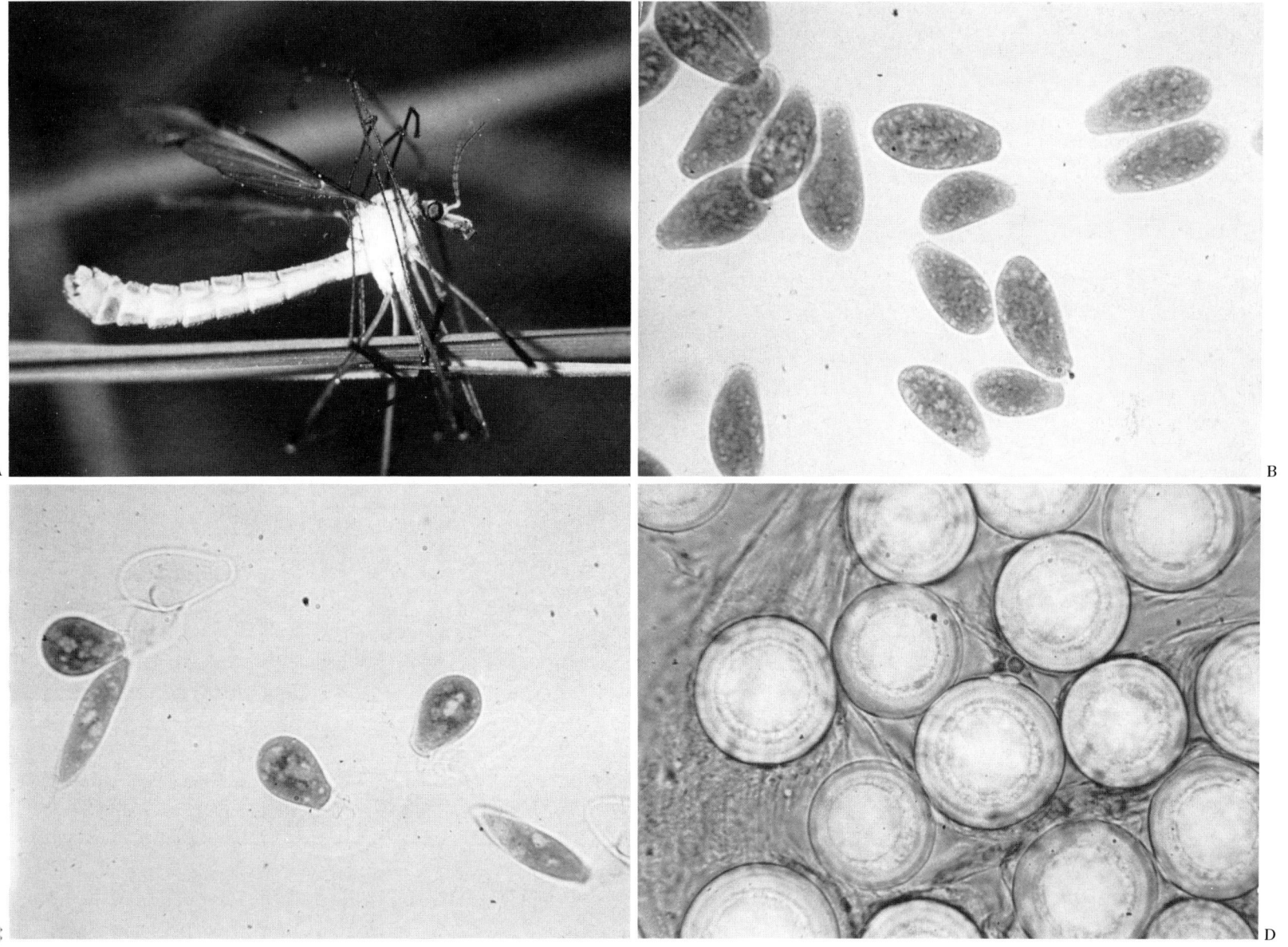

PLATE 16. *Entomophaga grylli*
A On grasshoppers, B spore formation with nuclei stained with aceto-orcein (×400), C resting spore (×1000), D primary spore (×1200), E secondary spore (×1200).

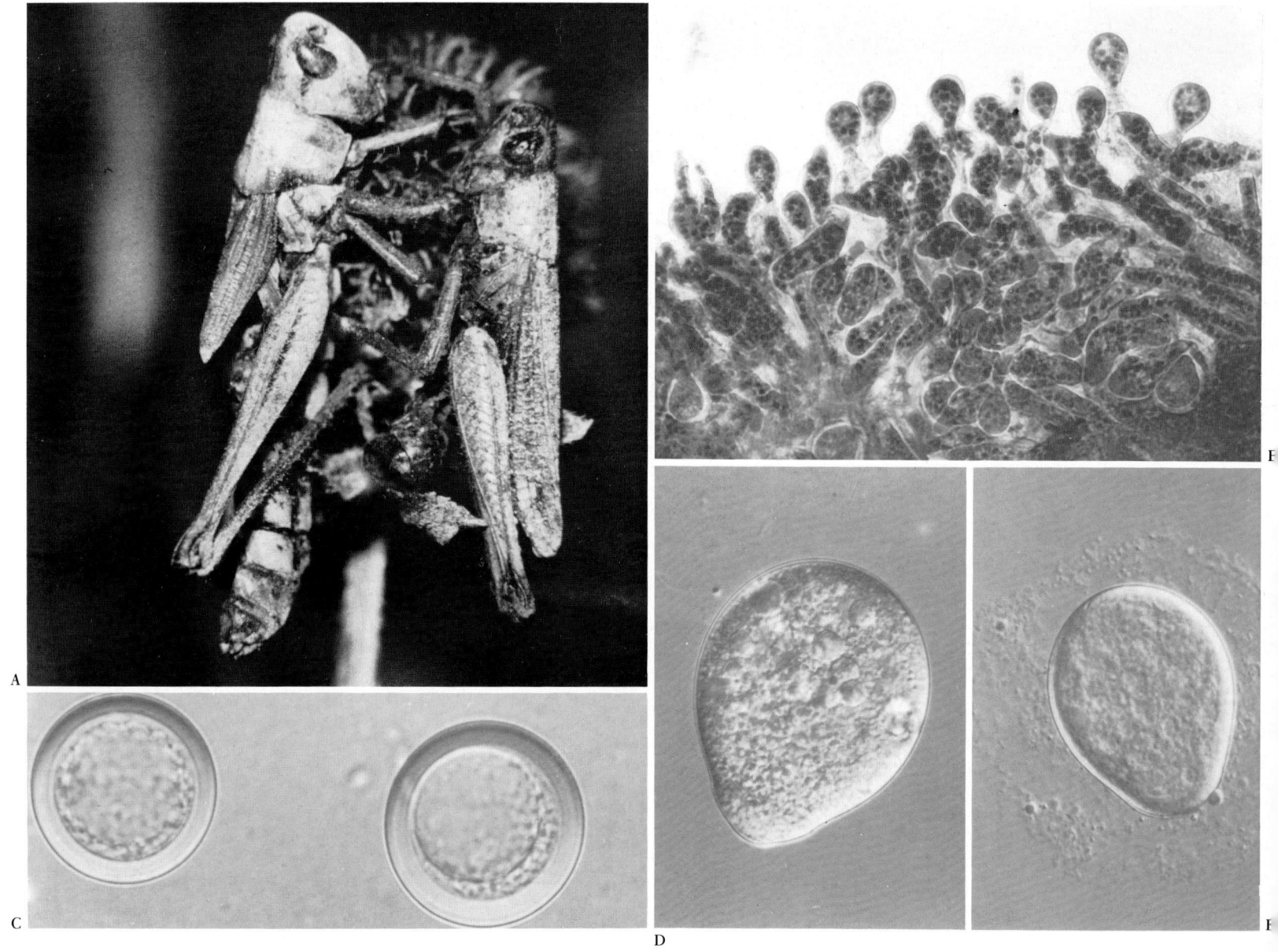

PLATE 17. *Entomophaga tenthredinis*
A On larva of sawfly, B primary spore (×500), C secondary spore formation (×500), D staining with aceto-orcein of the nuclei of the primary spores (×900), E resting spores (×600).

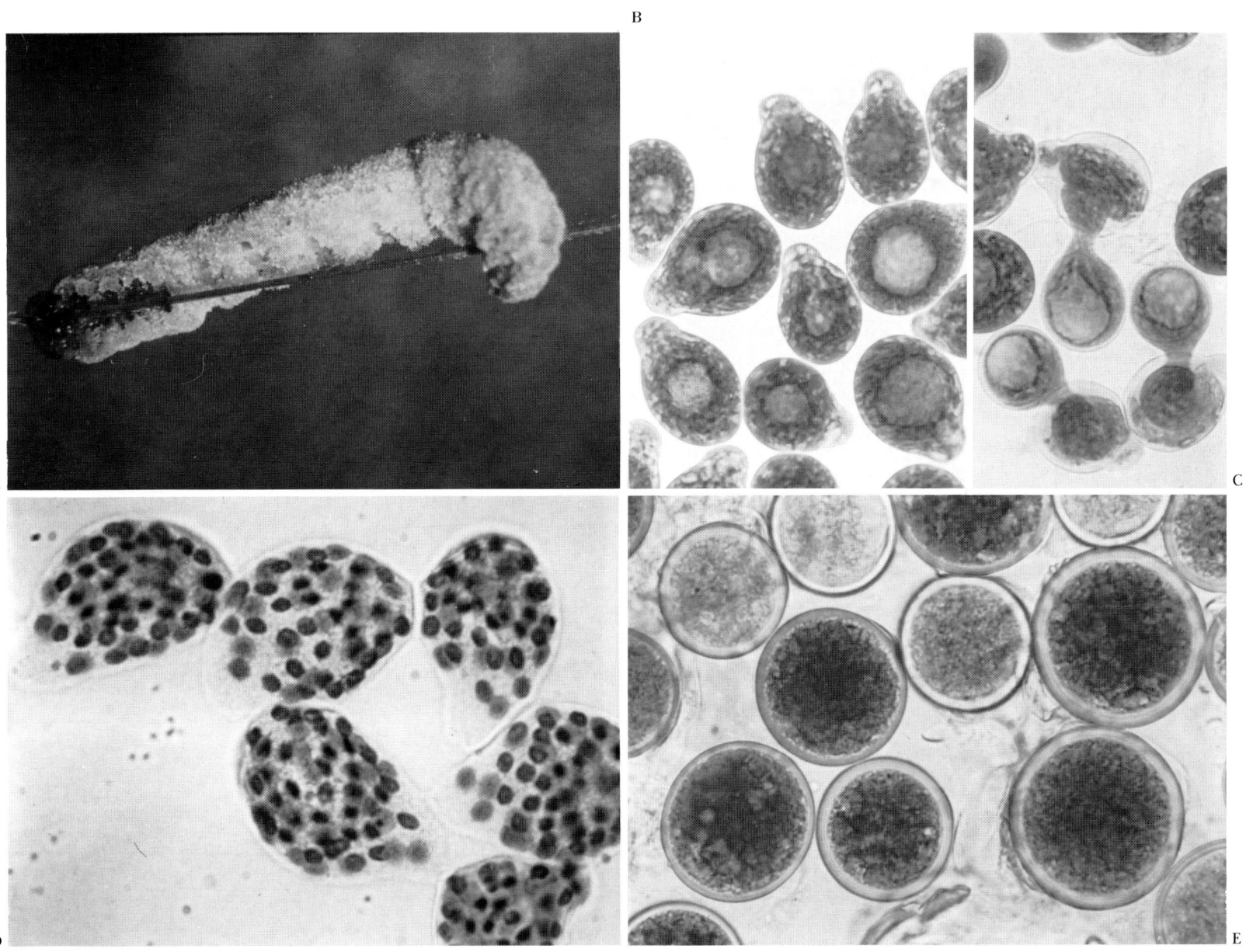

PLATE 18. *Entomophthora muscae*
A-B Discharged spores on flies, C primary spores (×1000), D-E different nuclei number in two species of the *E. muscae*-complex (×500).

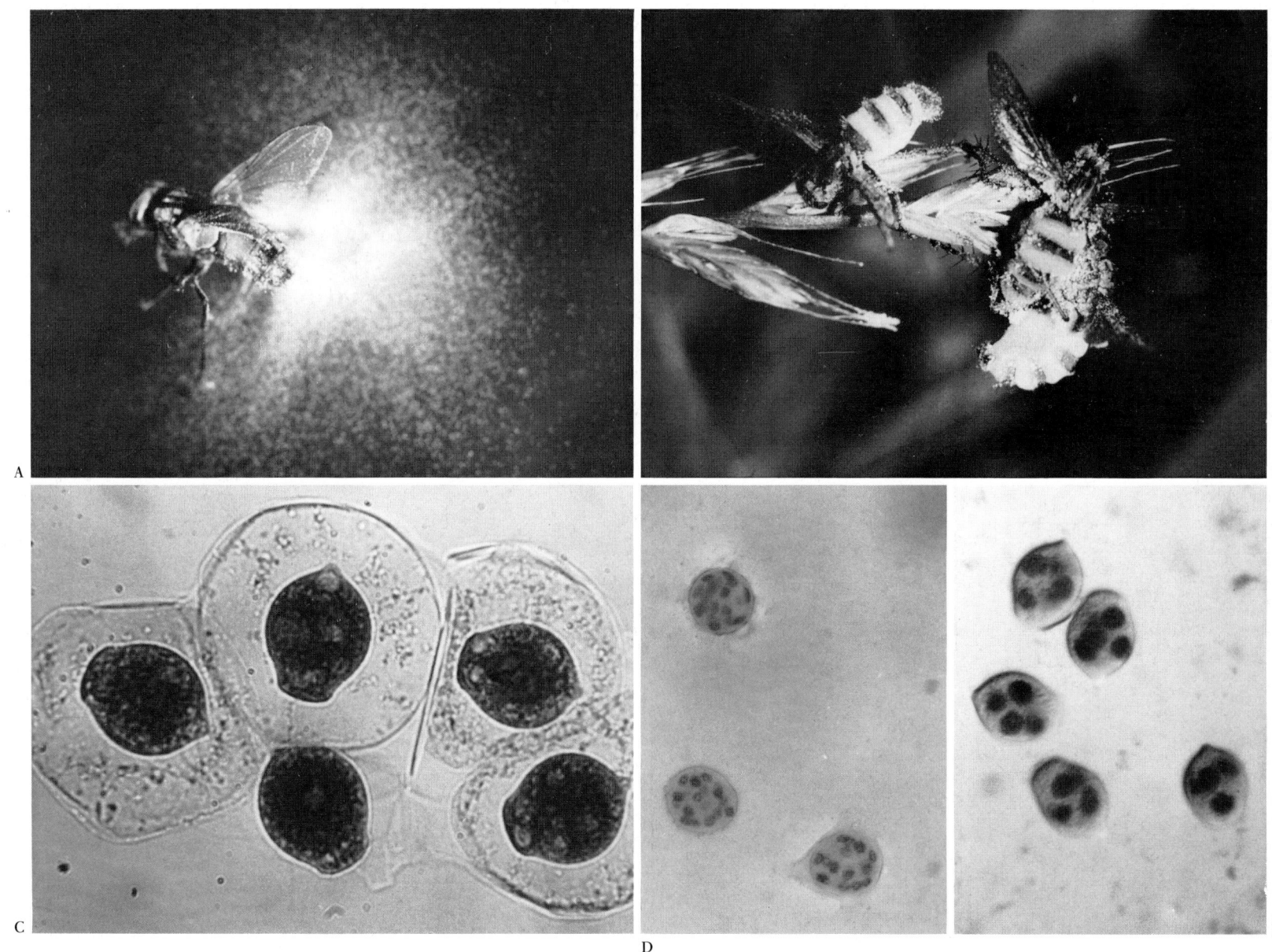

PLATE 19. *Entomophthora muscae*
A Secondary spore formation (×500), B resting spores (×700),
C–E protoplast stages from in vitro cultures of a strain attacking Psila
rosae (×2000).

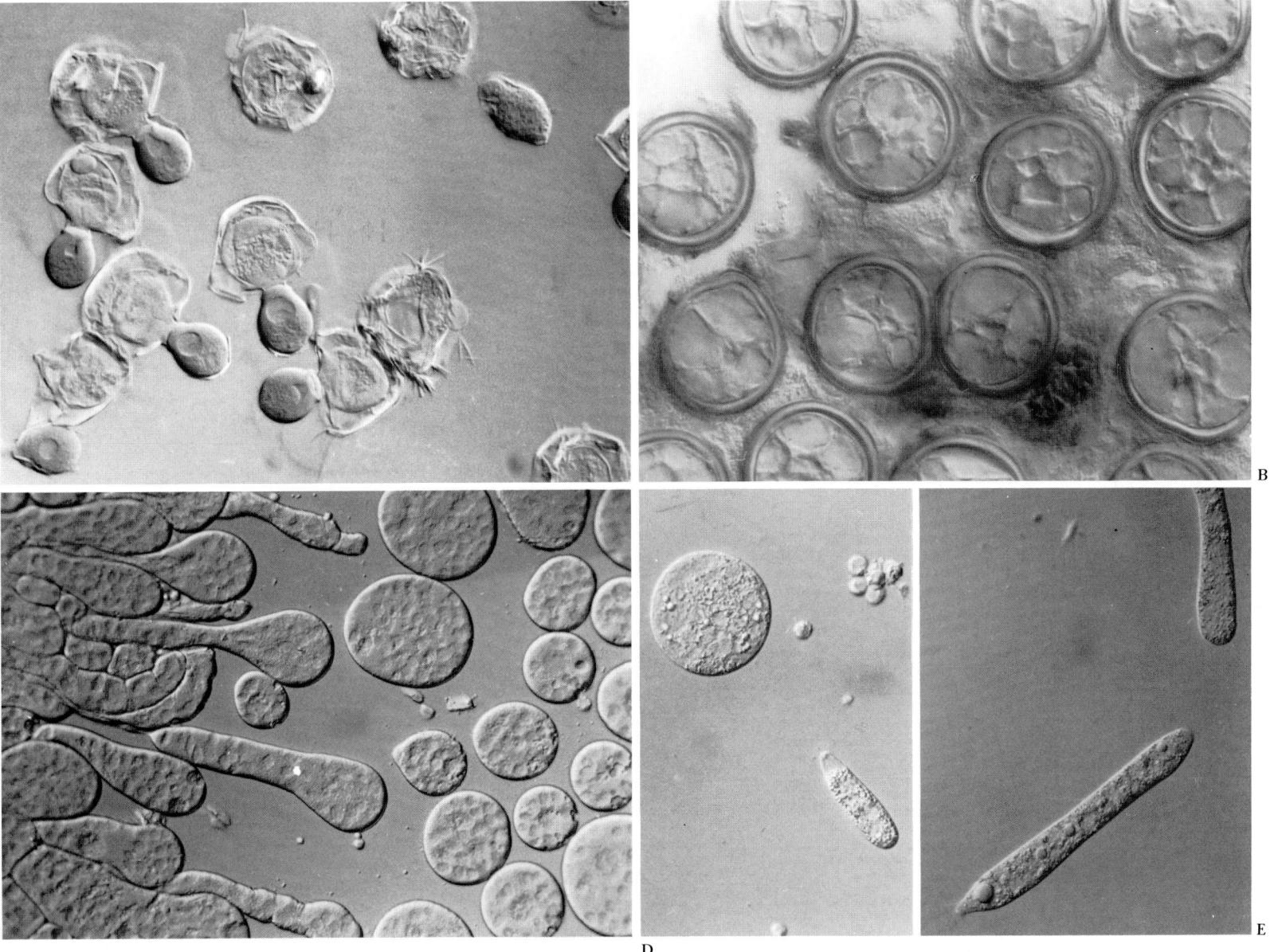

PLATE 20. *Entomophthora planchoniana*
A On aphids, B primary spore formation (×600), C primary spores (×700), D secondary spore formation (×700), E-F resting spores (×500).

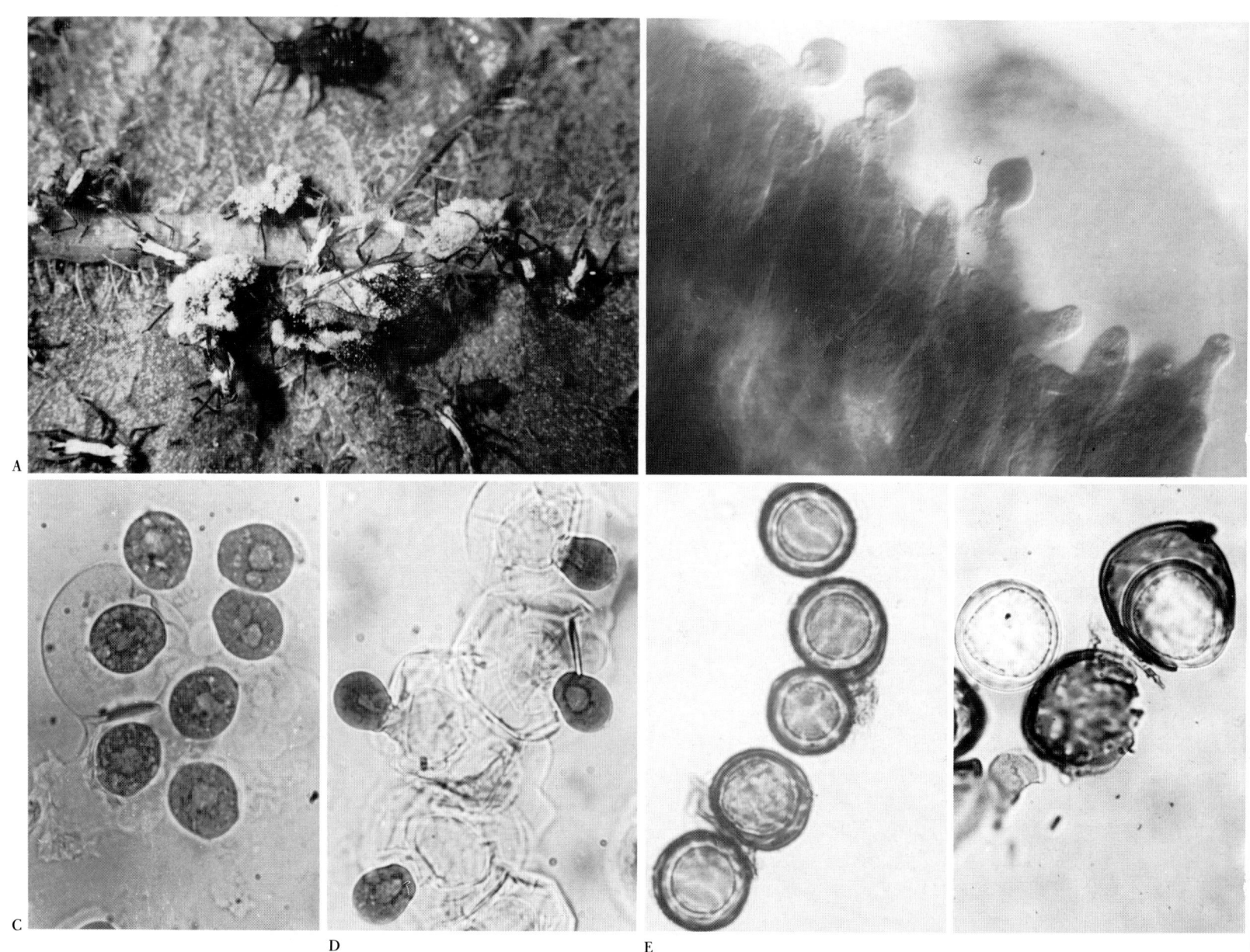

PLATE 21. *Erynia aquatica/ Entomophthora culicis*
A-B *Erynia aquatica*, A on diptera, B primary spores (× 1000);
C-E *Entomophthora culicis*, C on diptera, D-E primary spores (×900).

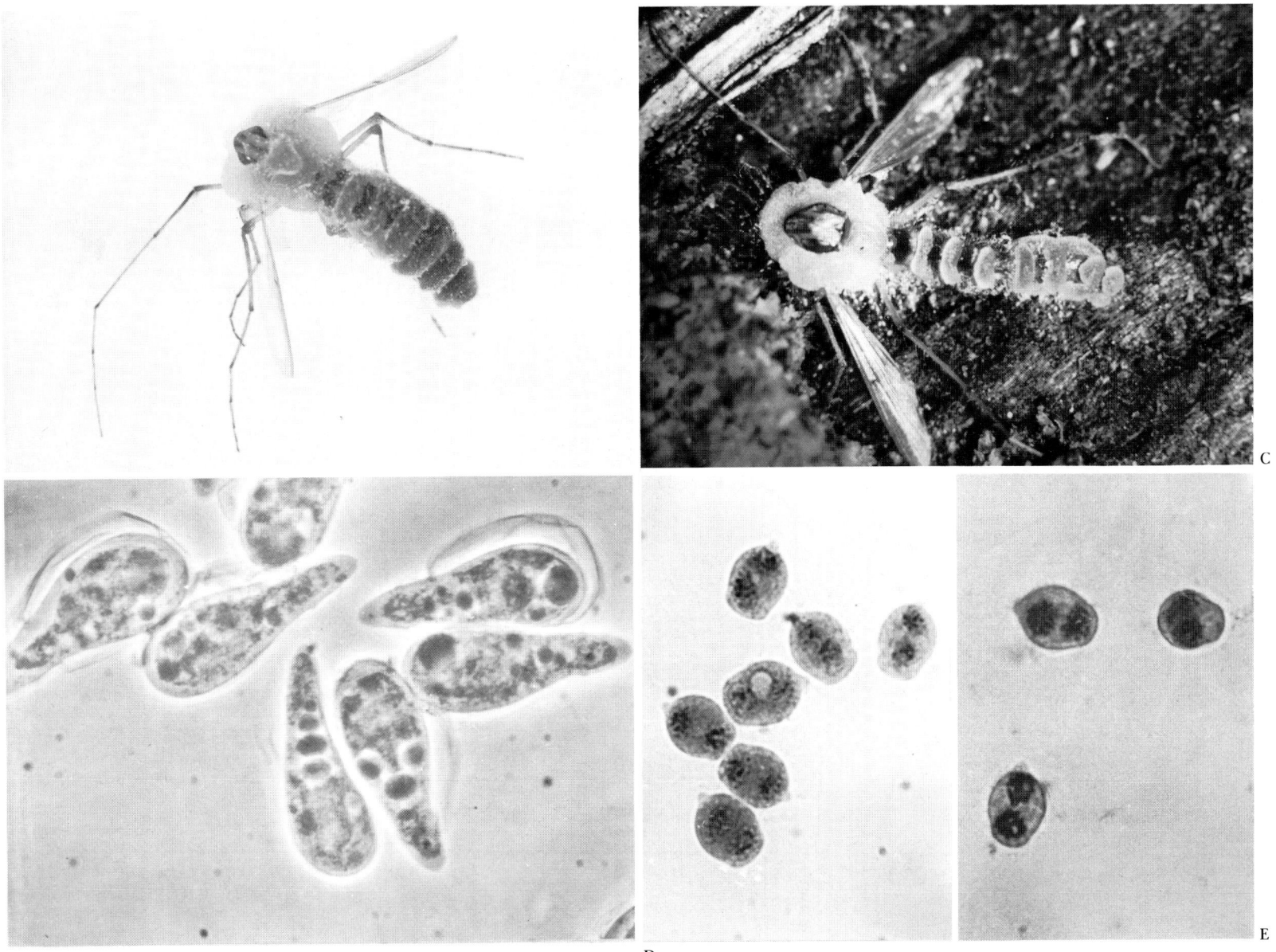

PLATE 22. *Erynia castrans/Massospora cicadina*
A-E *Erynia castrans*, A on flies with typical hole in the abdomen, B primary spores (×600) with double wall, C nuclear staining of the primary spores with one nucleus (×600), D resting spores (×800), E-G *Massospora cicadina*, E on Magicicada (Cicadidae), note open abdomen; F-G ornamented resting spores (×800).

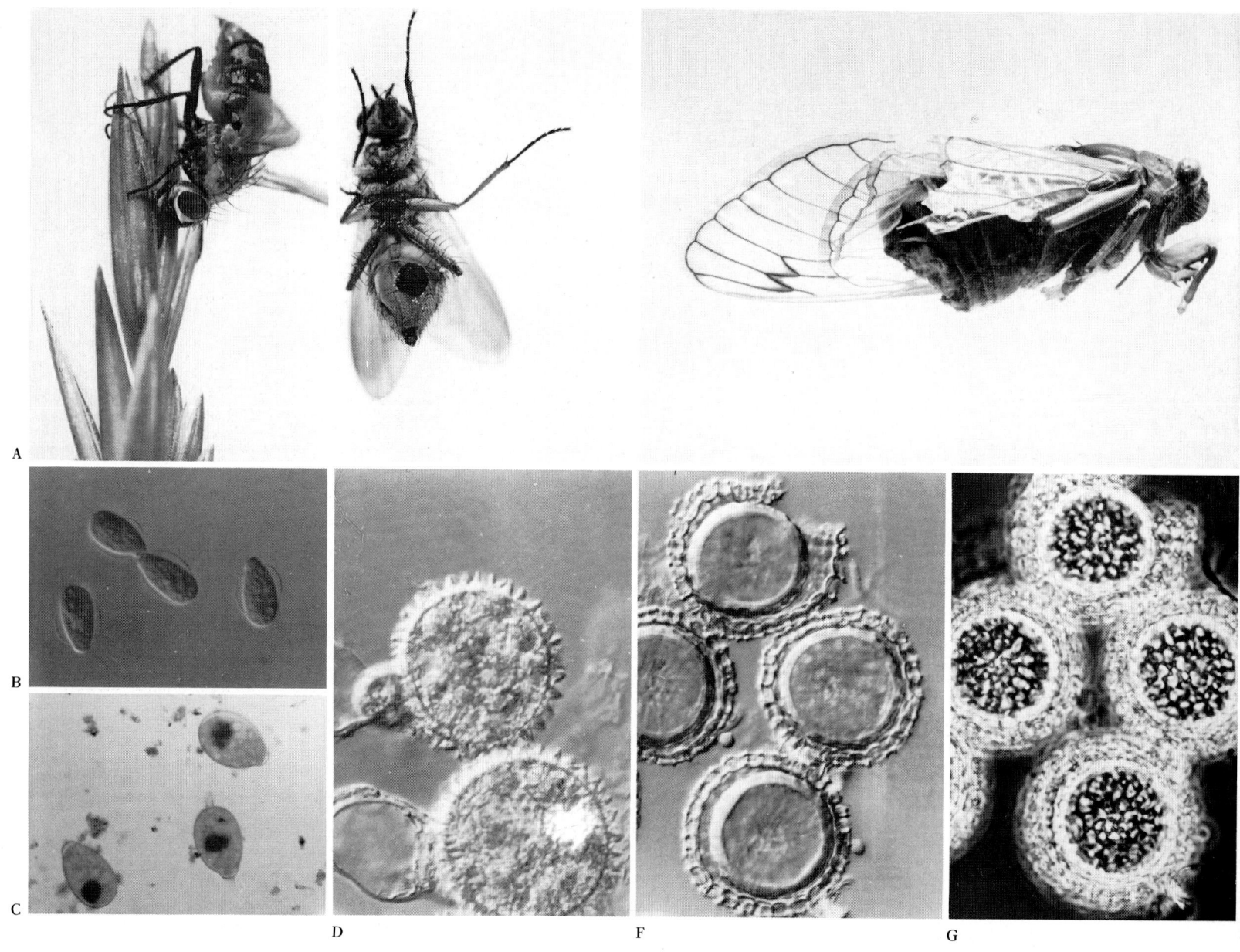

PLATE 23. *Erynia conica*
A On small aquatic-dwelling diptera, B aerial primary curved spores,
C-D aerial secondary spore formation, E aquatic primary coronate
spore, F aquatic secondary coronate spore (all × 400).

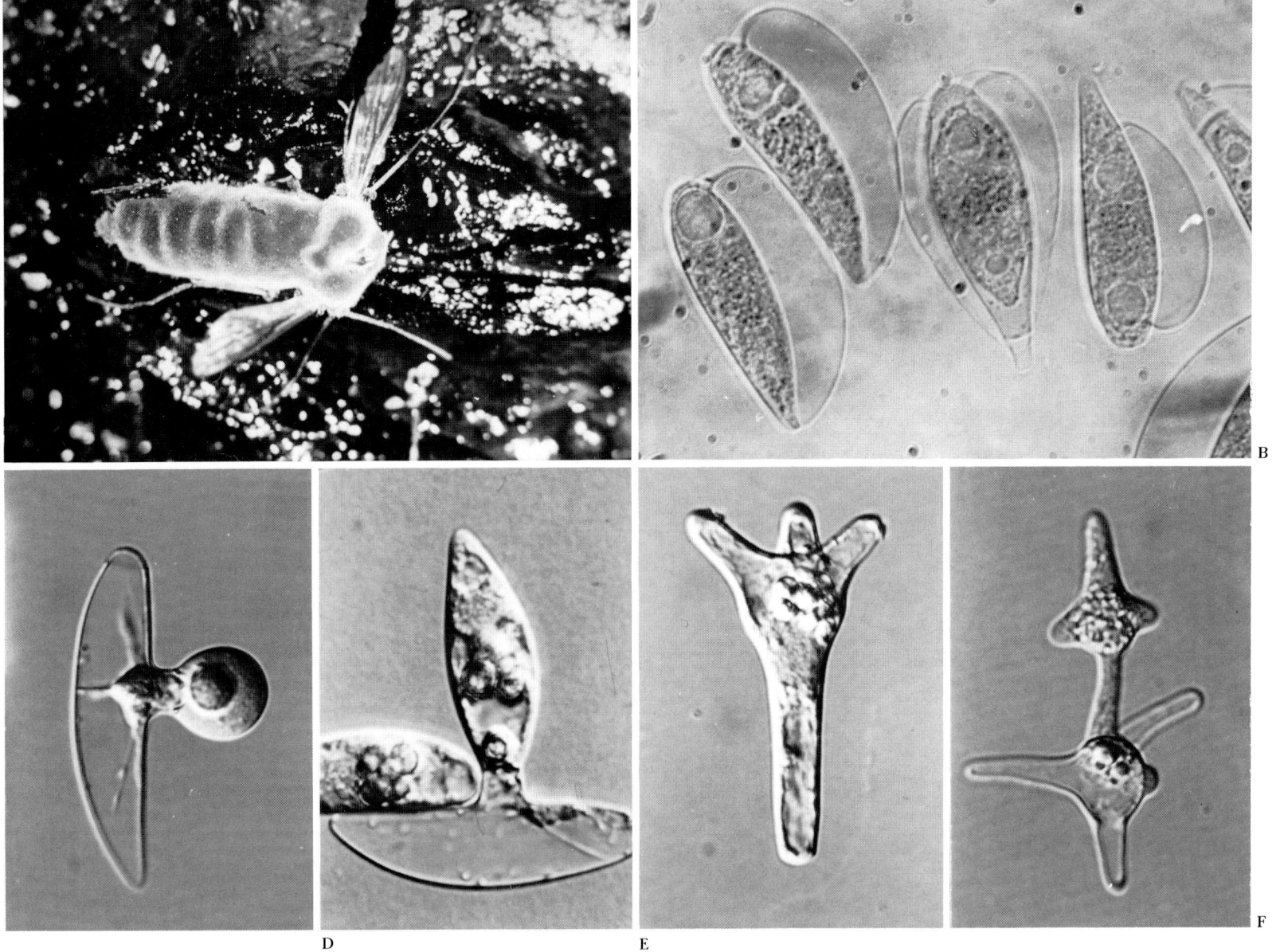

PLATE 24. *Erynia dipterigena / E. blunckii*
A–C *E. dipterigena*, A on small fly, B primary spores (×500), C secondary spore formation (×500), D–F *E. blunckii*, D on larva of *Athalia rosae*, E primary spores (×800), F resting spores (×800).

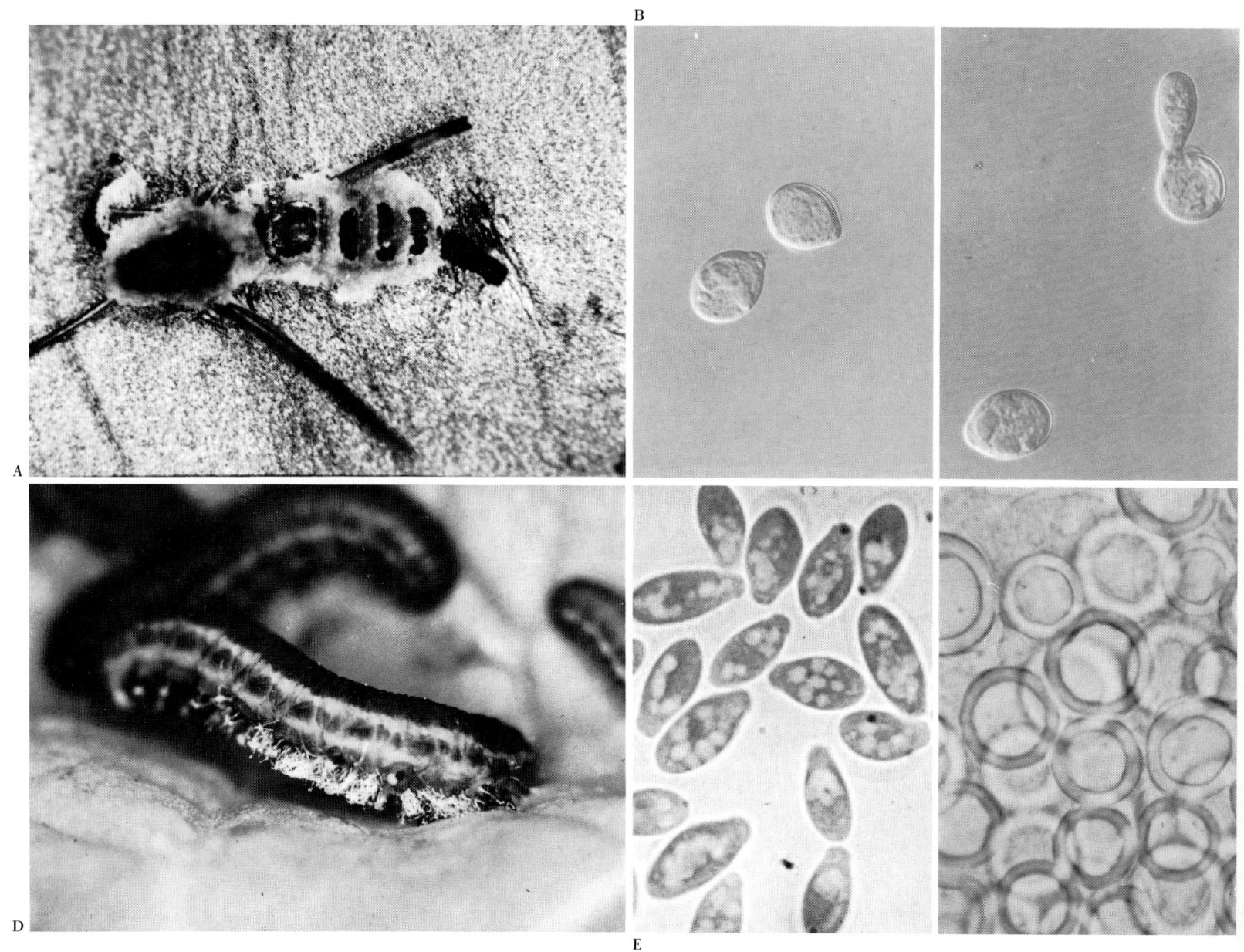

PLATE 25. *Erynia elateridiphaga*
A–B On *Agrotis sputator*, B host with rhizoids, C primary spores (×1200), D capillispore formation (×800), E capillispore (×1000), F resting spores (×900).

PLATE 26. *Erynia gammae*
A On moth larva, B primary spores with secondary spore formation (×800), C resting spores (×600).

PLATE 27. *Erynia neoaphidis*
A On *Acyrthosiphon pisum*, B rhizoids, C-D primary spores,
E secondary spores, F secondary spore formation (C-F × 1000).

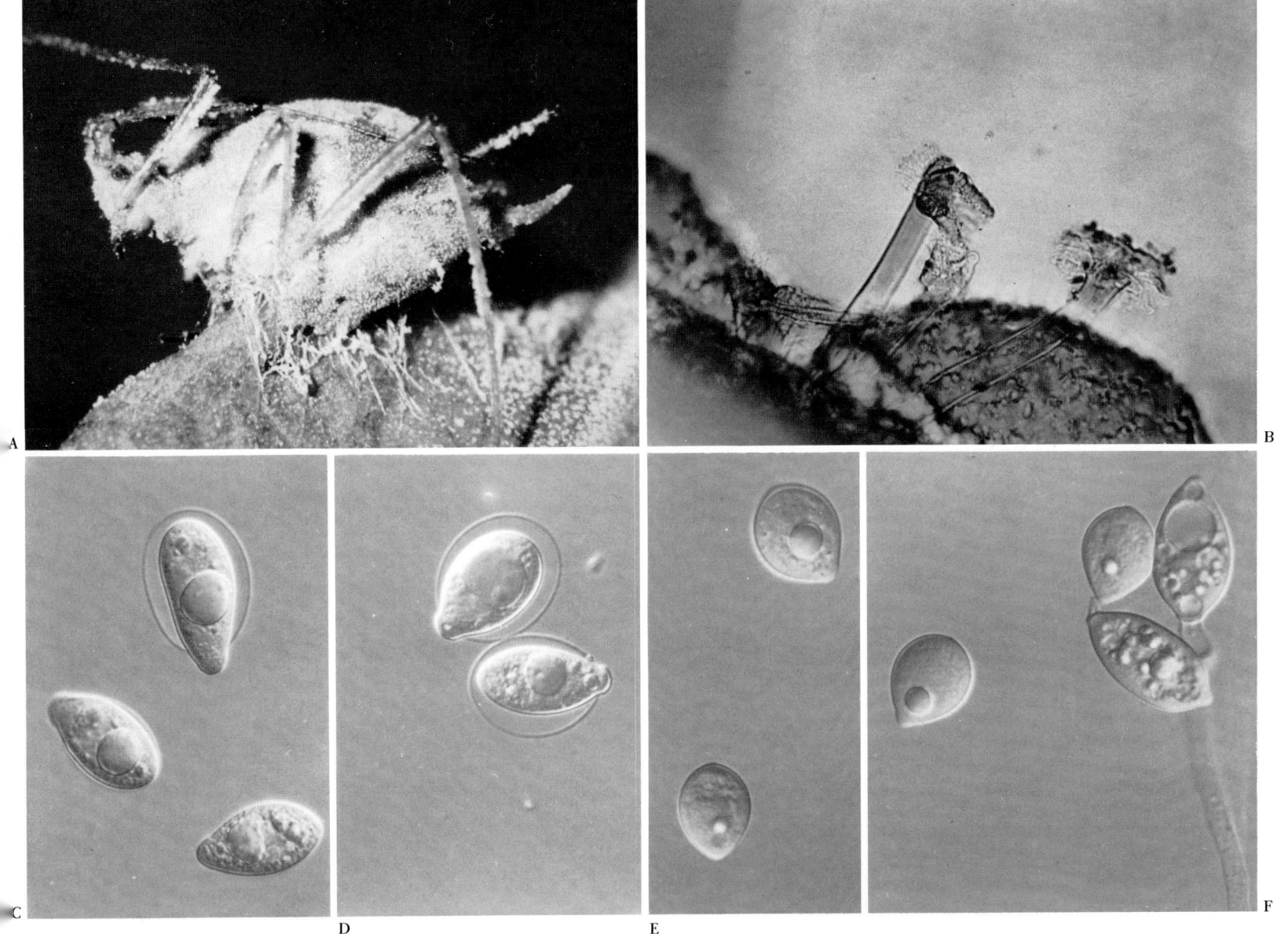

PLATE 28. *Erynia radicans*
A-B Two different hosts, C-D primary spores, E capillispores,
F resting spores (C-F ×900).

PLATE 29. *Erynia rhizospora/ E. plecopteri*
A–C *E. rhizospora*, A on caddis fly, B–C secondary spore formation
(×600 and ×800 resp.), D–F *E. plecopteri*, D on stone fly, E aerial
primary spores (×800), F aquatic coronate primary spore (×1500).

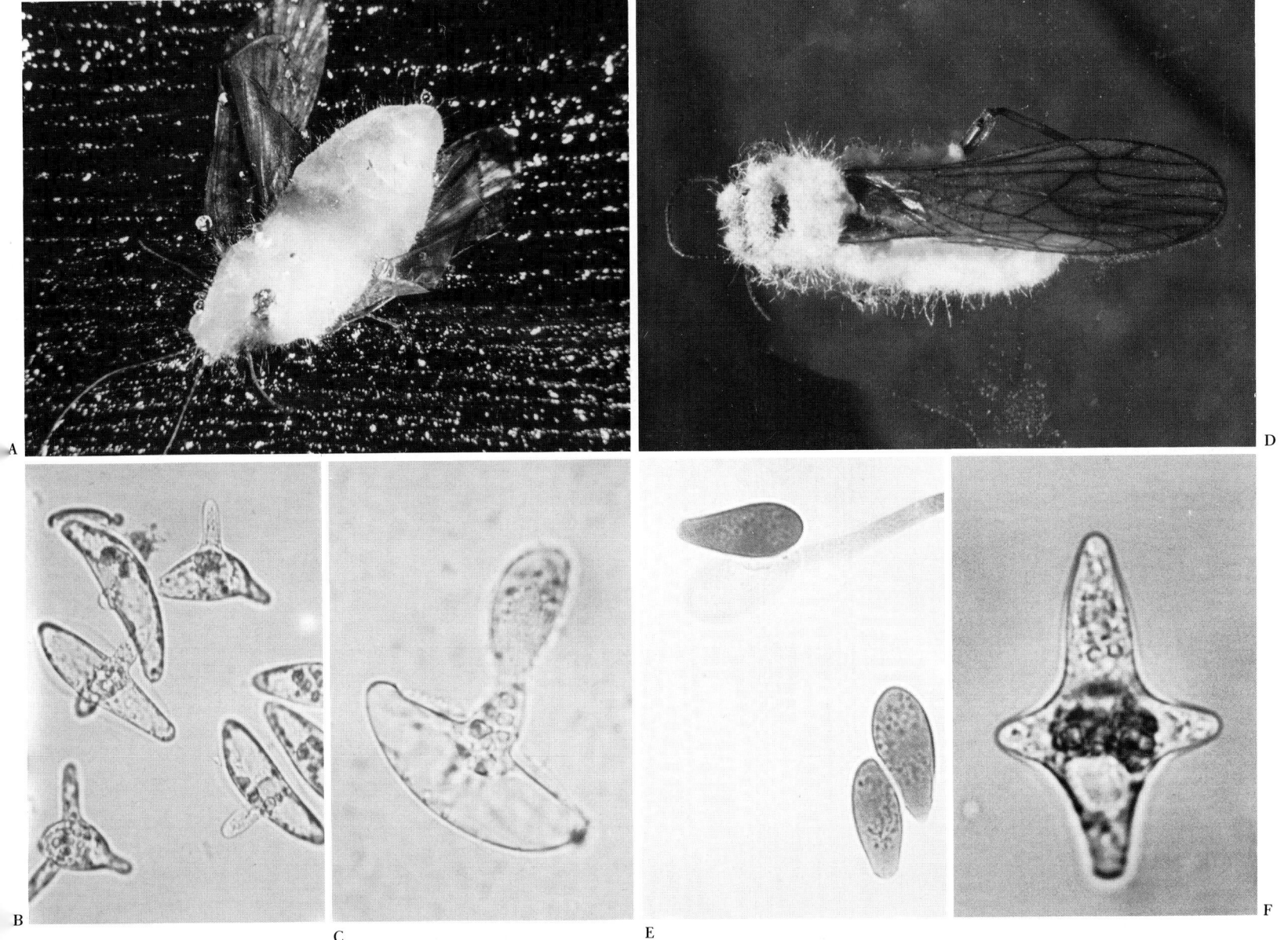

PLATE 30. *Erynia virescens*
A On larva of *Agrotis segetum* (Noctuidae), B primary spores, C nuclear staining of primary spores, D-E resting spores (all ×500).

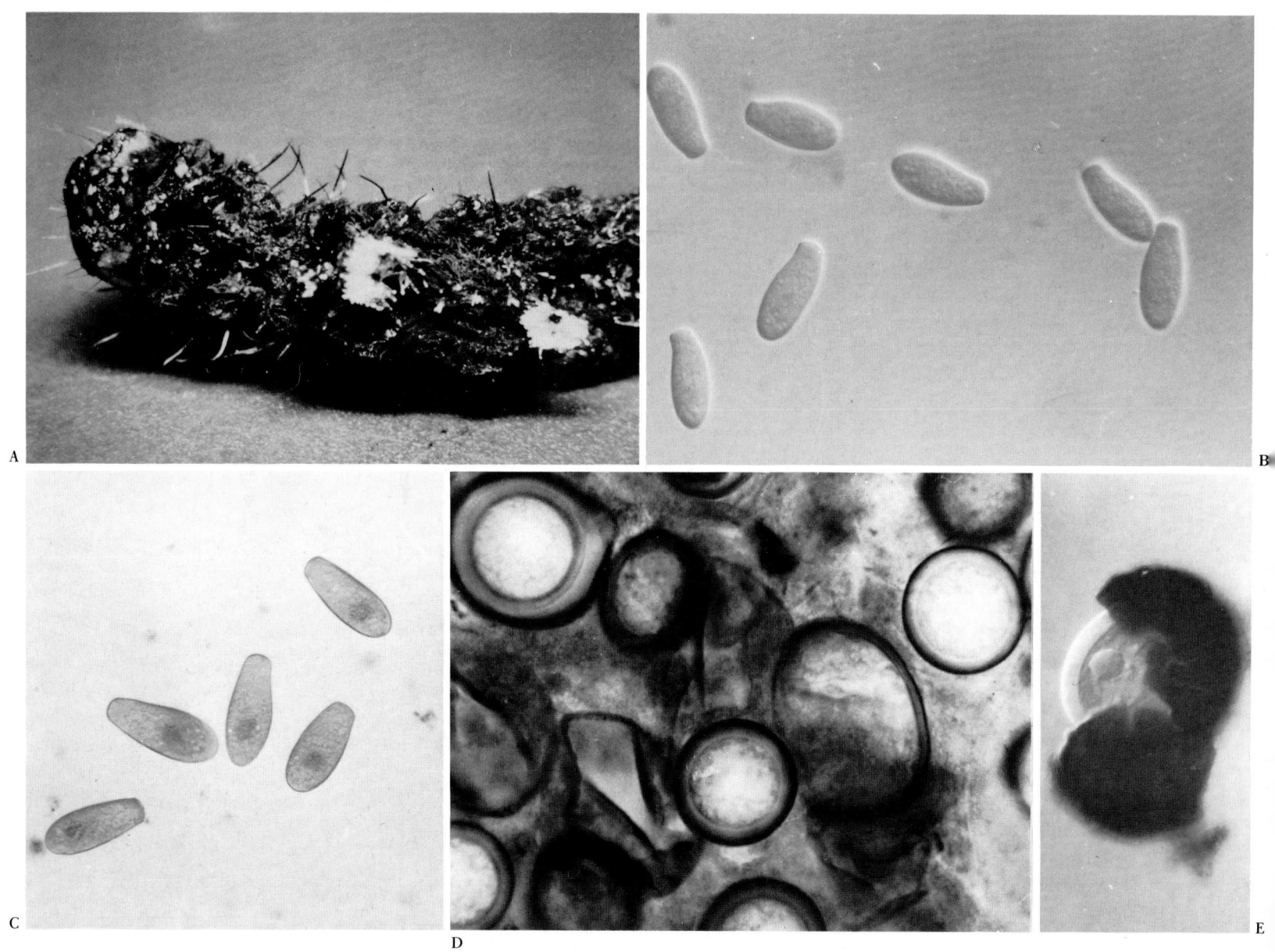

PLATE 31. *Neozygites adjarica/N. fumosa*
A On *Tetranychus urticae* with resting spores, B primary spores, C capillispore formation, D-G *N. fumosa*, D-E capillispore formation showing the curved capillisporophore (E-F) and the adhesive tip of the capillispore (D), G secondary capillispore formation (all × 800).

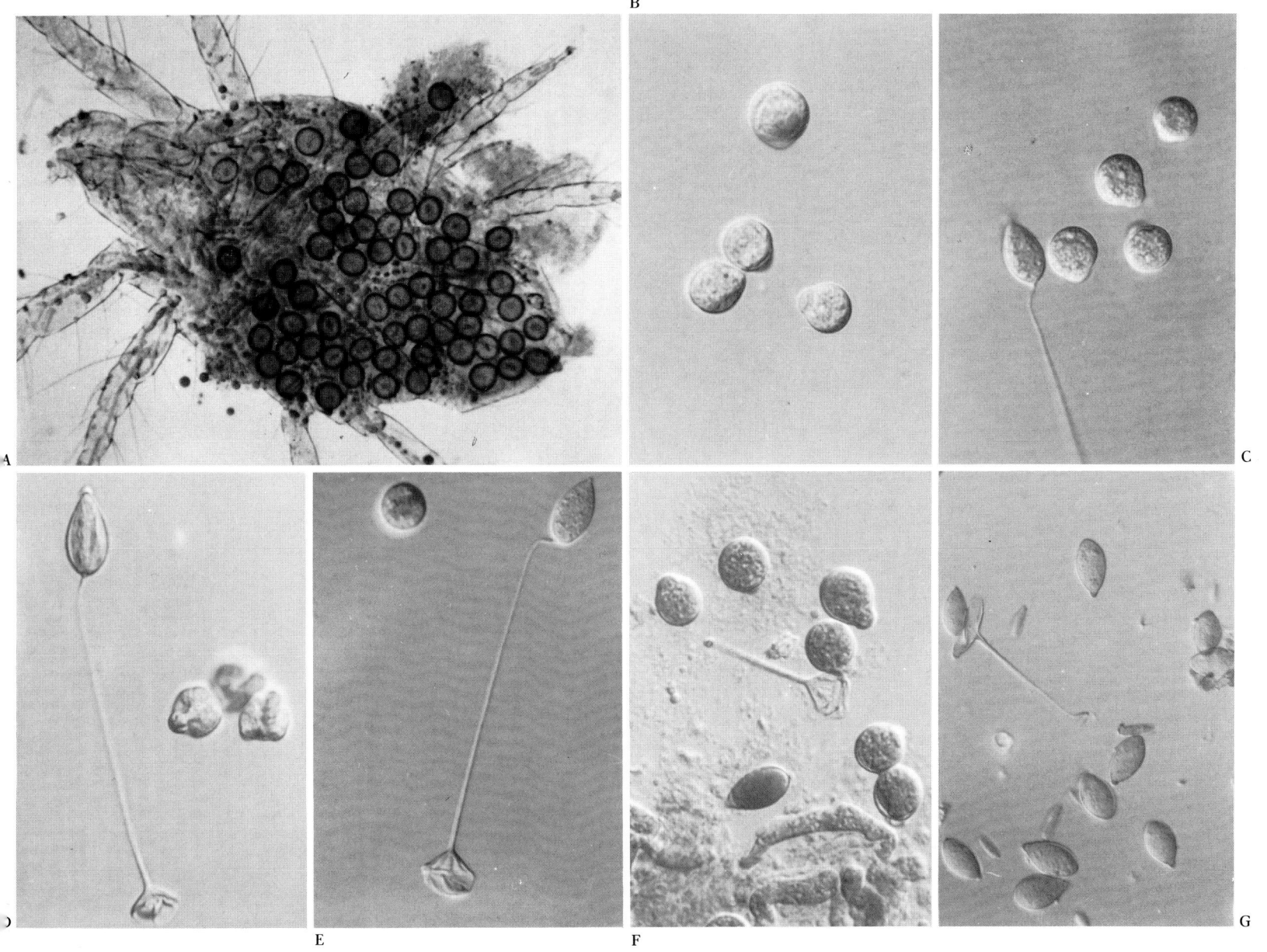

PLATE 32. *Neozygites fresenii*
A On *Aphis fabae*, B primary spores, C-E capillispore formation and capillispores, F resting spores (all ×500).

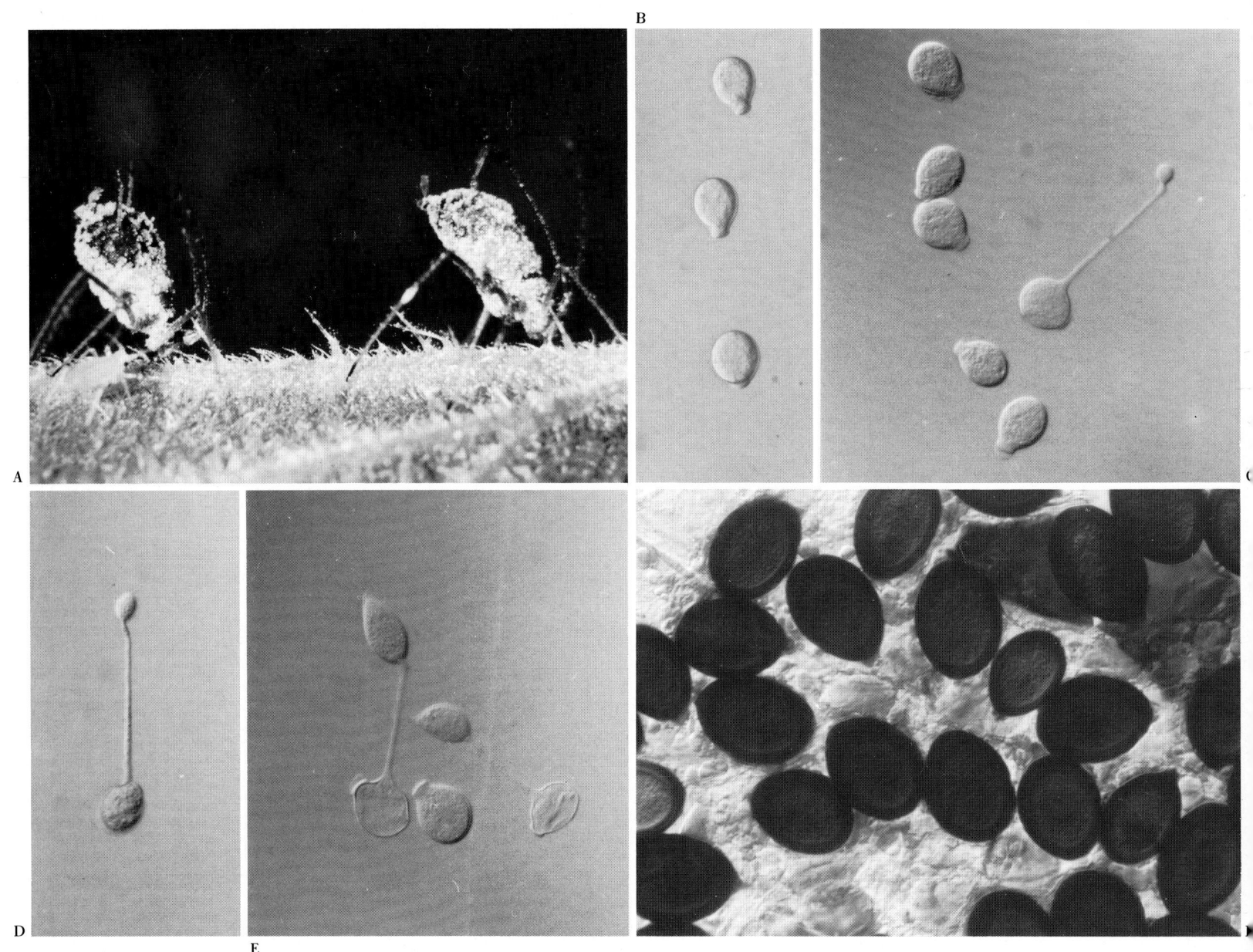

PLATE 33. *Sporodiniella umbellata*
A Sporangiophore formation on a membracid *(Umbonia* sp.), B sporangiophore (×75), C-D sporangia and spores (×750), E zygospores (×200).

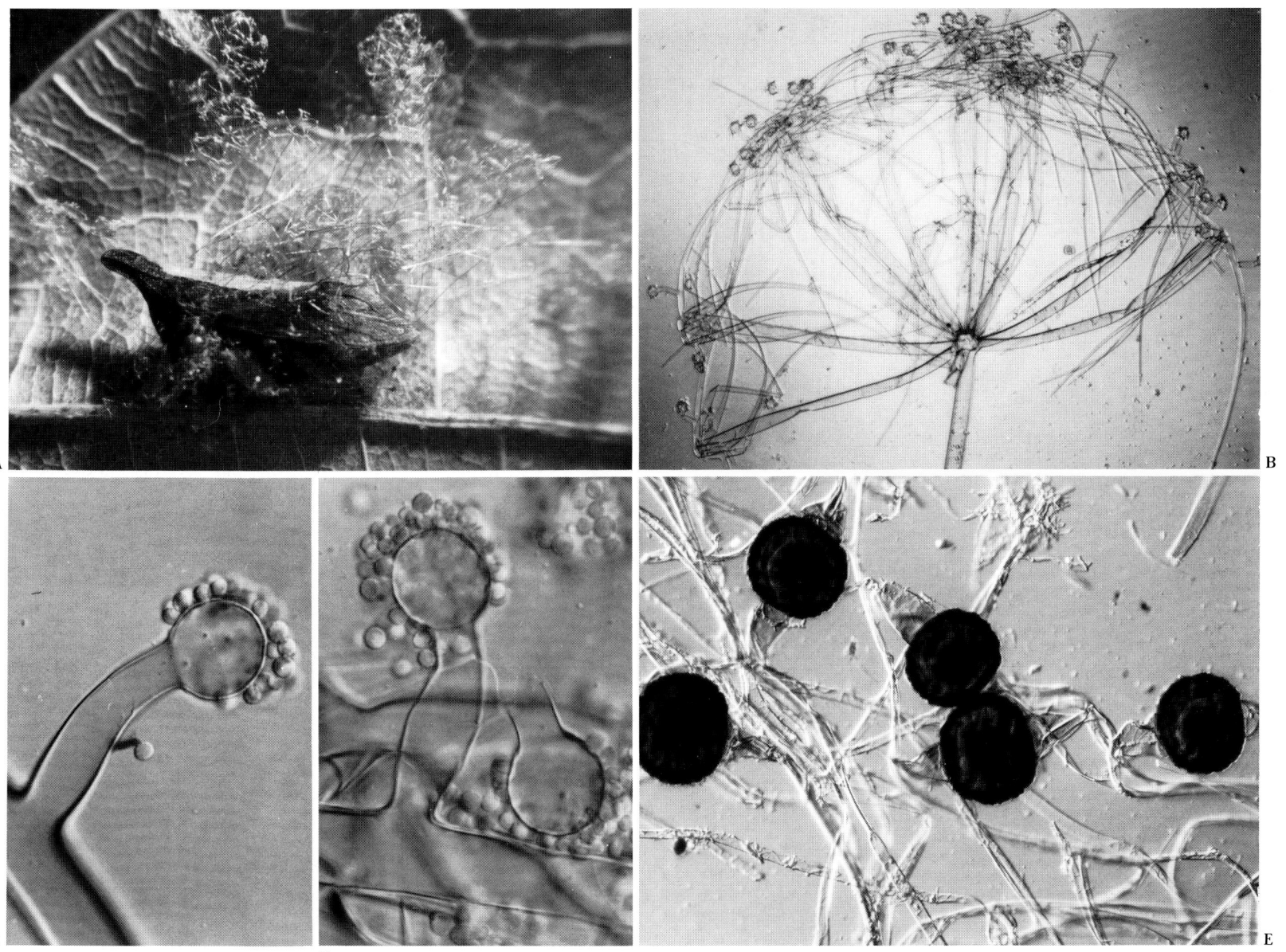

PLATE 34. *Ascosphaera aggregata*
A On larvae of *Megachile rotundata*, B-C ascospores occurring in characteristic balls (× 750), D ascospores (× 1875).

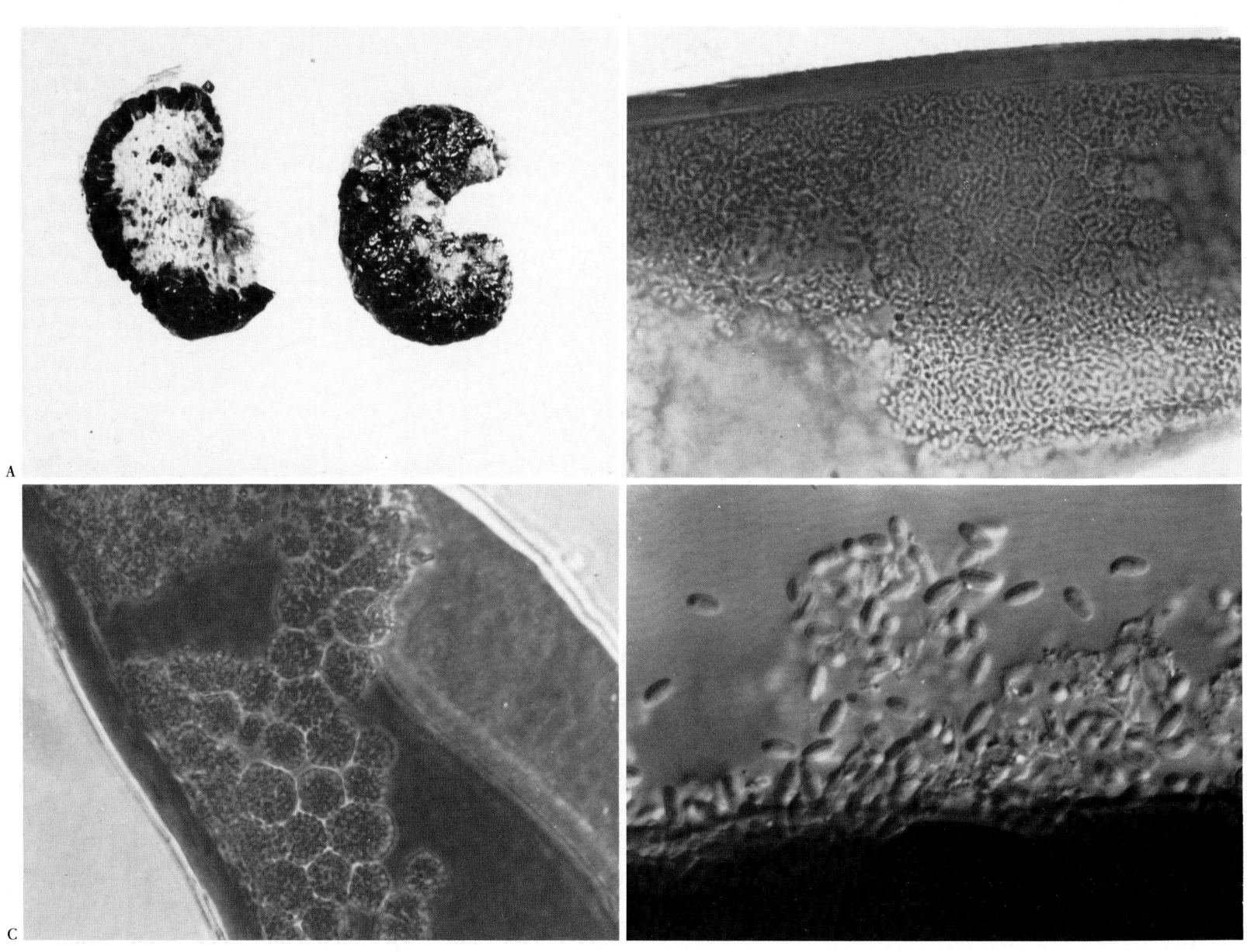

PLATE 35. *Ascosphaera apis*
A On larvae of an *Apis* sp., B-C globose cyst-like ascoma containing ascospores (×750), D ascospores (×1875).

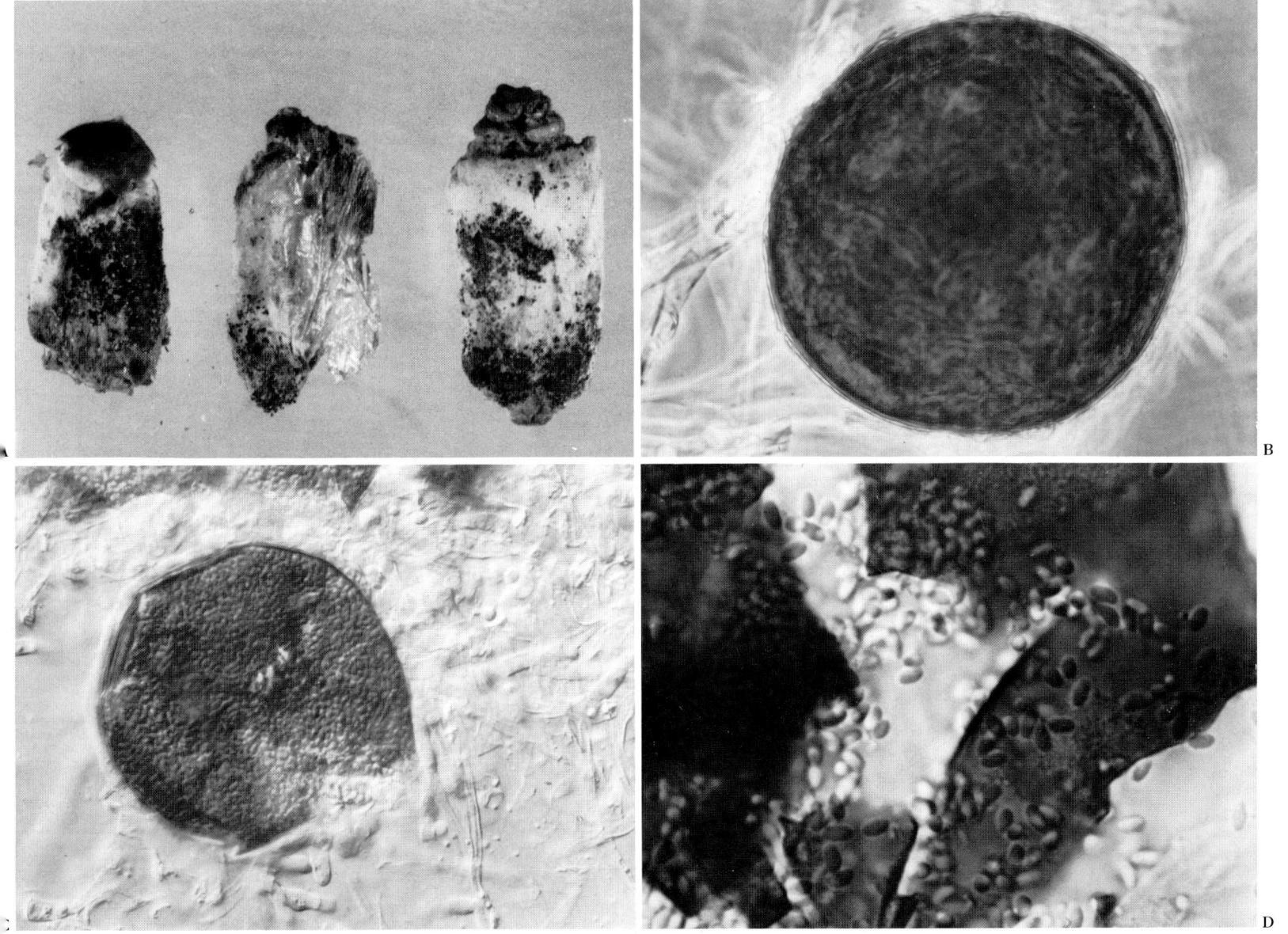

PLATE 36. *Calonectria pruinosa*
A Ascomata and mycelial development on adult homopteran (Cicadellidae), B ascoma (× 200), C asci (× 750), D ascospores (× 750).

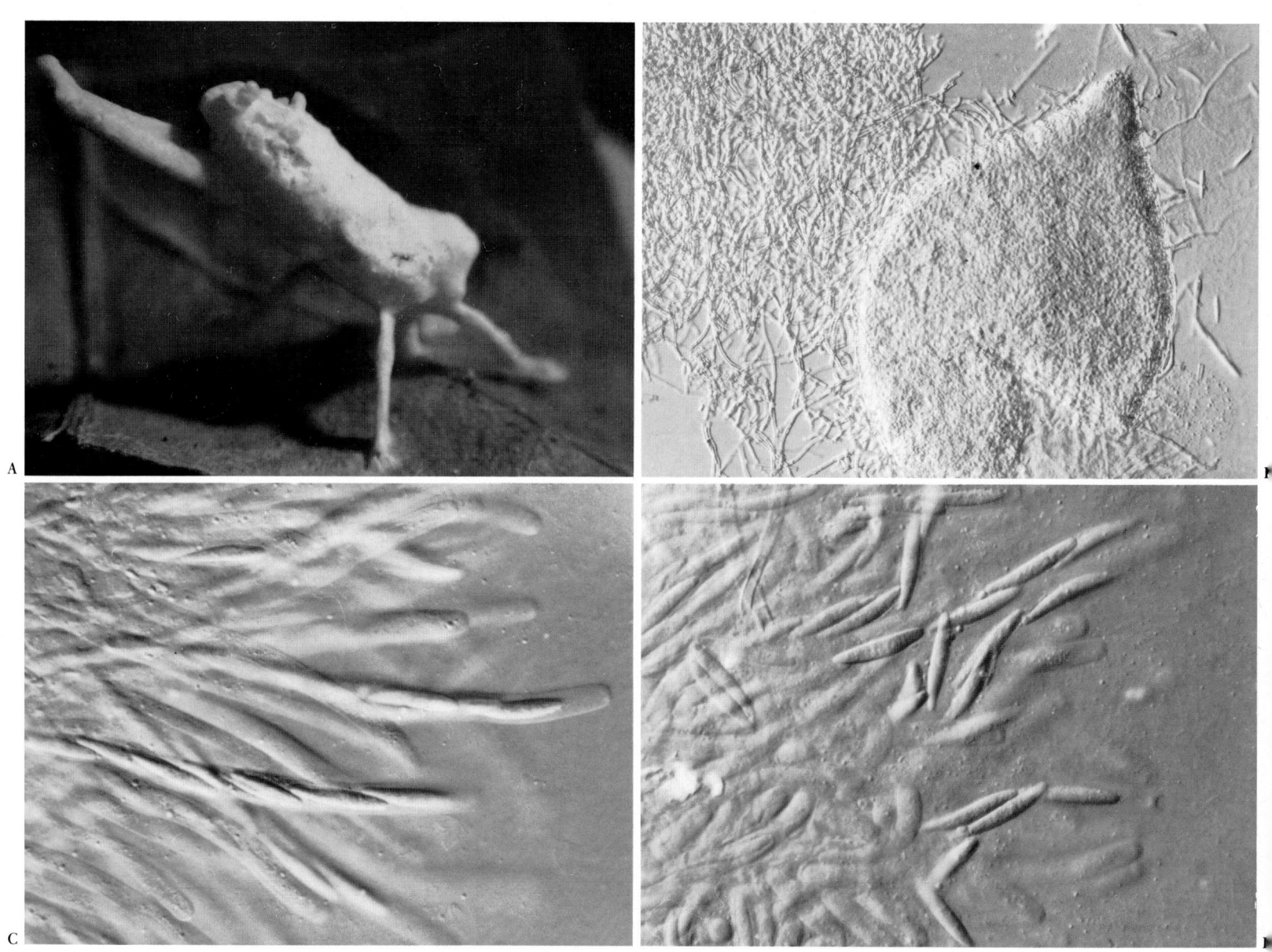

PLATE 37. *Cordycepioides octosporus* / *C. bisporus* A-D *C. octosporus*, A on termite (*Tenvirostitermes* sp.) (× 8), B ascus development in ascoma (× 450), C ascus (× 320), D ascospores (× 550), E *C. bisporus*, ascospores (× 800).

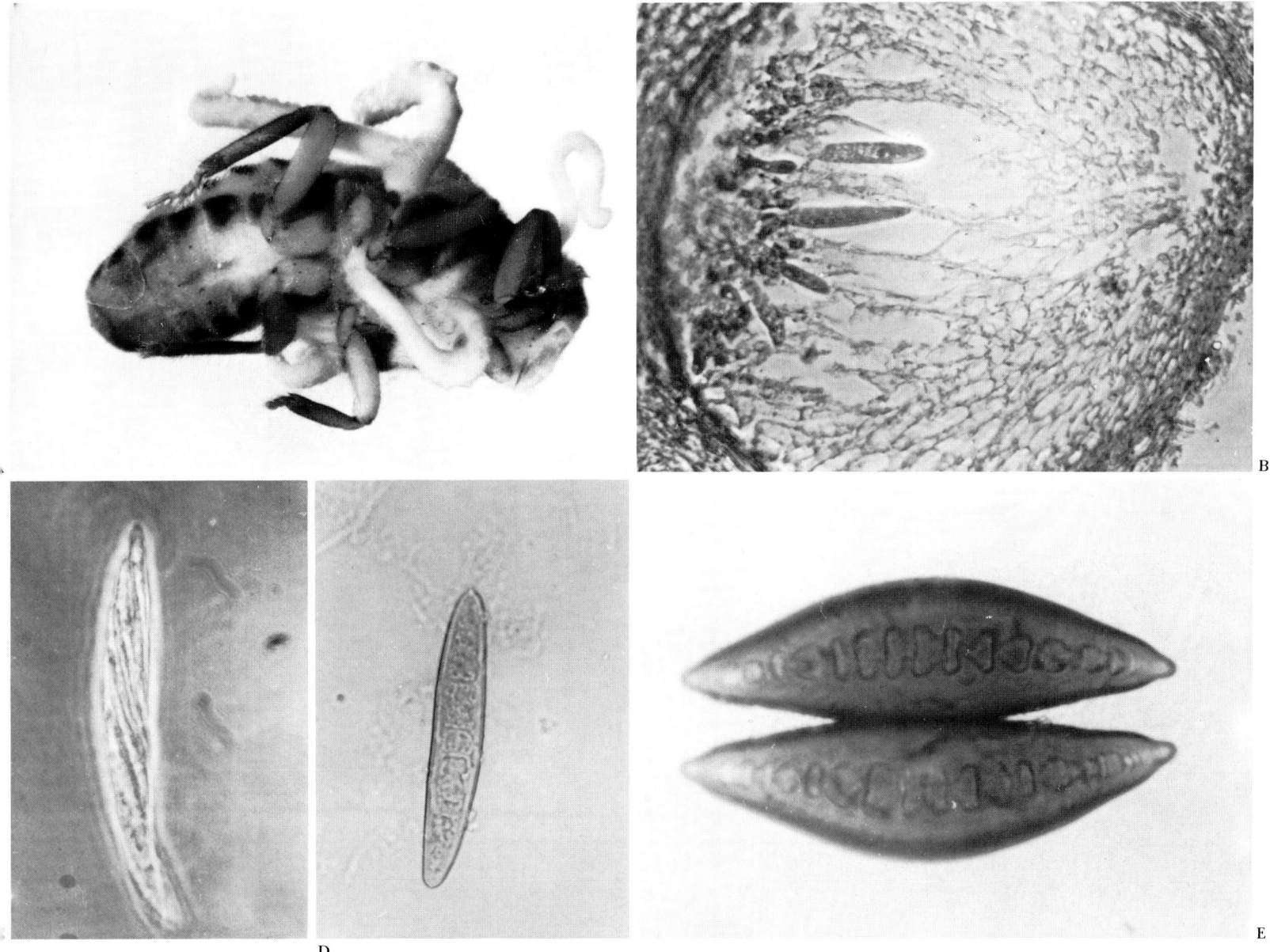

PLATE 38. *Cordyceps australis*
A Clava on ponerine ant *(Paltothyreus tarsatus)*, B fertile head,
C stroma showing ascomata (×75), D ascoma with asci (×200),
E ascospores disarticulating into part spores (×750).

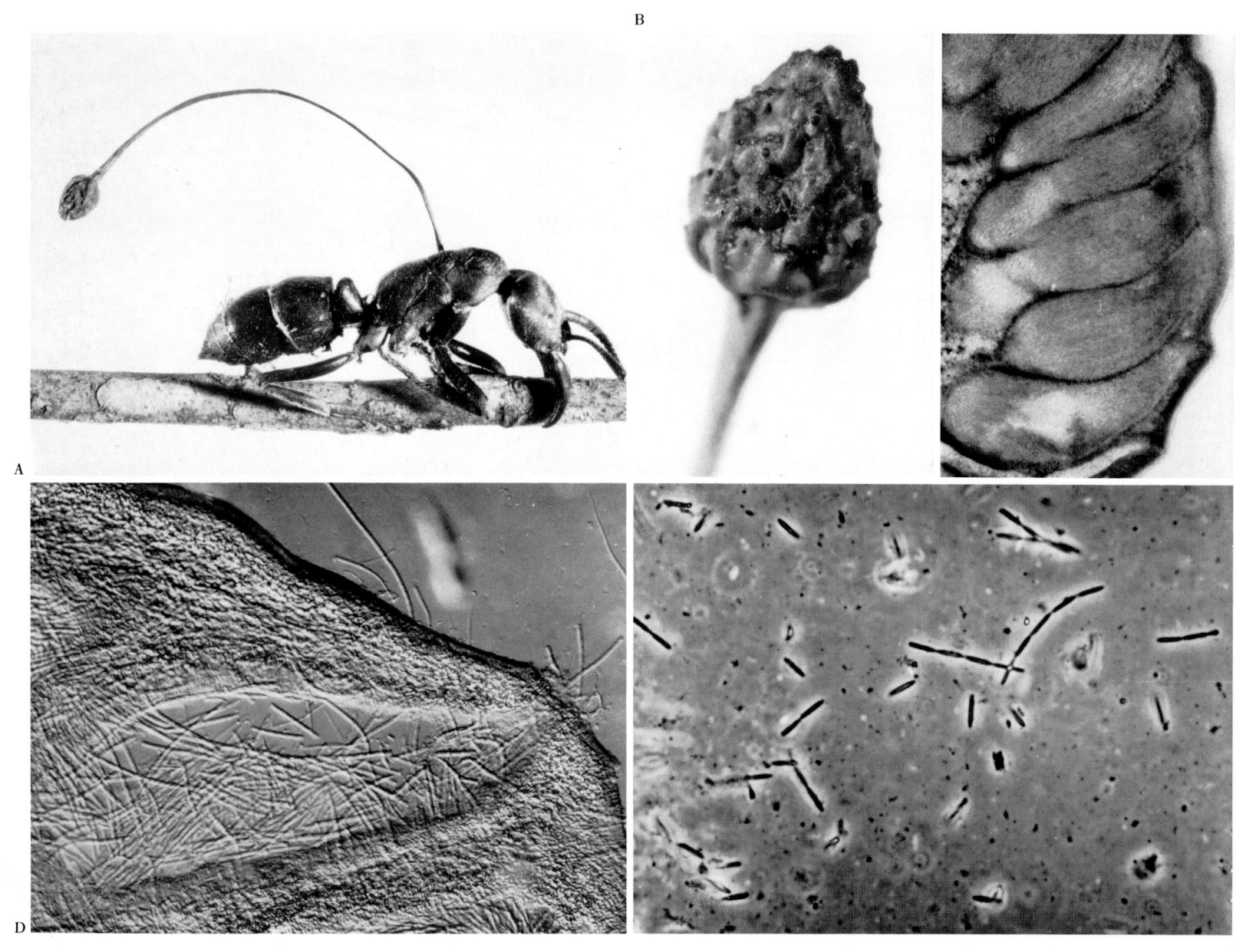

PLATE 39. *Cordyceps calocerioides*
A Two clavae arising from a large trapdoor spider (Mygalidae),
B stroma showing ascomata (× 75), C ascoma with asci (× 200),
D asci and ascospores (× 200).

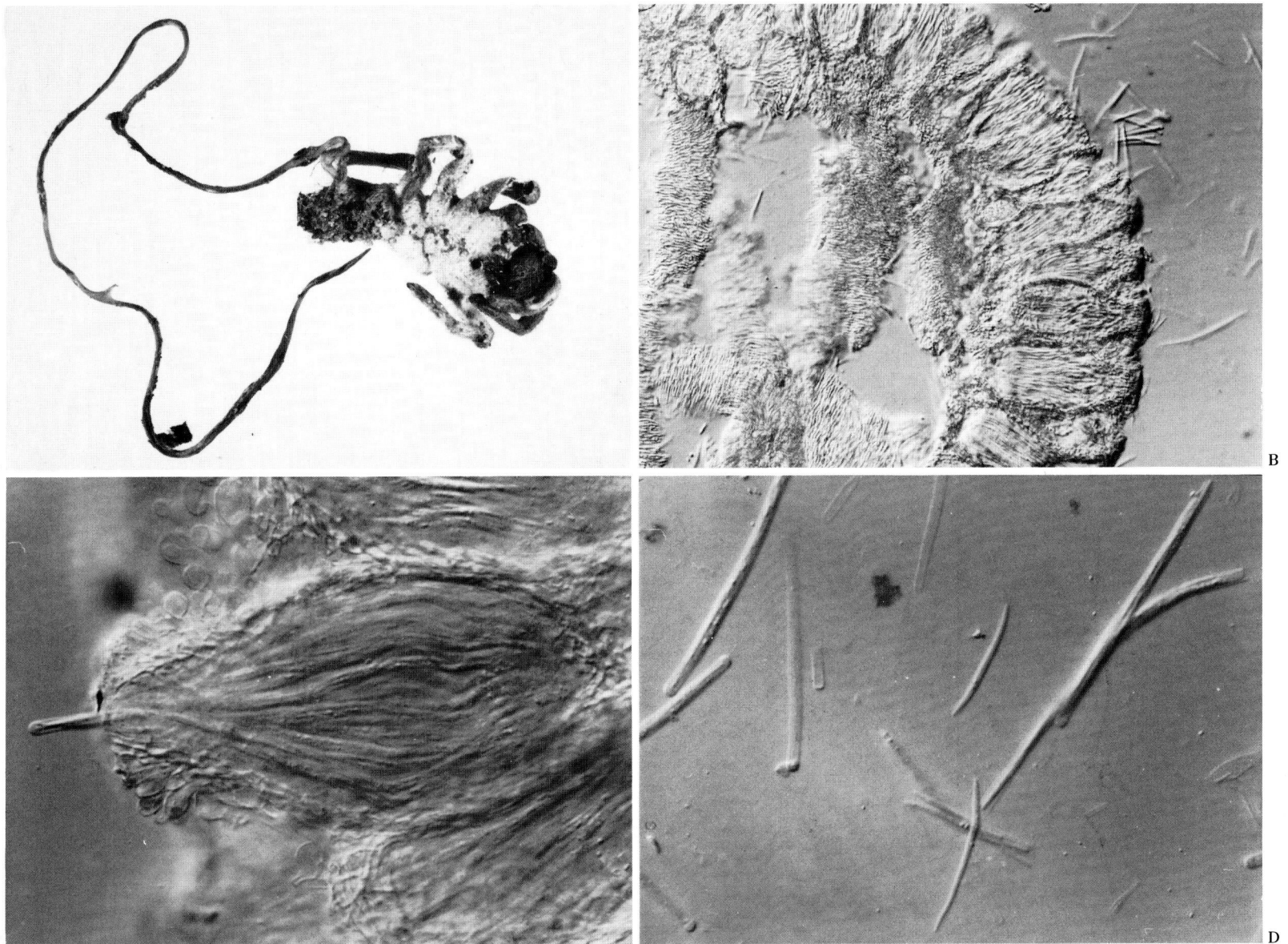

PLATE 40. *Cordyceps gunnii*
A Stroma removed from burrowing lepidopteran larva *(Neleus* sp.),
B ascoma containing asci (× 200), C–D multiseptate ascospores
(× 750).

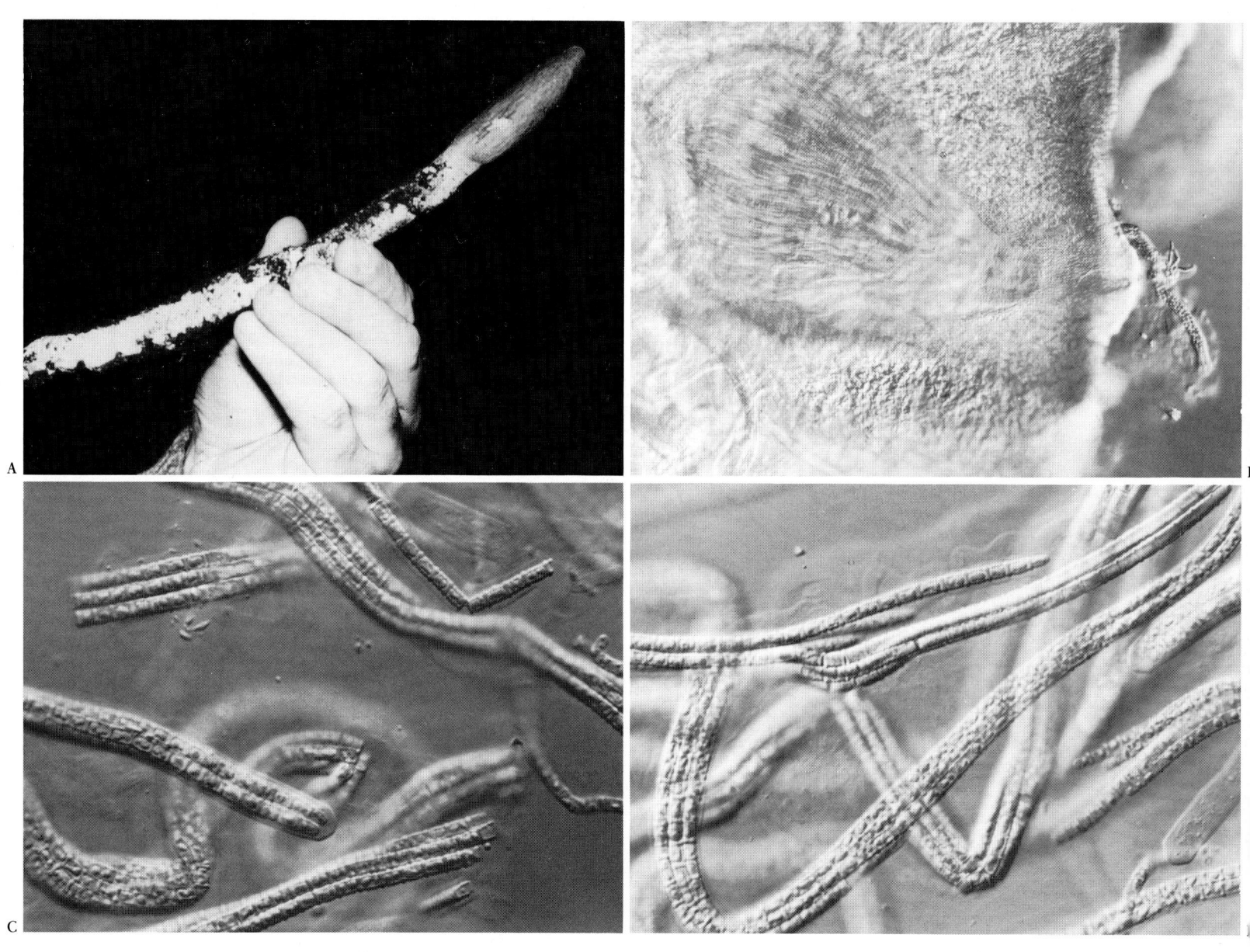

PLATE 41. *Cordyceps lloydii*
A Clava and synnema formation on formicine ant *(Camponotus* sp.), B clavae formation of the var. *binata* on ant host, C-D ascomata in stroma (×75 and ×200 resp.), E asci (×750), F hyphal bodies within host (×750).

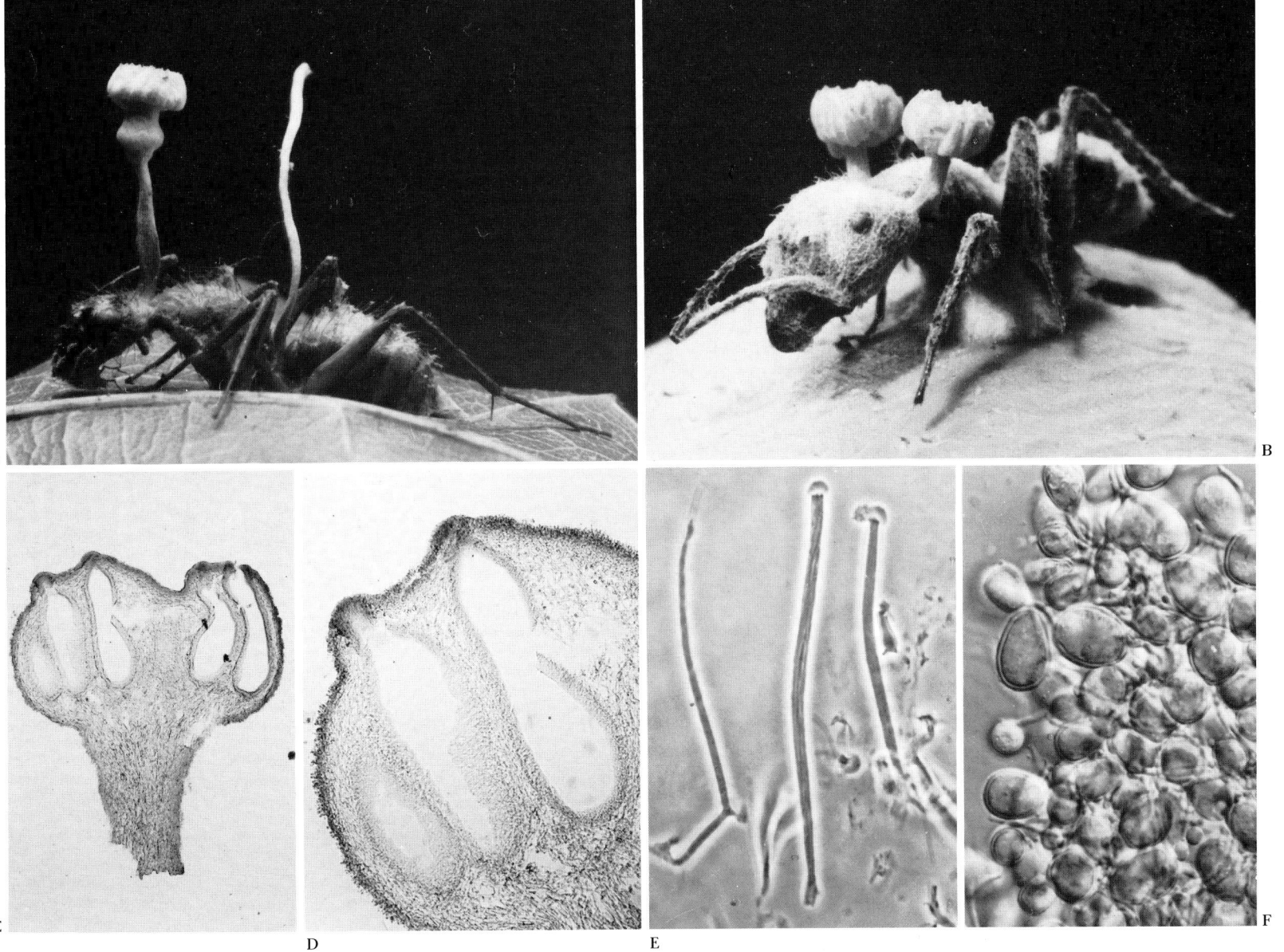

PLATE 42. *Cordyceps martialis*
A Clavae on coleopteran larvae (Elateridae), B structure of stroma (×750), C asci (×750), D ascospores which have been disarticulated into part spores (×750).

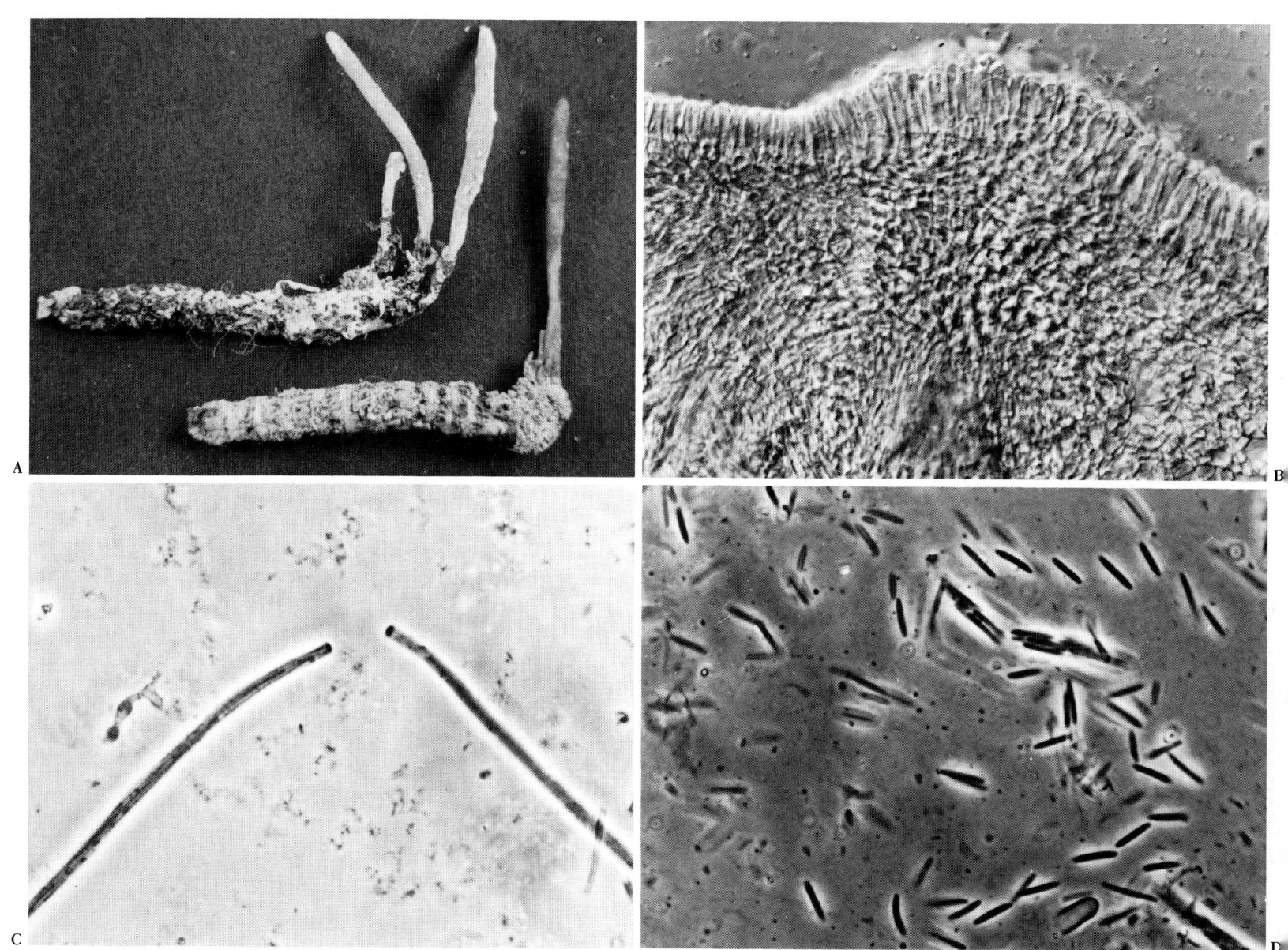

PLATE 43. *Cordyceps militaris*
A Clavae on lepidopteran pupae, B ascomata in squash preparation (×75), C ascoma (×200), D-E asci (×750).

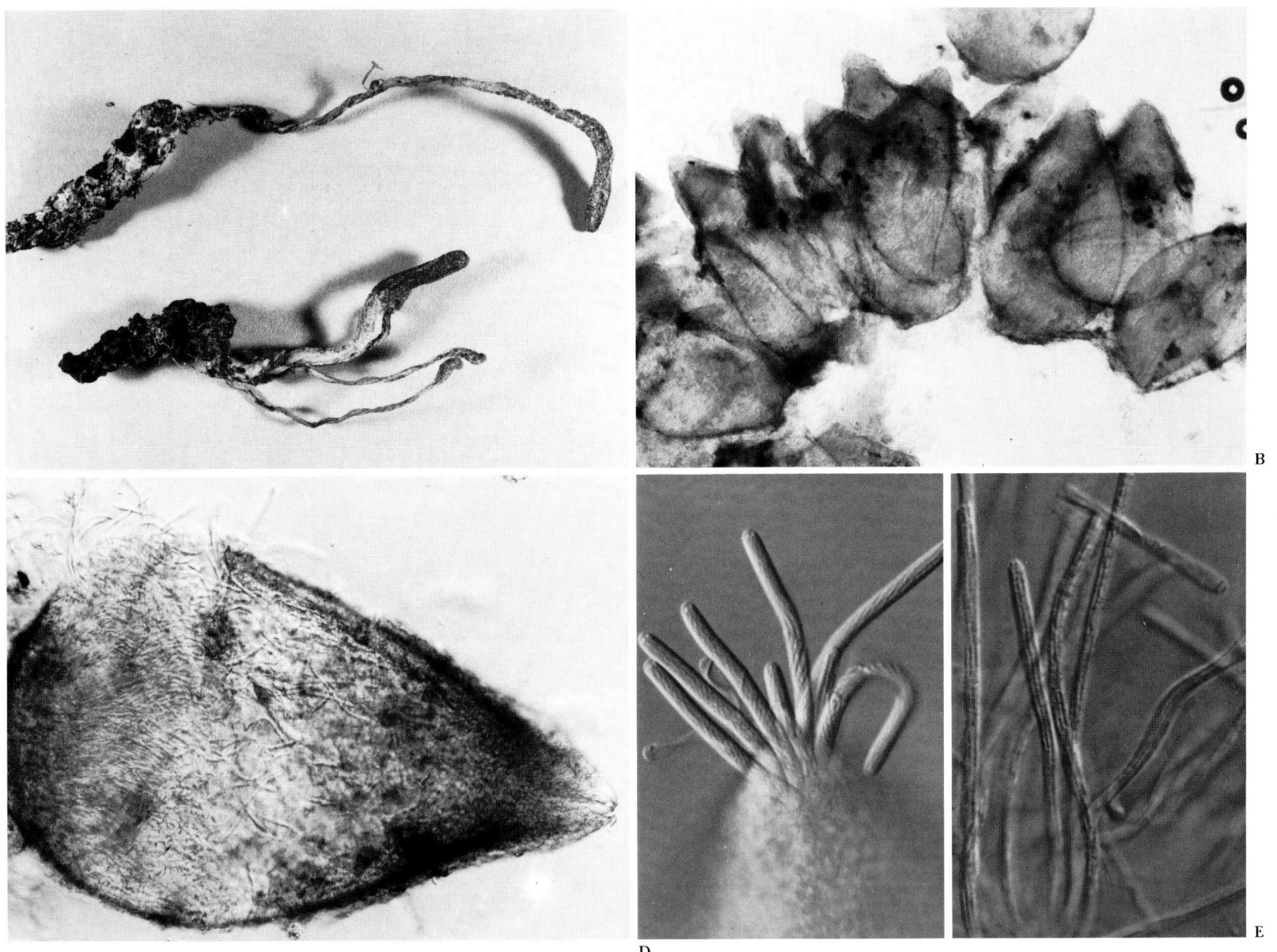

PLATE 44. *Cordyceps nutans*
A Clavae on pyrrhocorid *(Callibaphus* sp.), B ascomata in stroma (×200), C asci (×750), D ascospores which have disarticulated into part spores (×750), E hymenium-like layer of conidiogenous cells of *Hymenostilbe* anamorph (×750).

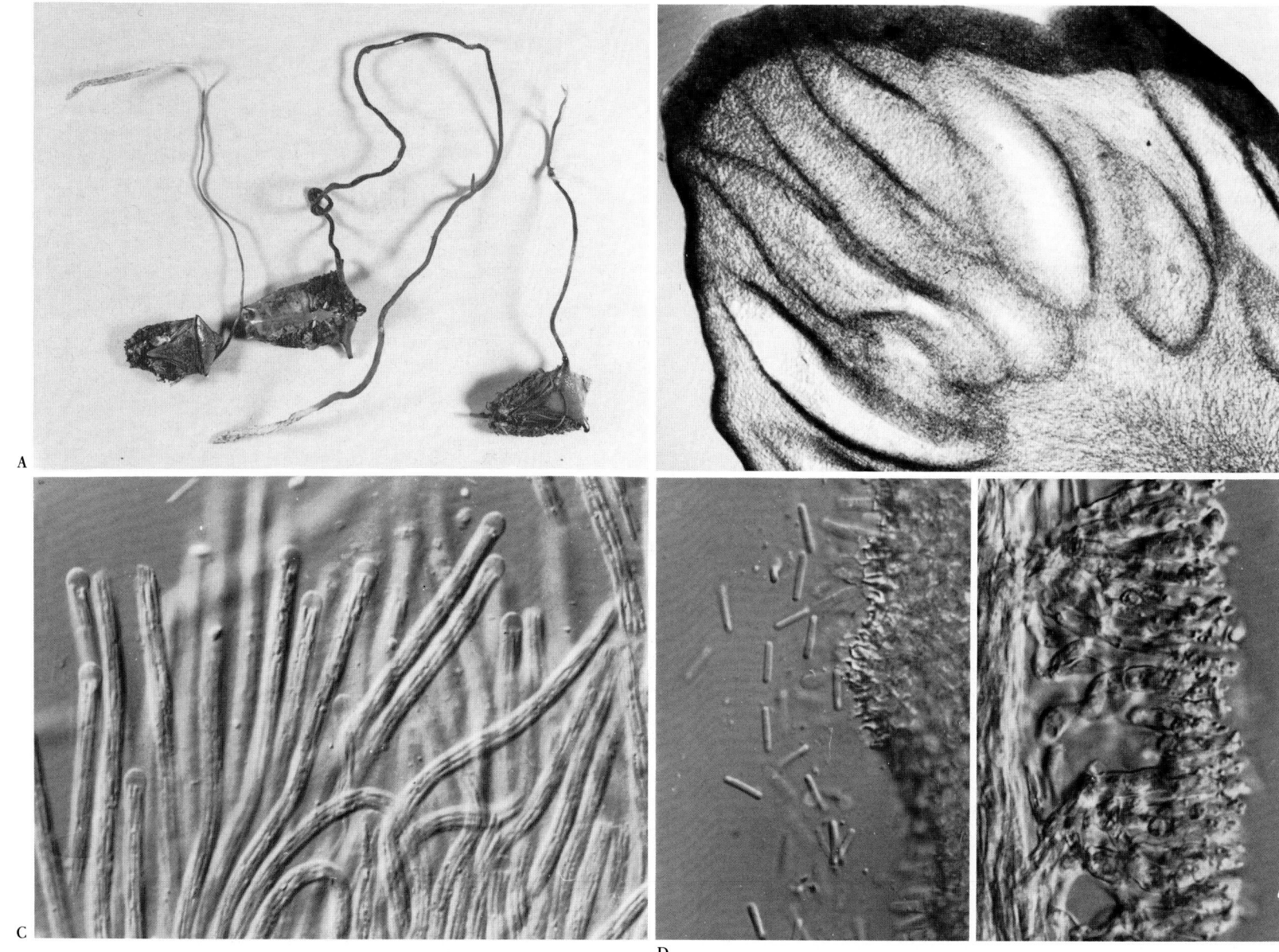

PLATE 45. *Cordyceps polyartha*
A Clavae on lepidopteran pupa, B ascomata in stroma (× 200), C asci protruding from ascoma (× 750), D asci (× 750).

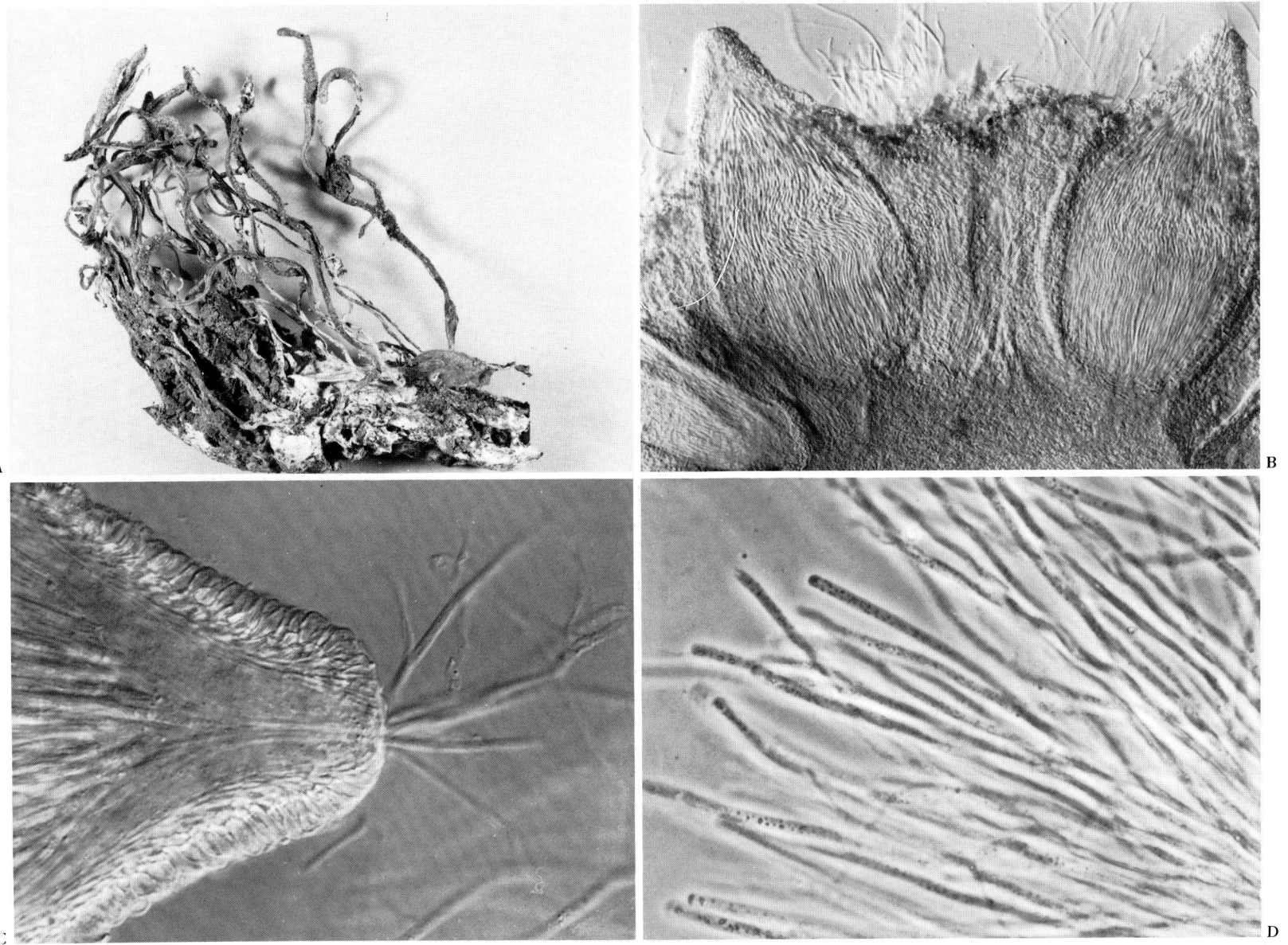

PLATE 46. *Cordyceps sobolifera*
A Clavae on cicada nymph (Cicadidae), showing teleomorph (ascomata) in darker apical ends and *Paecilomyces* anamorph (conidial state) in lighter apical parts (see also Plate 92), B-C ascomata in stroma (× 75 and × 200 resp.), D-E asci (× 750 and × 1875 resp.).

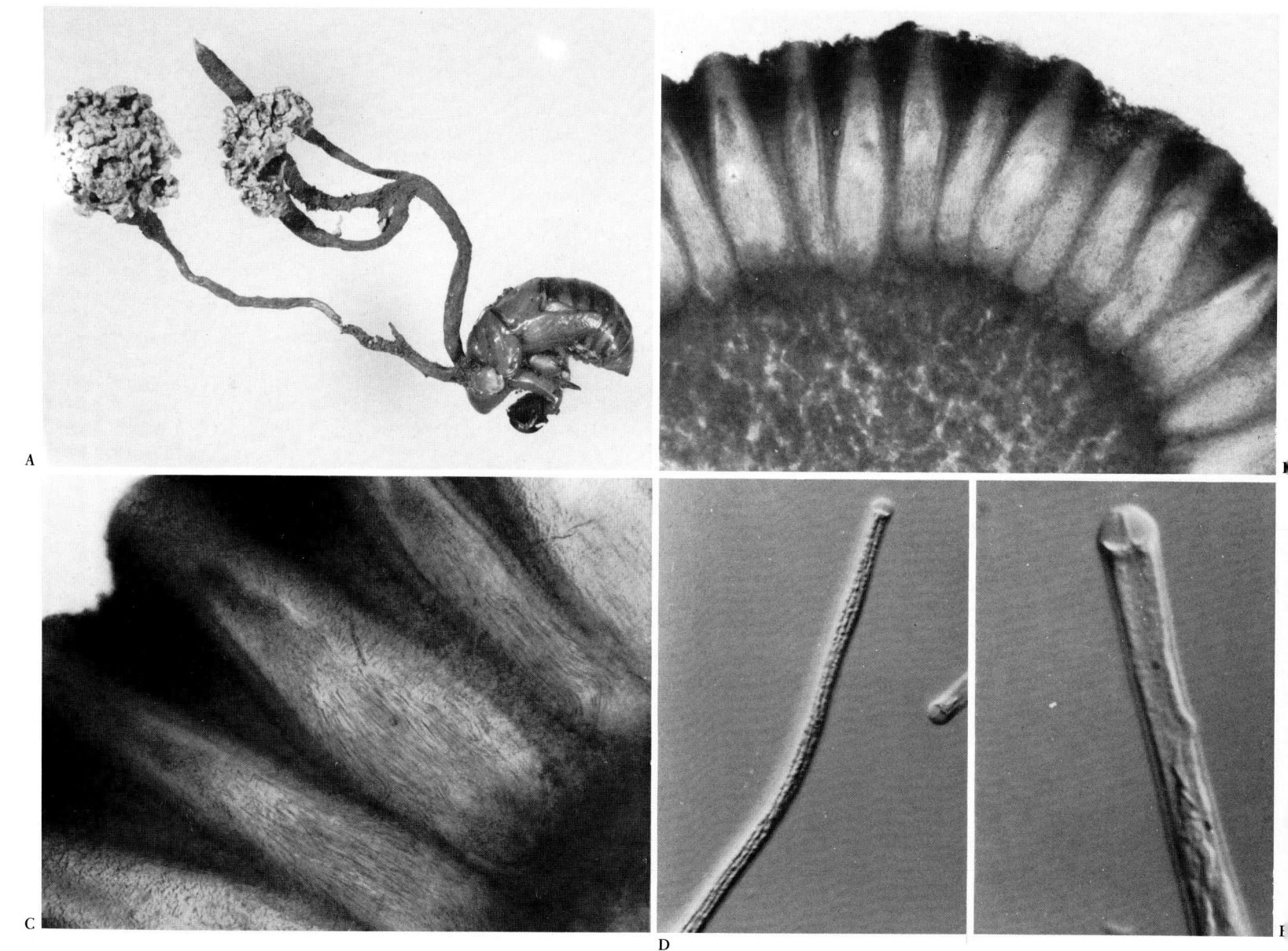

PLATE 47. *Cordyceps tuberculata*
A-B Clavae formed on tropical moths, C ascomata in stroma (× 200),
D asci (× 750).

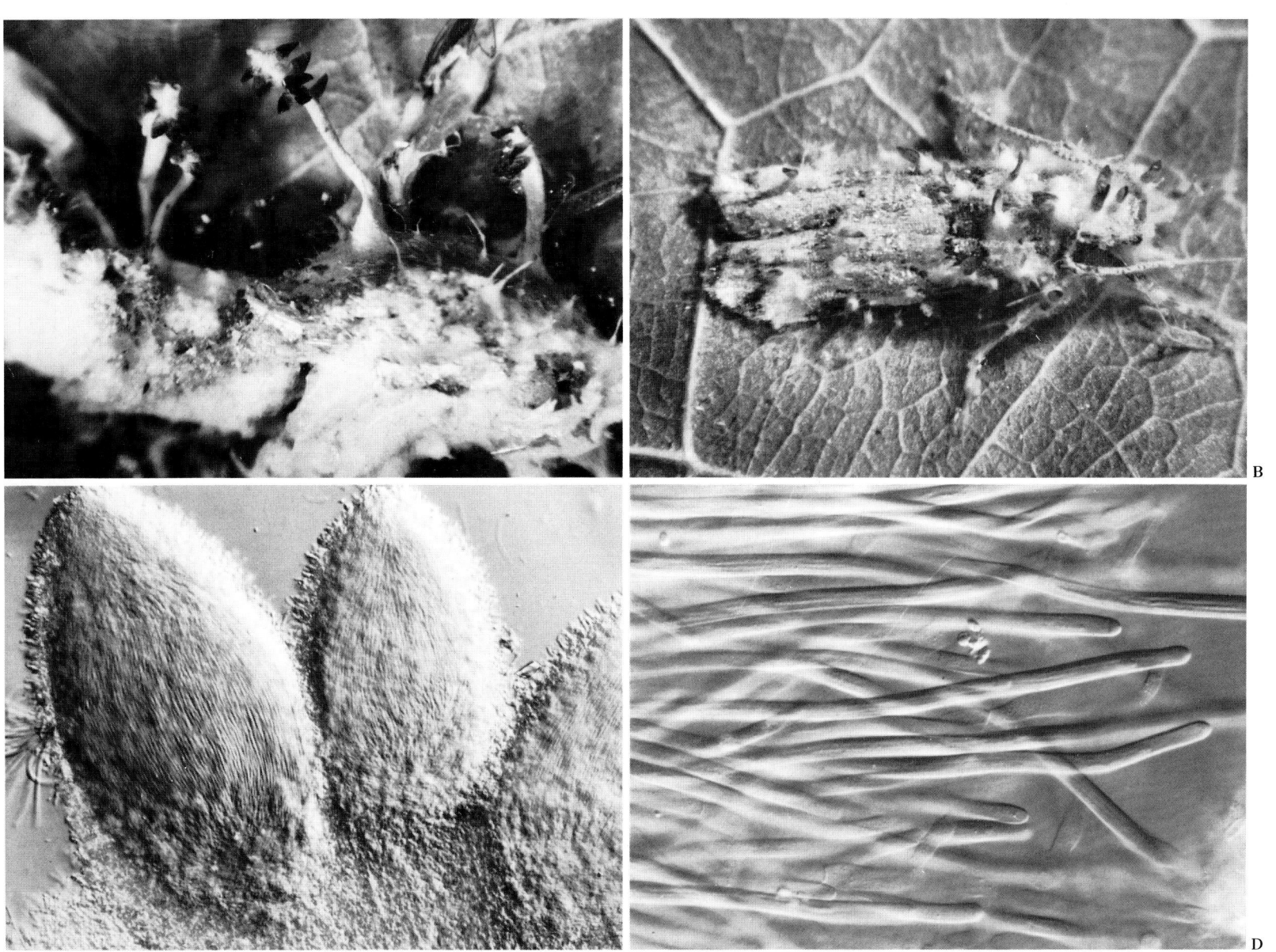

PLATE 48. *Cordyceps unilateralis*
A–D Different clavae structures formed on formicine ants *(Camponotus* spp. and *Polyrhachis* spp.).

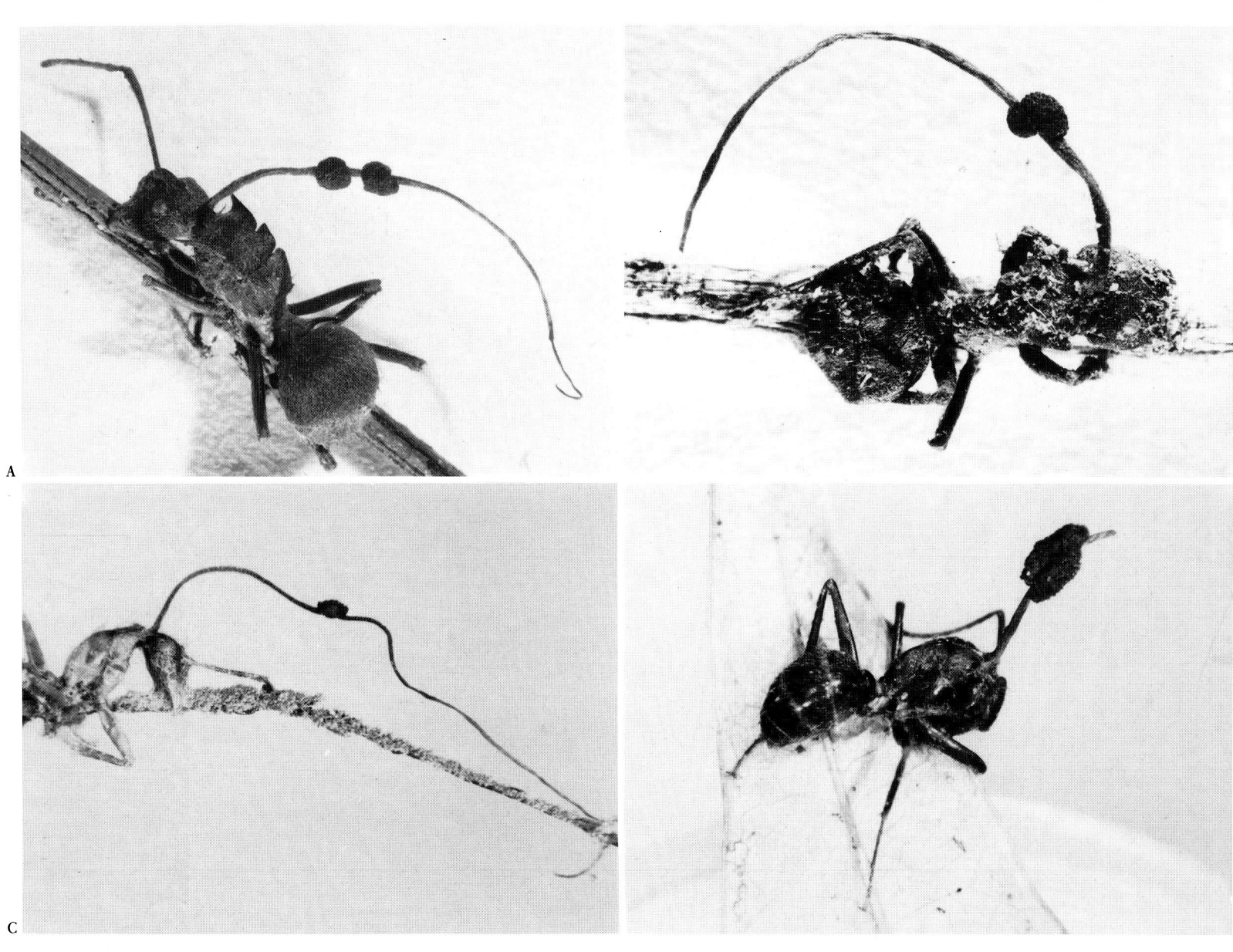

PLATE 49. *Cordyceps unilateralis*
A Stroma, B section through stroma showing ascoma (×75), C ascomata (×200), D ascospores (×750), E apical part of clavae showing hymenium-like layer of conidiogenous cells of *Hirsutella formicarum* (×200), F phialides of *Hirsutella* anamorph (×750).

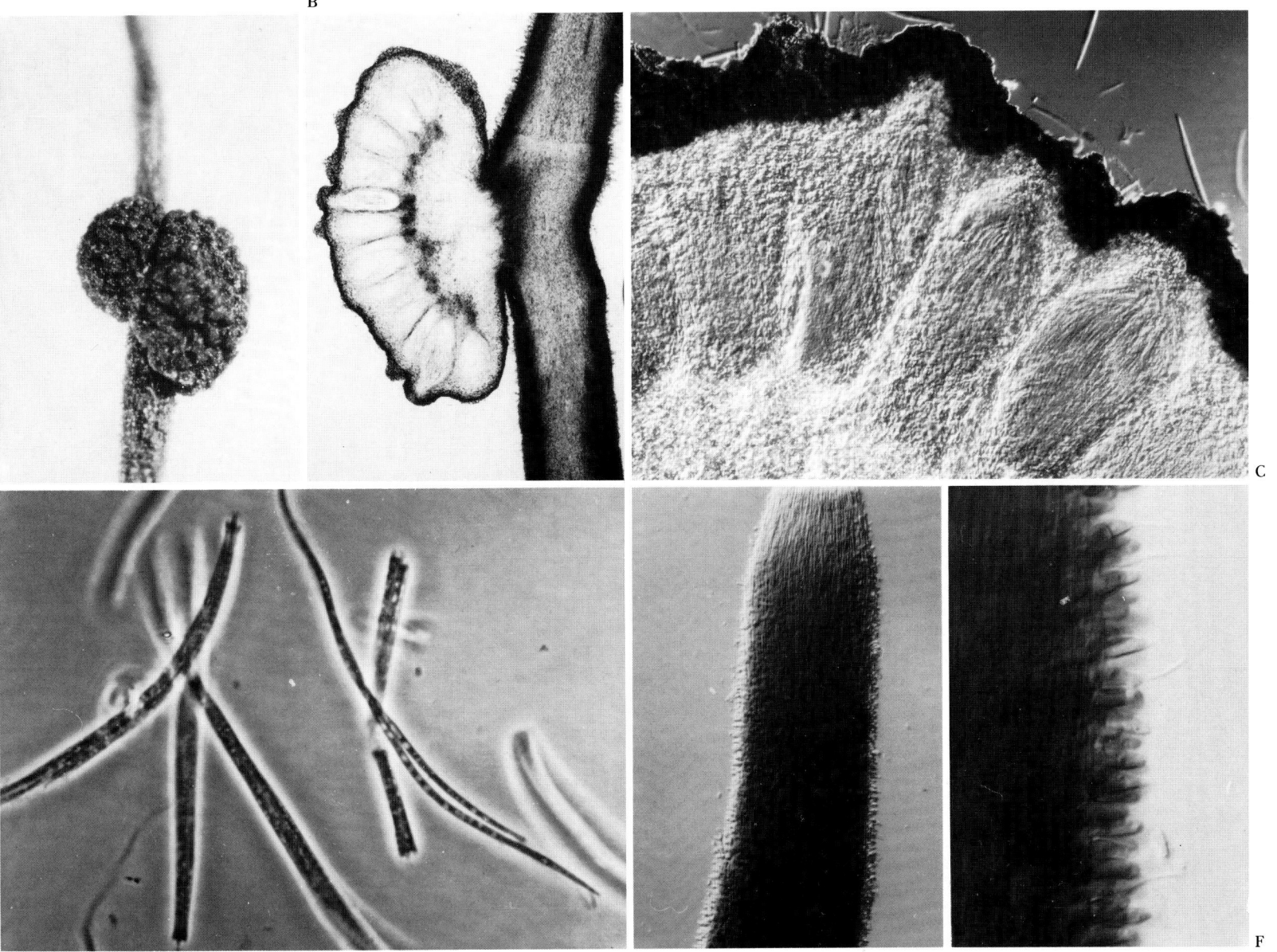

PLATE 50. *Hypocrella amomi*
A Stromata op tropical scales (Lecaniidae), B section through stroma showing two ascomata (× 75), C ascoma (× 200), D-E asci containing fusiform ascospores (× 750).

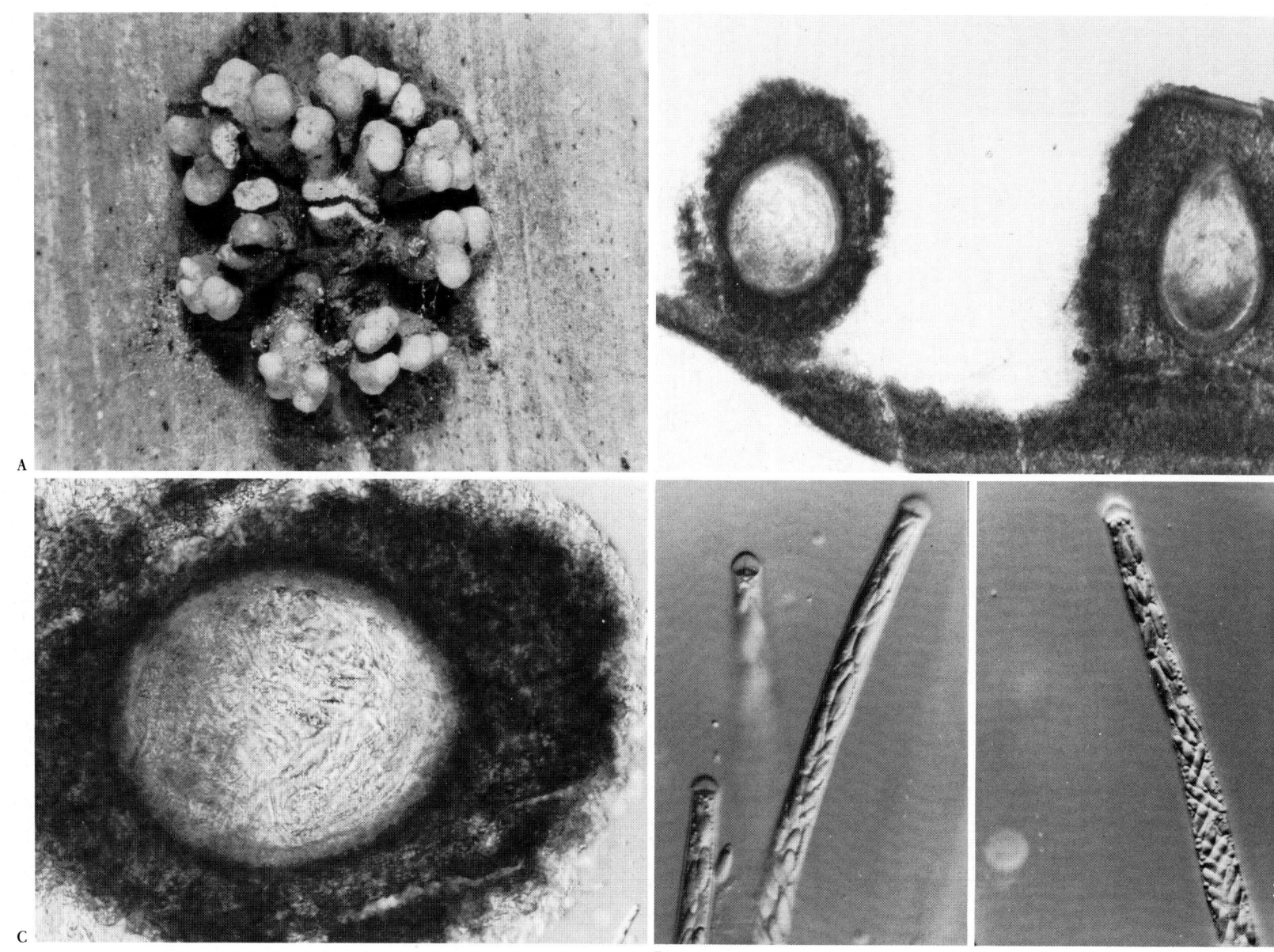

PLATE 51. *Myriangium duriaei*
A Stromata on willow scales (Lecaniidae), B section through stroma showing cavities containing asci (×75), C–D asci (×750), E–F asci and multiseptate ascospores (×750).

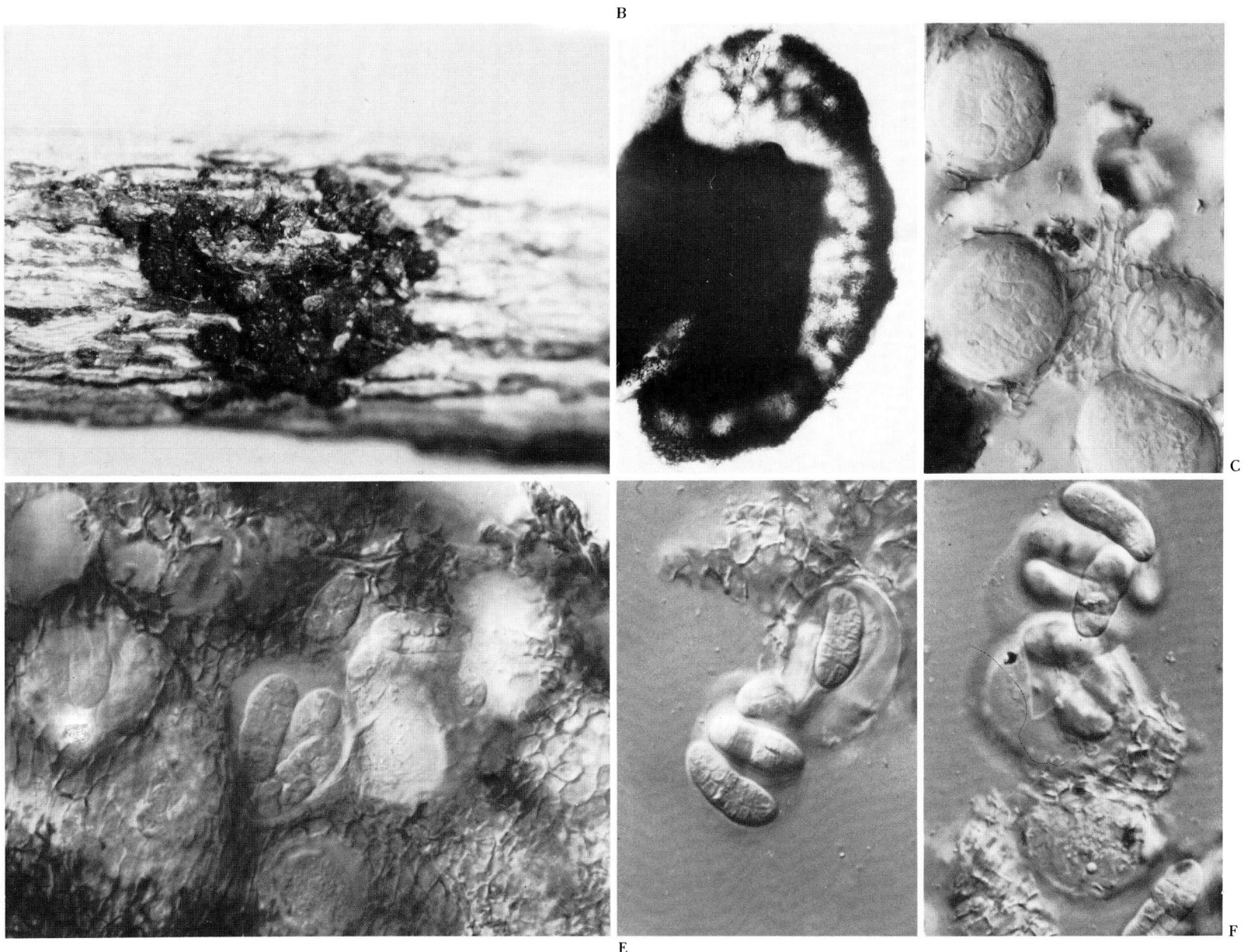

PLATE 52. *Nectria flammea*
A Ascomata on Diaspid scale *(Ischnaspsis* sp.), B-C ascomata (× 200),
D-E asci containing two-celled ascospores (× 750).

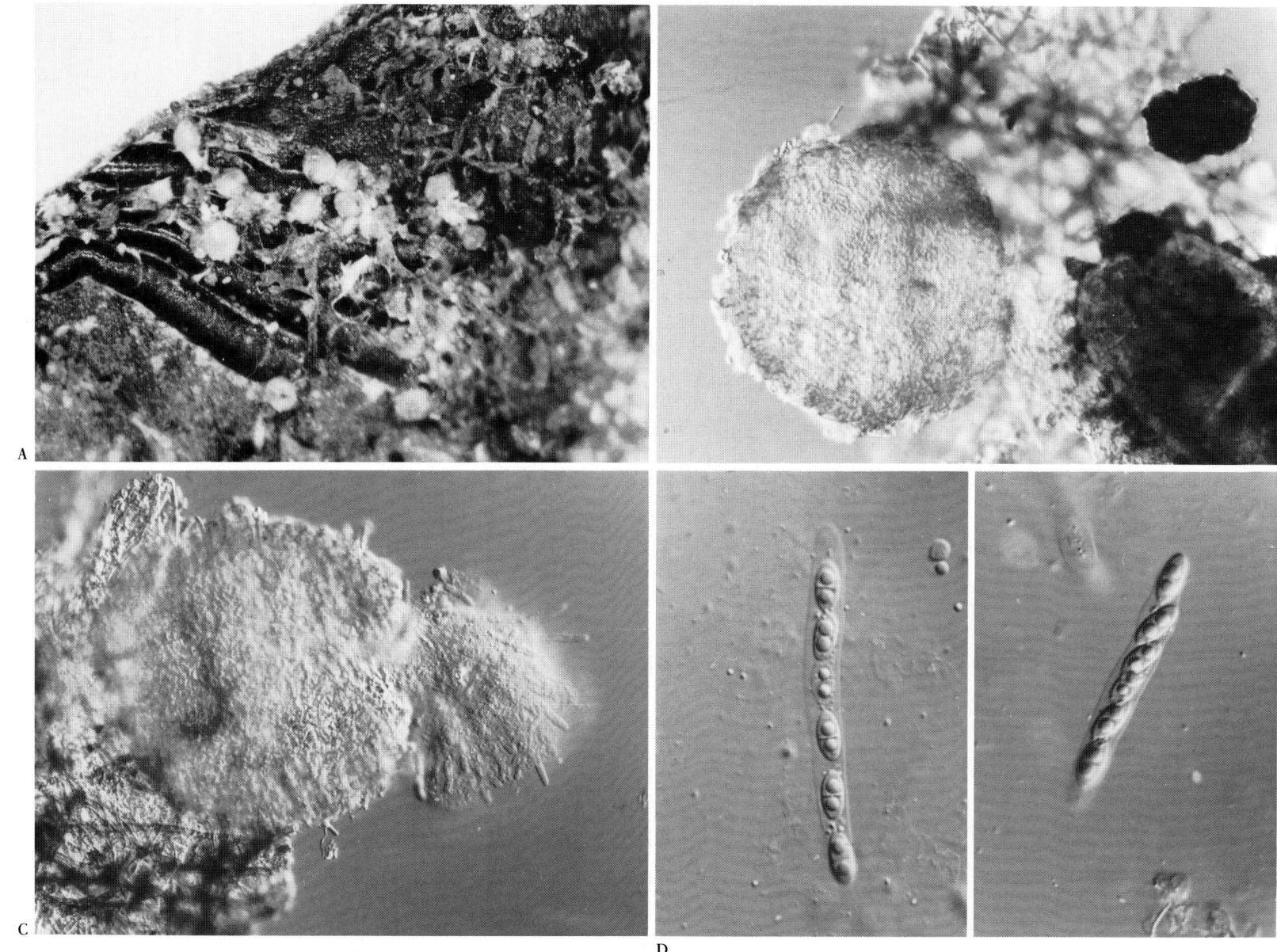

PLATE 53. *Podonectria coccicola*
A Ascomata (smooth, globose structures) of *P. coccicola* and sporodochia (cushion-shaped) of *Tetracrium* anamorph on citrus scales (*Lepidosaphes*), B squash preparation of an ascoma (× 200), C asci (× 750). D multiseptate ascospores (× 750).

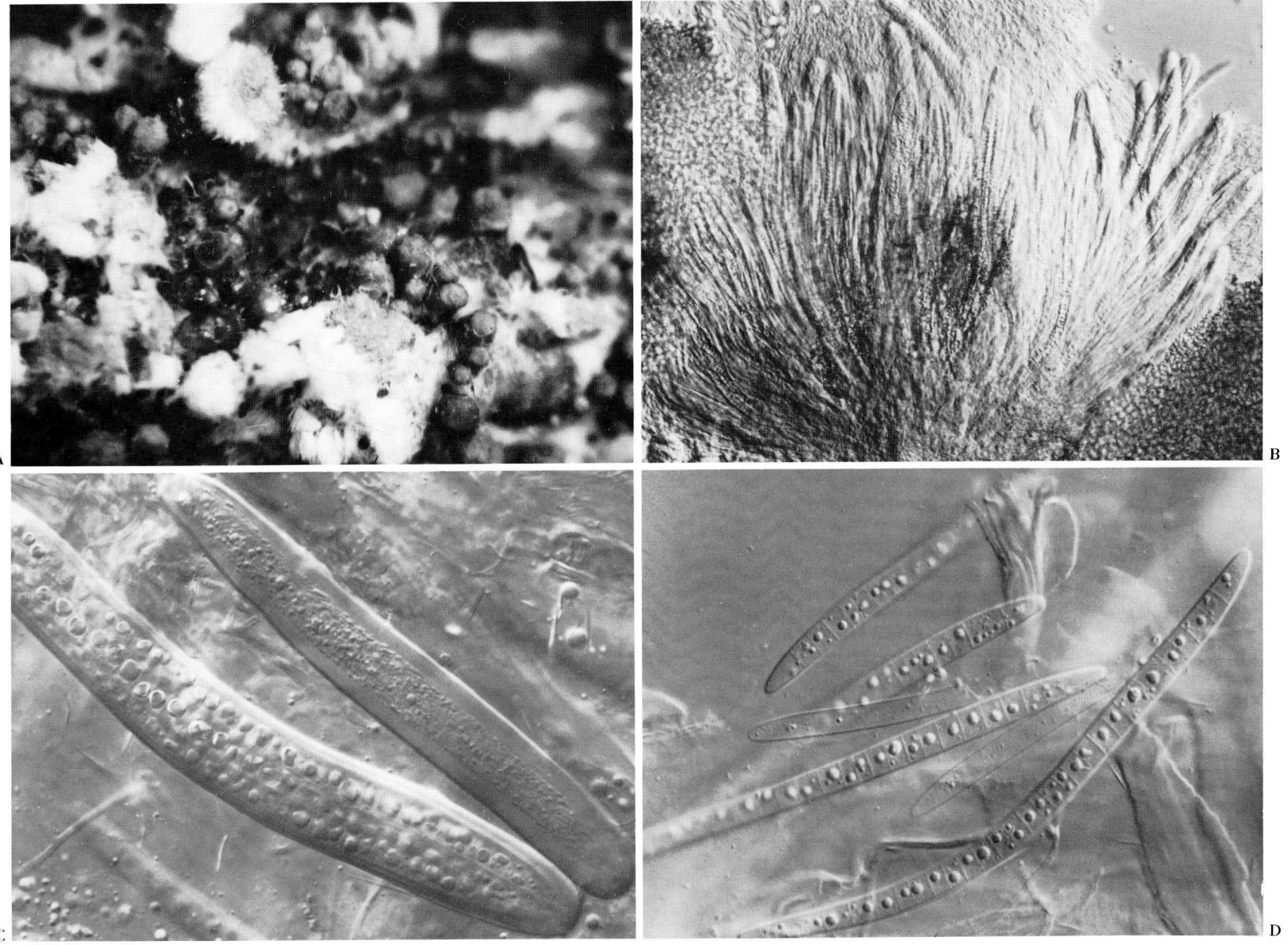

PLATE 54. *Torrubiella arachnophila*
A-B Ascomata on small salticid spiders, C-D ascomata (×75 and ×200 resp.), E ascoma wall showing the presence of the *Granulomanus* anamorph (×750), F-G asci (×750).

PLATE 55. *Torrubiella carnata*
A Ascomata on ponerine ant (*Paltothyreus tarsatus*), B ascomata (×200), C ascoma with protruding asci (×750), D asci (×750).

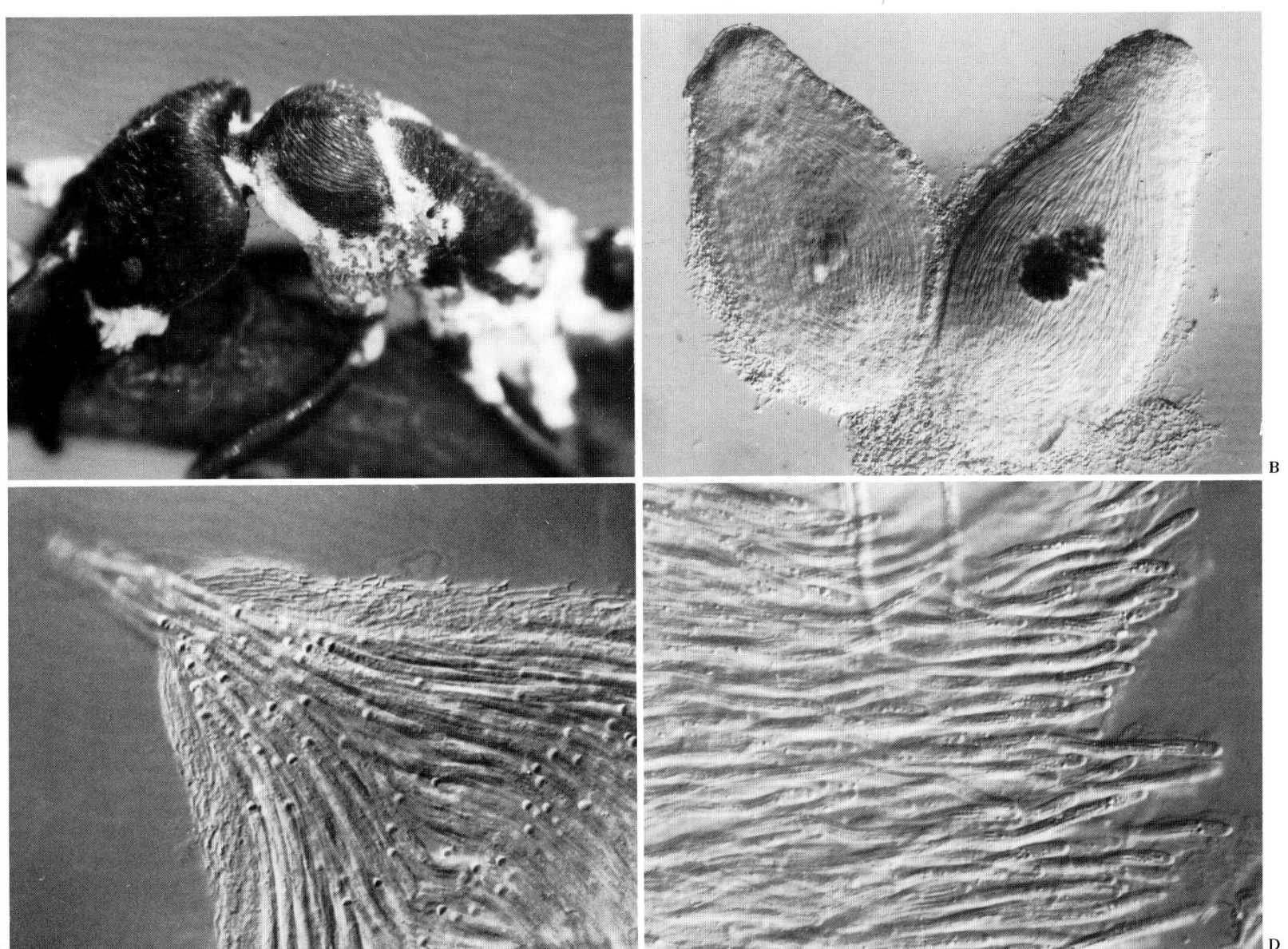

PLATE 56. *Torrubiella rubra*
A Ascomata on coccids, B-C darkly pigmented ascomata (× 75 and × 200 resp.), D asci (× 750).

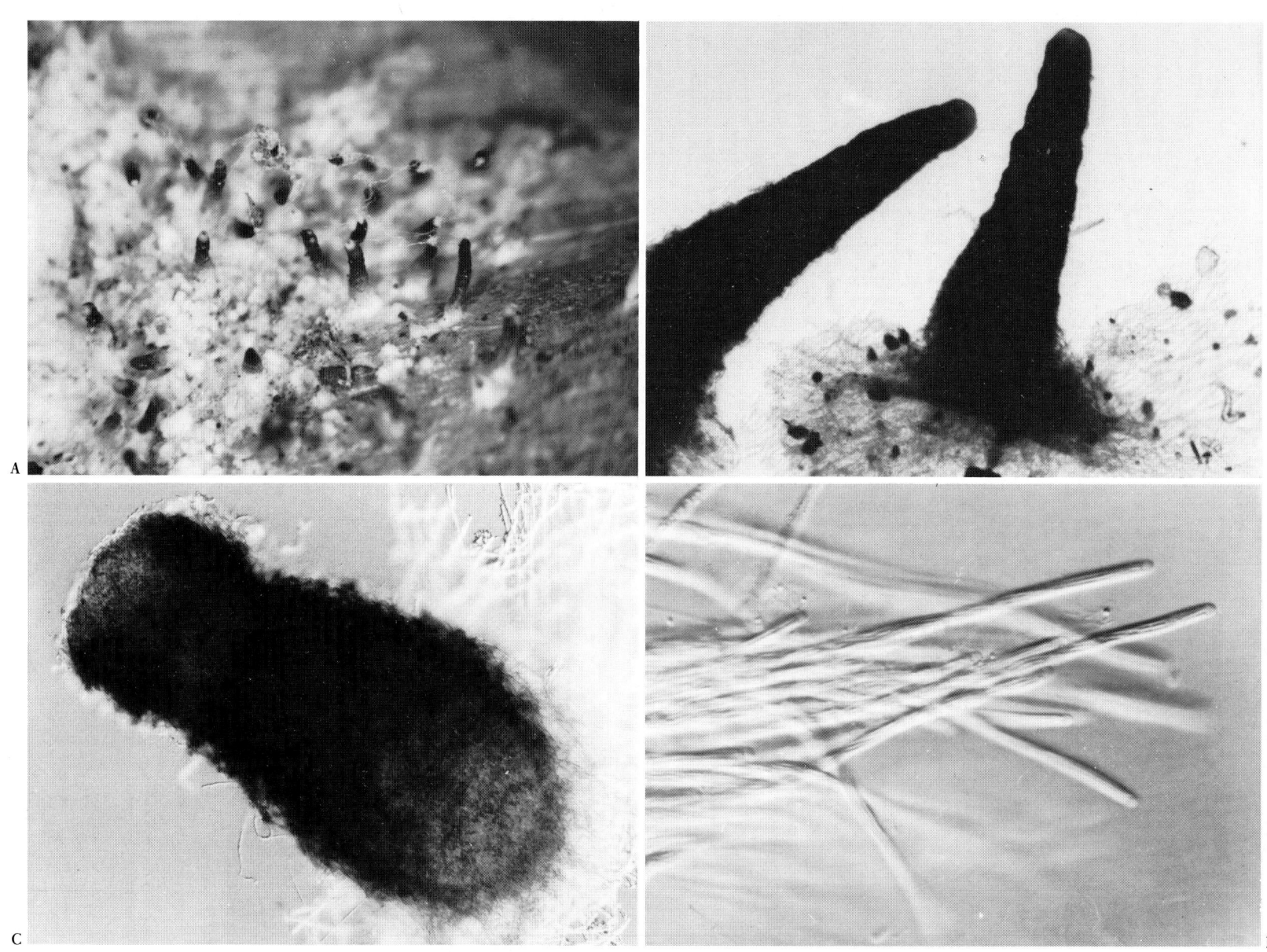

PLATE 57. *Akanthomyces aculeatus*
A Synnemata on a tropical moth (Noctuidae), B-D hymenium-like layer of phialides and conidia (B × 750, C and D × 1875).

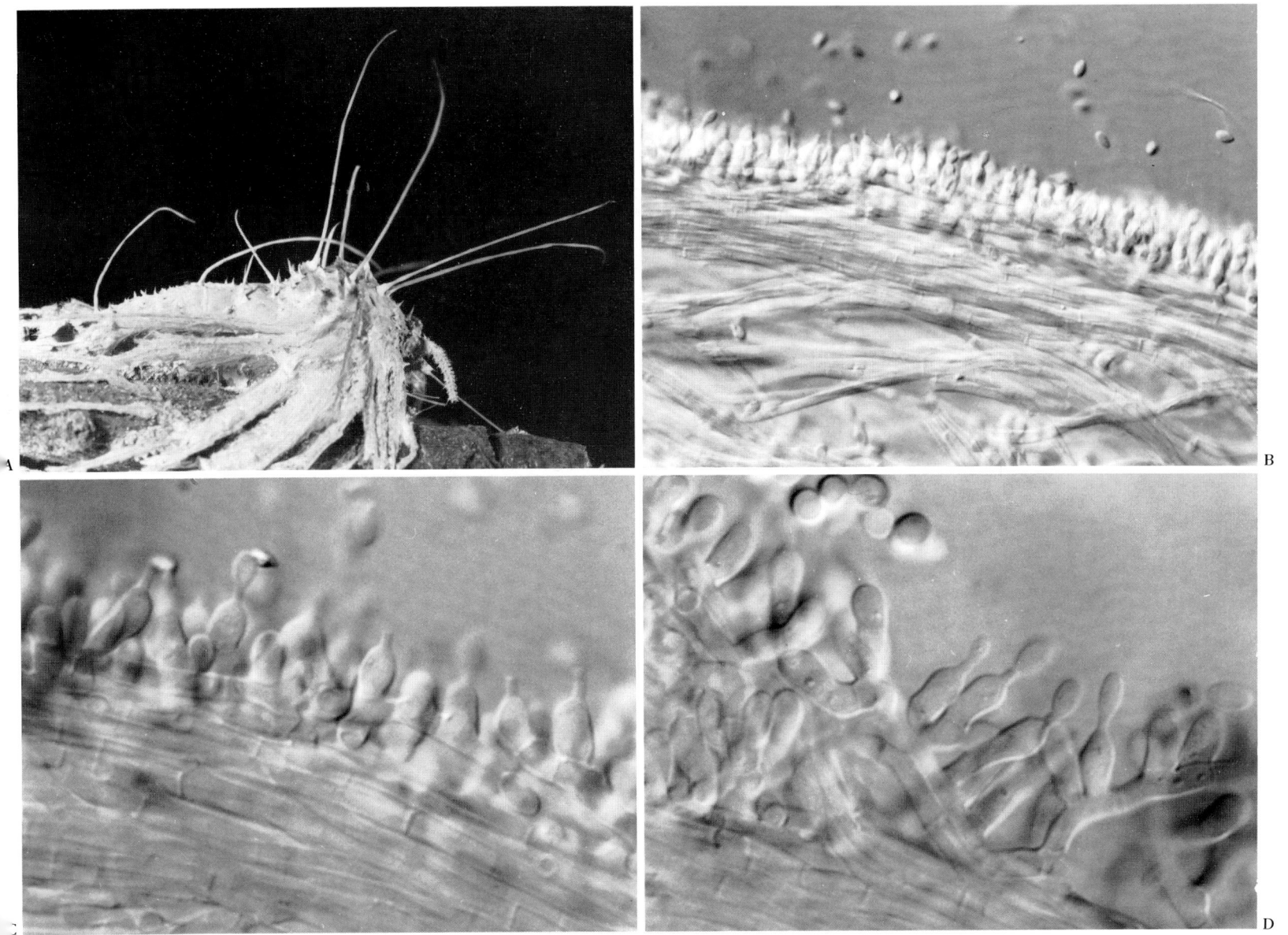

PLATE 58. *Akanthomyces gracilis*
Synnemata on A *Callibaphus* sp. (Pyrrhocoridae) and B ponerine ant *(Paltothyreus tarsatus)*; C-D hymenium-like layer of phialides along the synnema (× 750 and × 1875 resp.), E conidia (× 1875).

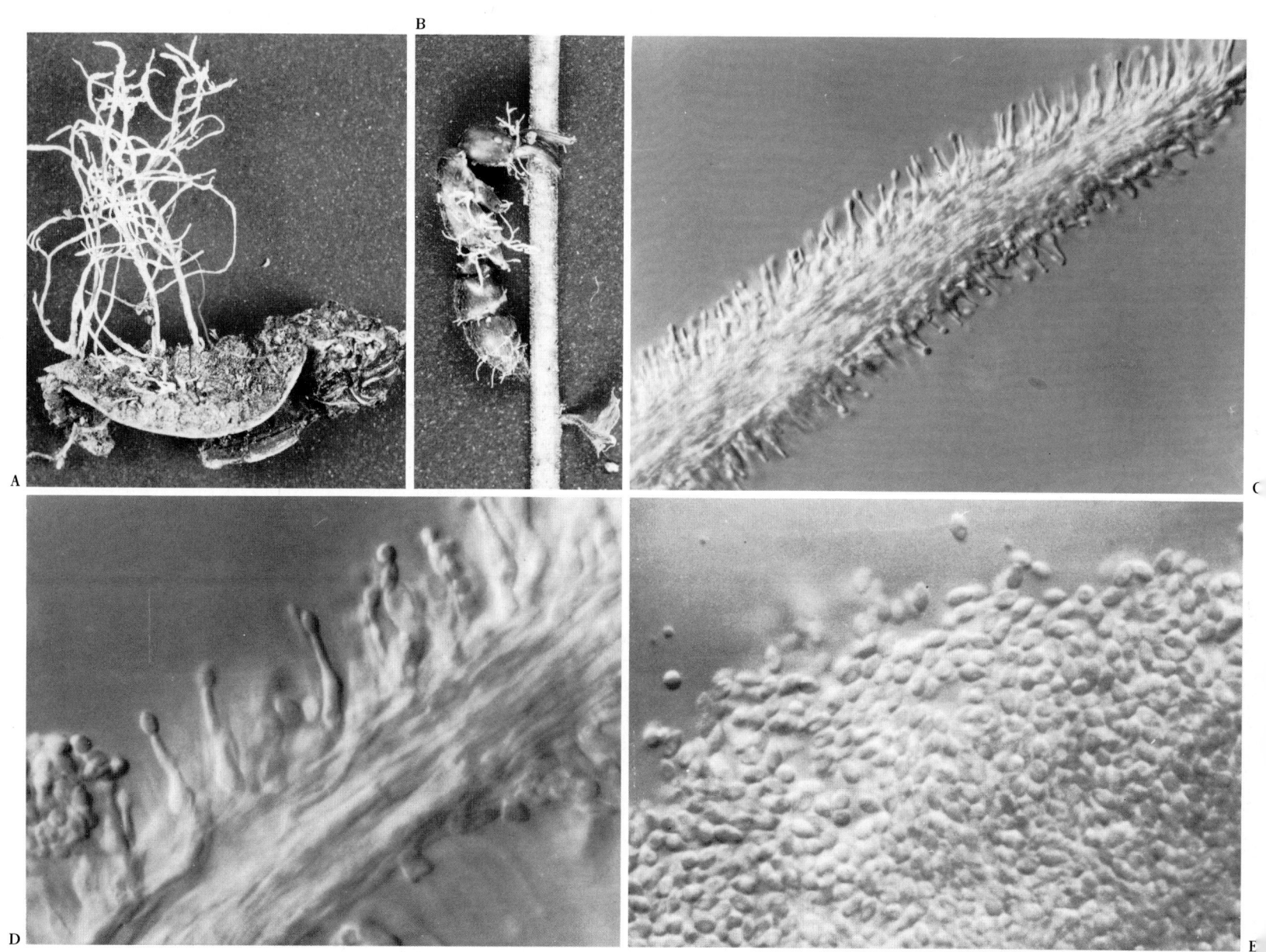

PLATE 59. *Akanthomyces pistillariiformis*
A Synnemata on a tropical moth (Sphingidae), B-D different parts of synnema showing phialides (B × 200, C × 750, D × 875), E conidia (× 1875).

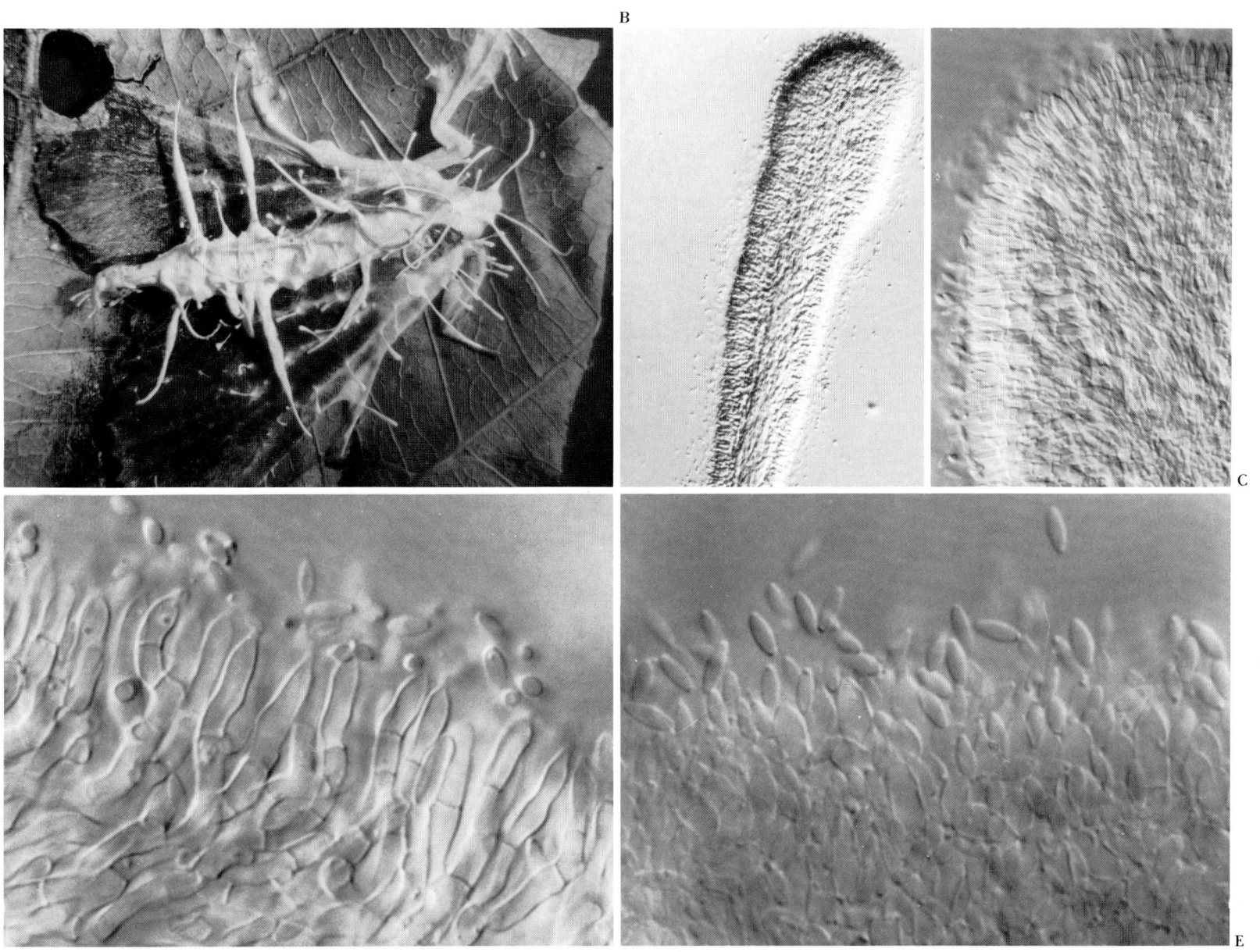

PLATE 60. *Aschersonia aleyrodis*
A White fly nymphs (Aleyrodidae) with mycelial and pycnidial development, B part of pycnidium (× 200), C phialides within the pycnidium (× 750), D conidia (× 1875).

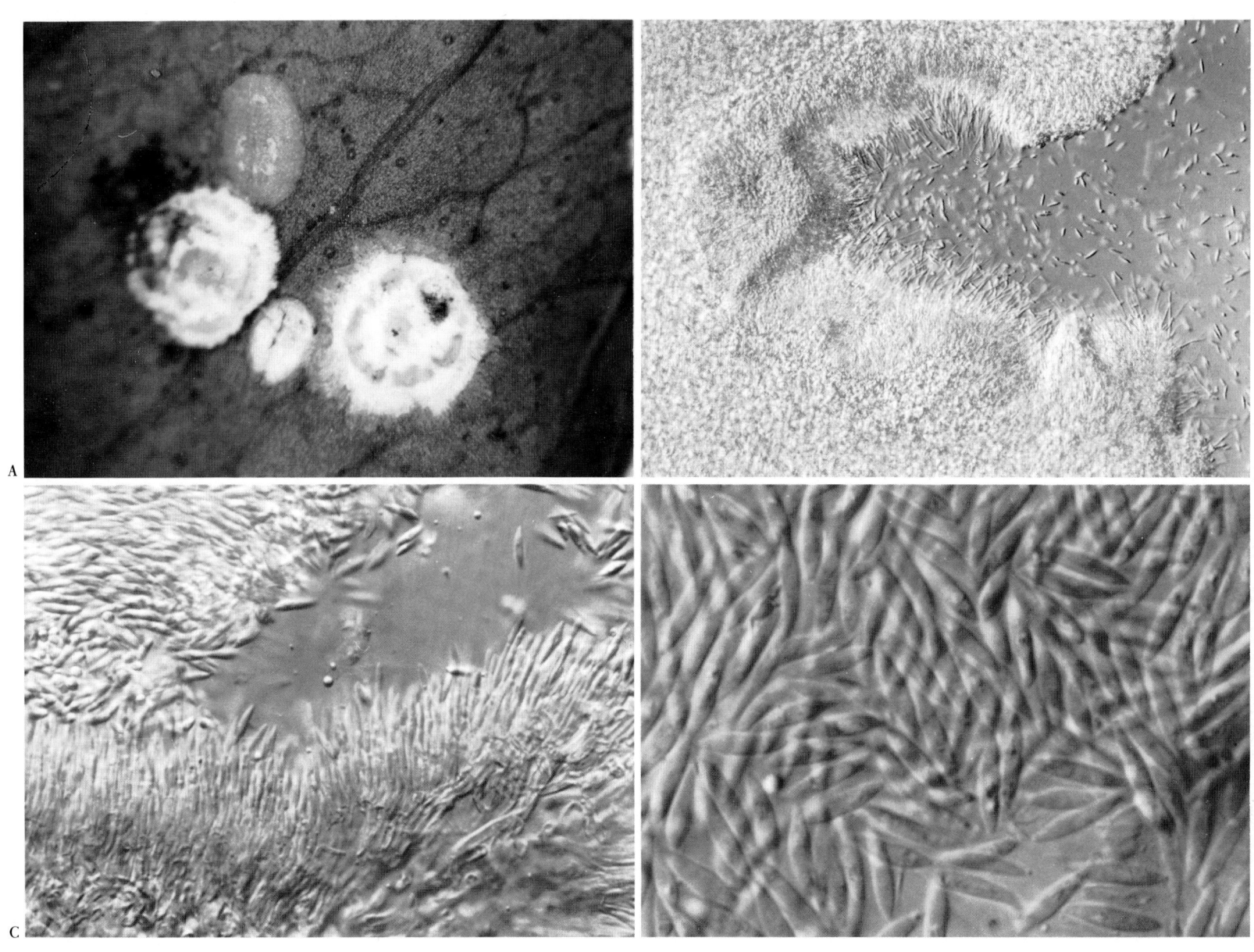

PLATE 61. *Aschersonia cubensis*

A Fruitbodies on soft scales (Lecaniidae), B part of pycnidium with phialides (×750), C thick-walled, curved hyphae of vegetative mycelium (×750), D conidia (×1875).

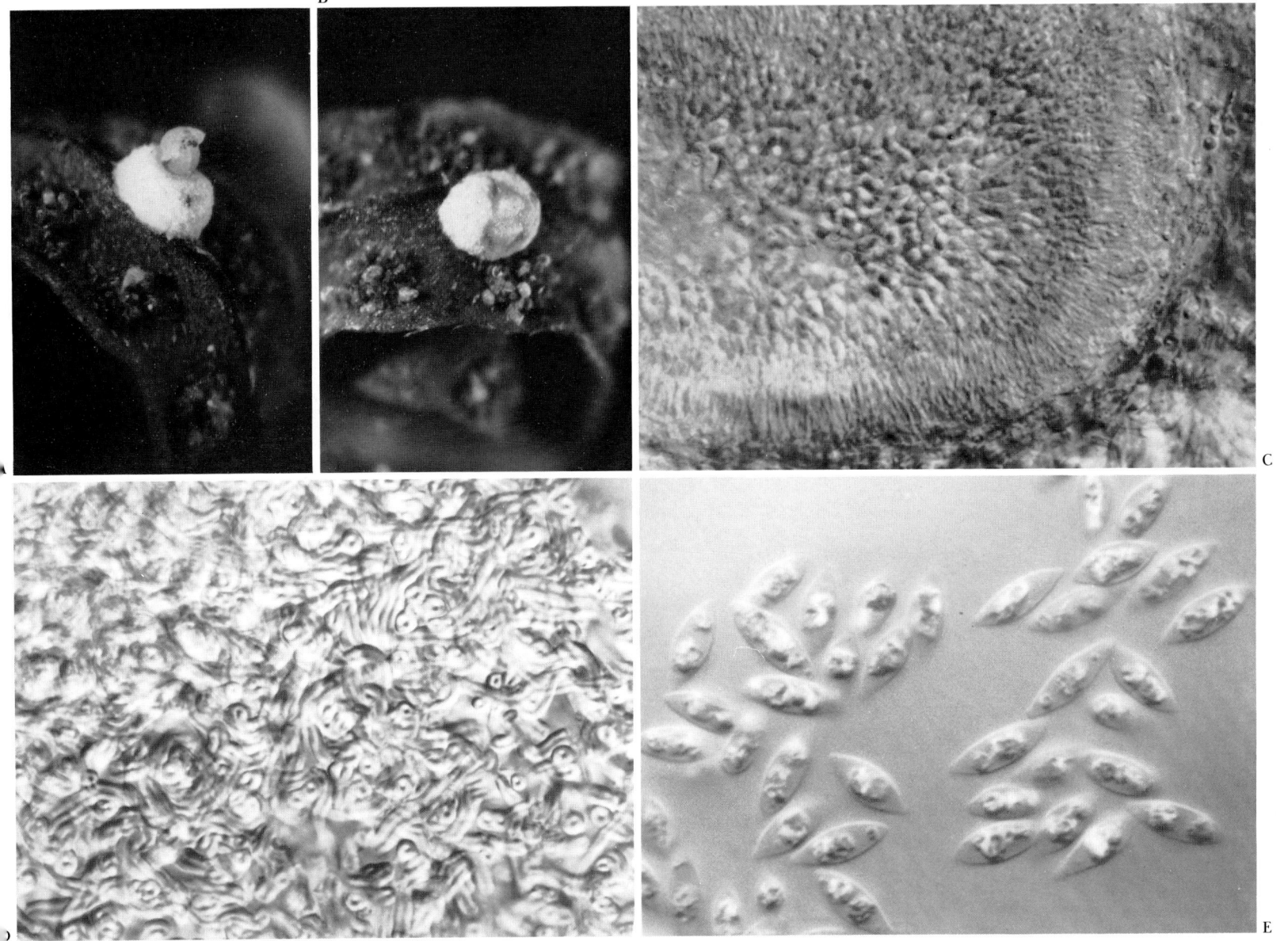

PLATE 62. *Aschersonia turbinata*
A-B Typical cup-shaped fruitbodies on citrus scale (Lecaniidae),
C part of pycnidium showing phialidic conidiogenous cells (×750),
D-E conidia (×1875).

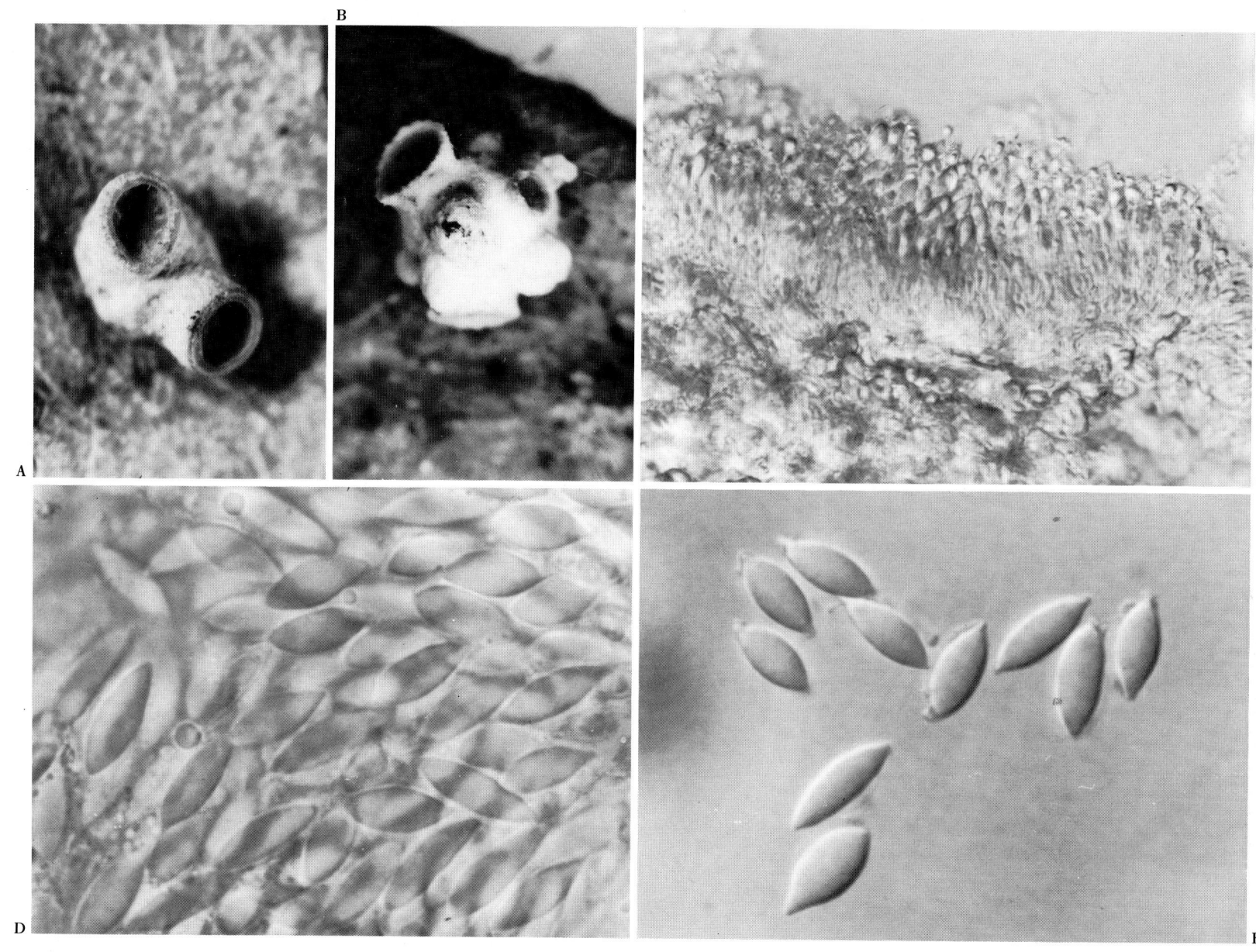

PLATE 63. *Aspergillus parasiticus*
A Conidiophores on *Encarsia formosa*, B-C conidiophores showing stipe, vesicle, phialide and conidial development (× 750 and × 200 resp.).

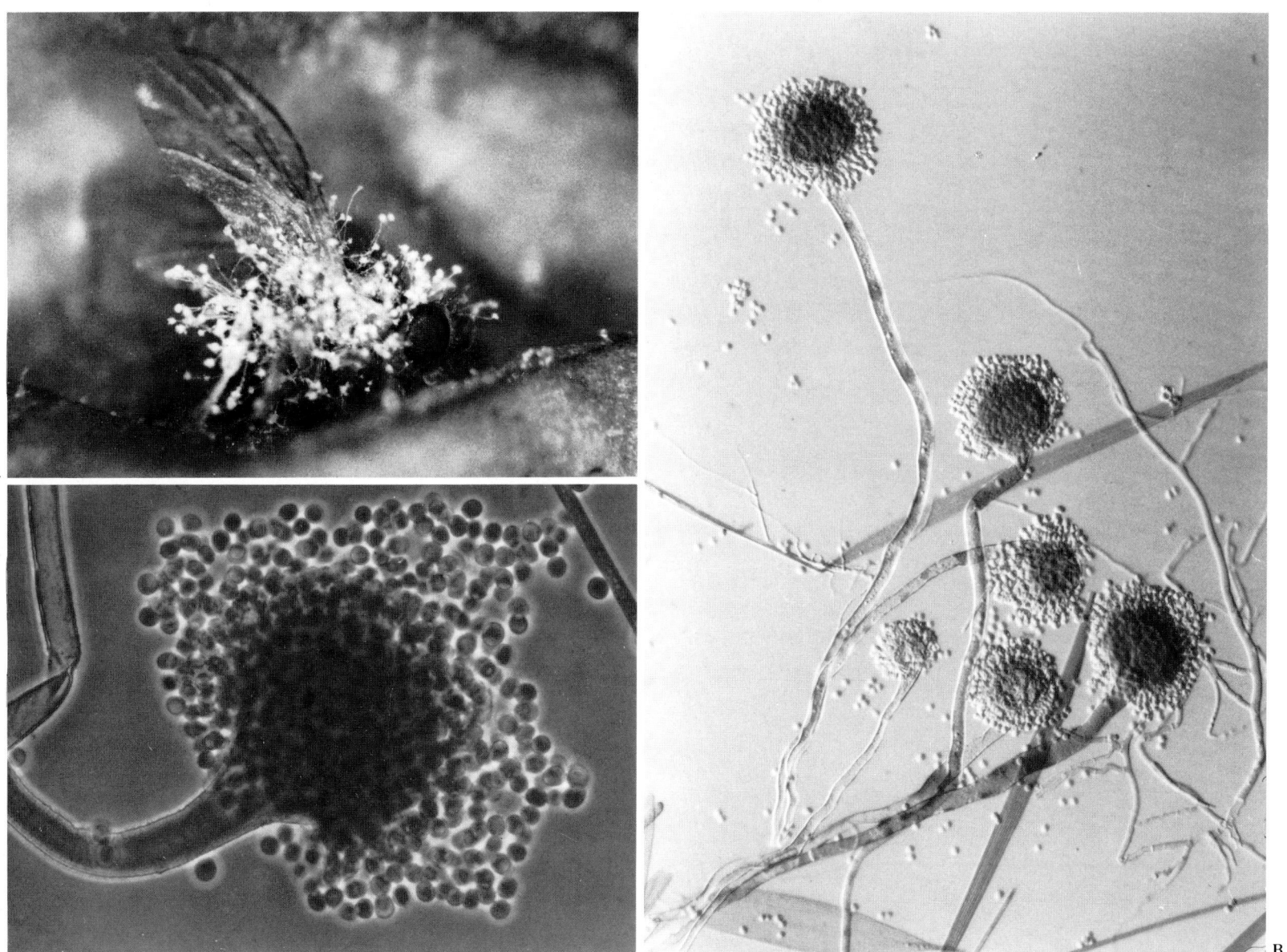

PLATE 64. *Beauveria bassiana*
Colonies on A weevils, B wasp, C lepidopteran larvae and D adult cicada.

PLATE 65. *Beauveria bassiana / B. brongniartii*
A-C *B. bassiana*, conidiophores, conidiogenous cells and conidia (A × 750, B-C × 1875), D-E *B. brongniartii*: conidiophores, conidiogenous cells and conidia (D × 750, E × 1875).

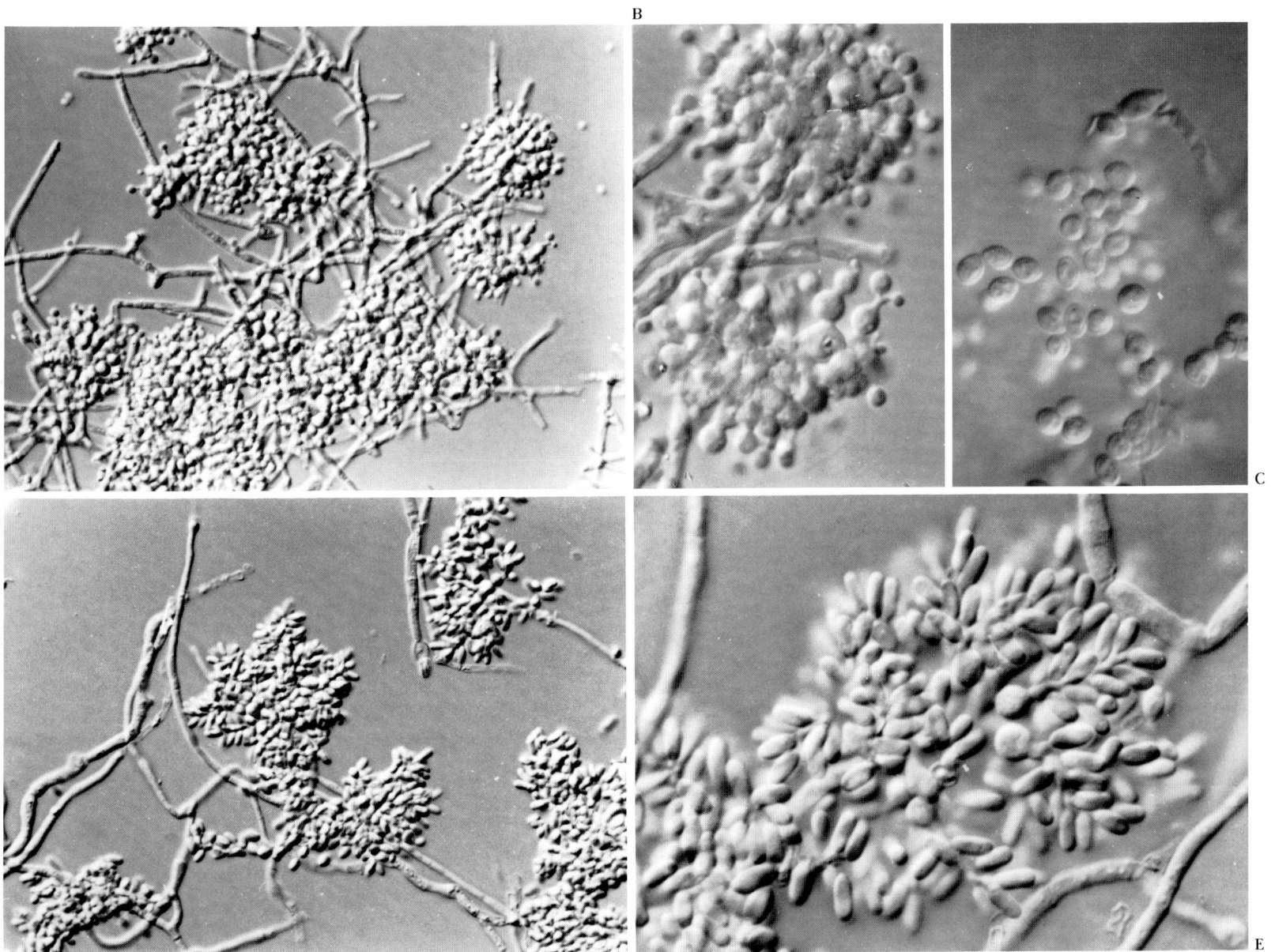

PLATE 66. *Beauveria amorpha / B. velata*
A-C *B. amorpha:* A on coleopteran adult, B-C conidiogenous cells and conidia (× 1875); D-E *B. velata:* D on lepidopteran larva, E conidia (× 1875).

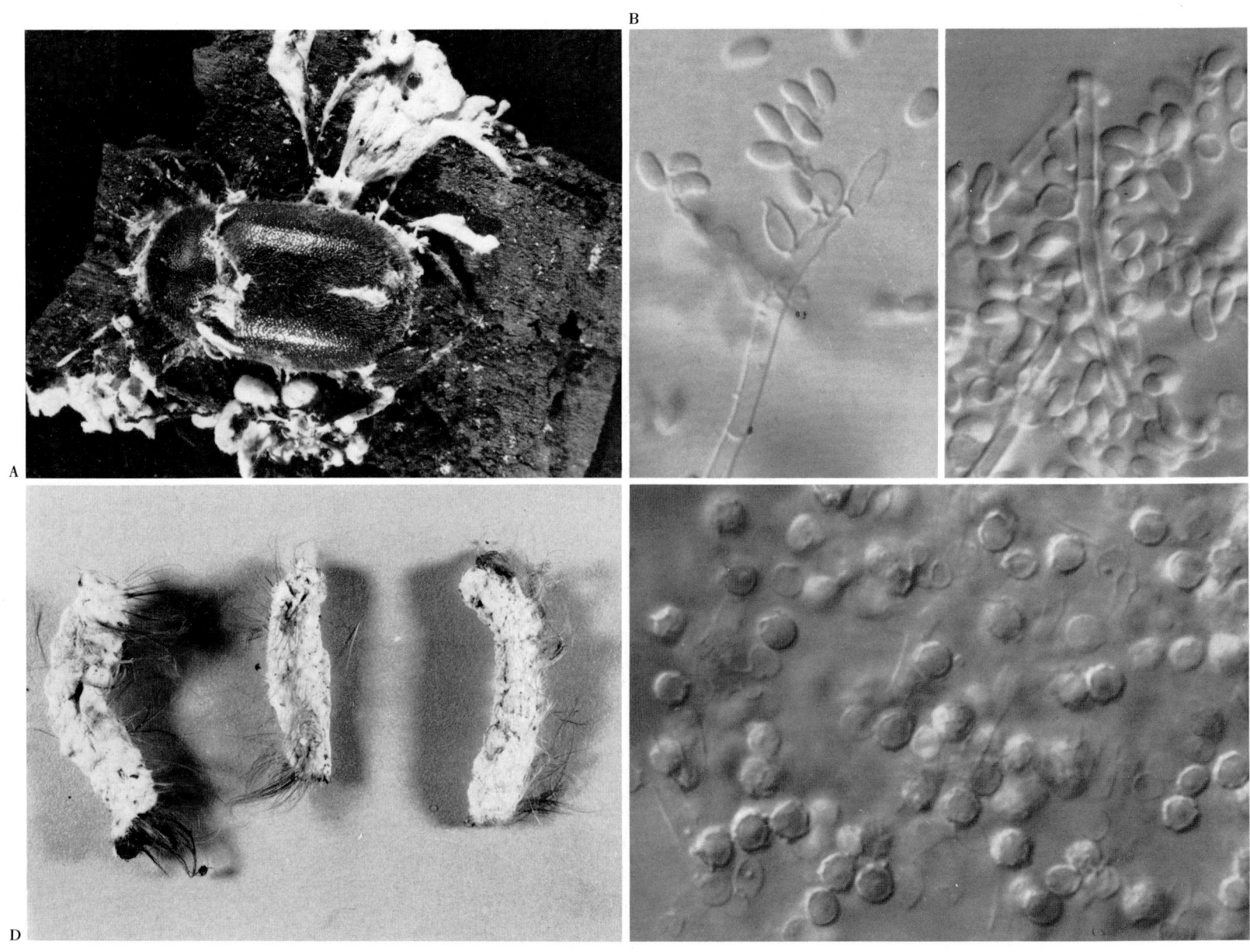

PLATE 67. *Culicinomyces clavisporus*
A Colonies on mosquito larva, B conidiophores arising from host (× 750), C-E phialides and conidia (× 1875).

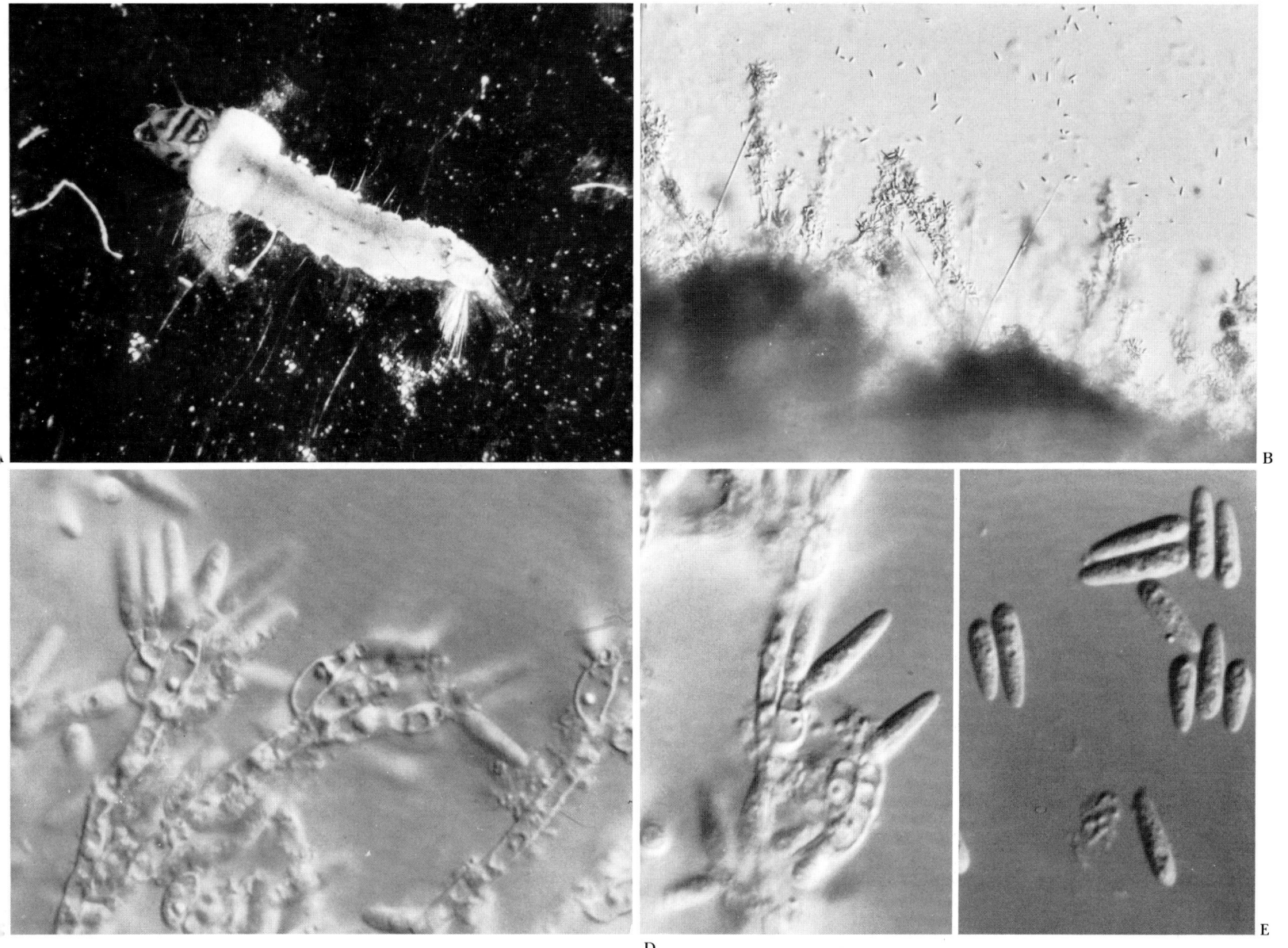

PLATE 68. *Engyodontium aranearum*
A Colonies on arachnids (Opilionidae), B conidiophores on legs of host (× 200), C–D phialidic and sympodial conidiogenous cells (× 750), E conidia (× 750).

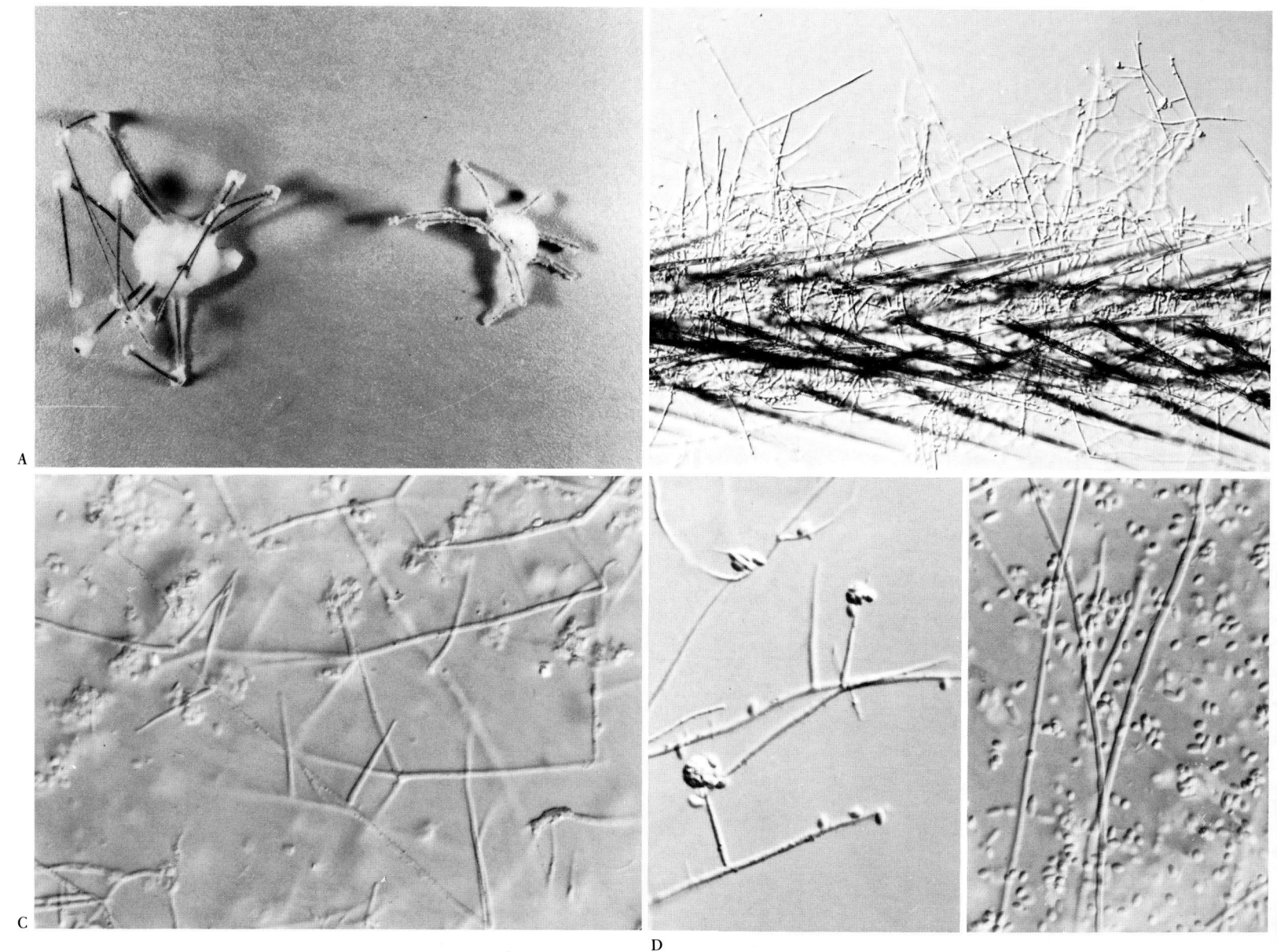

PLATE 69. *Fusarium coccophilum*
Sporodochia of *F. coccophila* on A pine needle scales and B diaspid scale; C sporodochium (×75), D phialides (×200), E multiseptate conidia (×750).

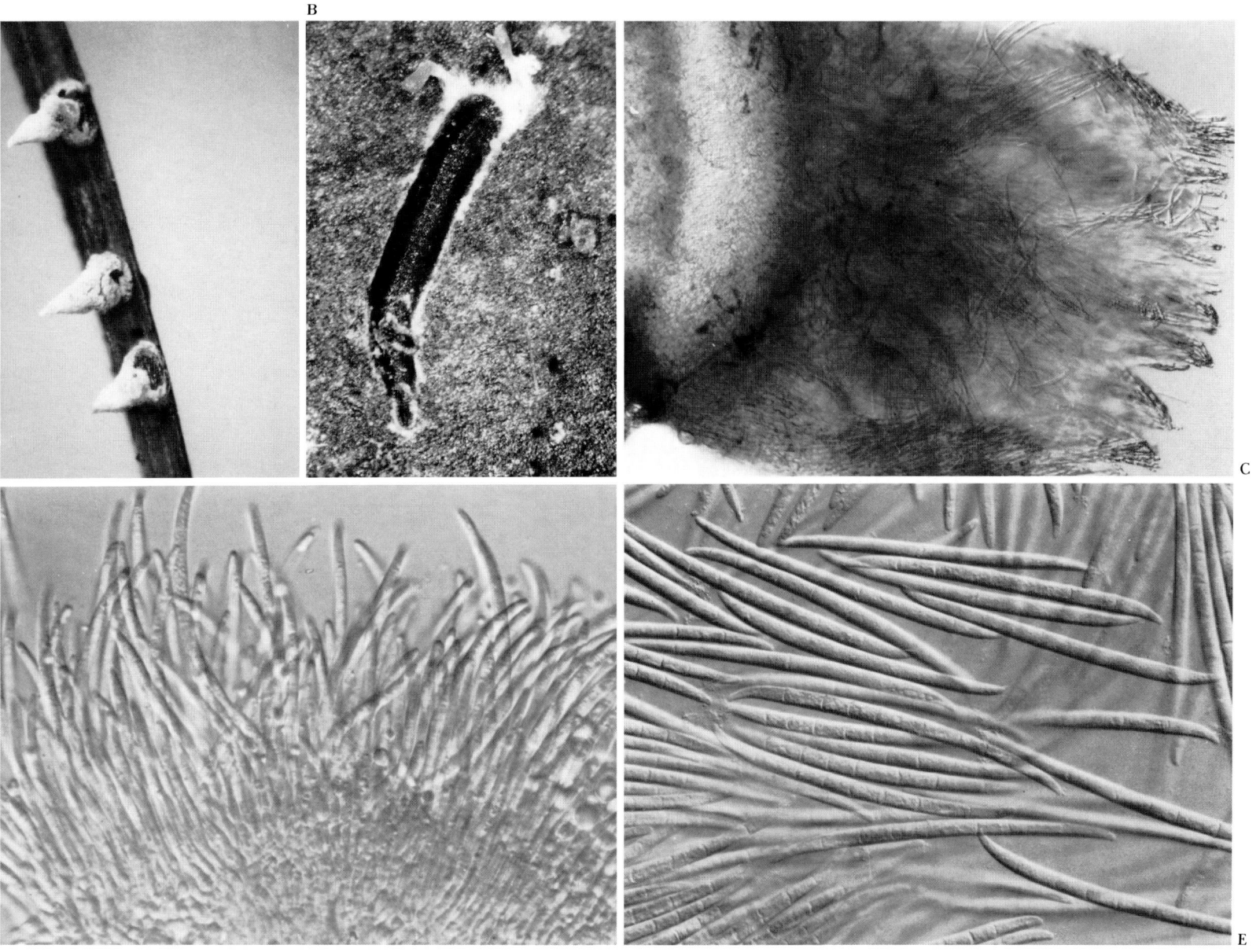

PLATE 70. *Gibellula alata*
A Synnema with distinct wing-like structure on a small tropical spider, B part of synnema showing wing-like structure and conidiophores (×75), C conidiophores (×200), D conidiophores with vesicle, phialides and conidia (×750).

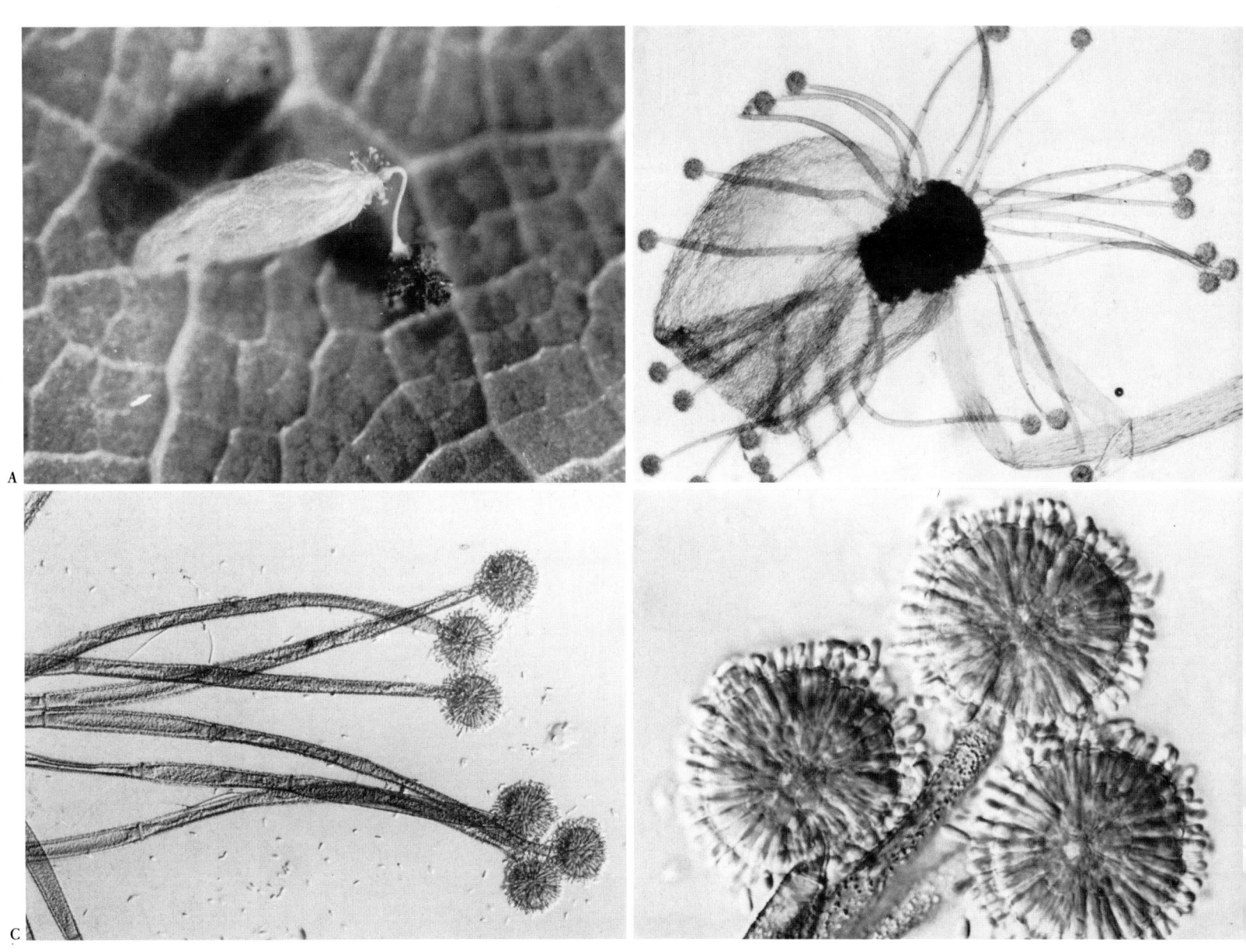

PLATE 71. *Gibellula leiopus*
A Synnemata on a tropical spider (Salticidae), B-D part of synnema showing conidiophores on short stipes, phialides and conidia (B × 750, C-D × 1875).

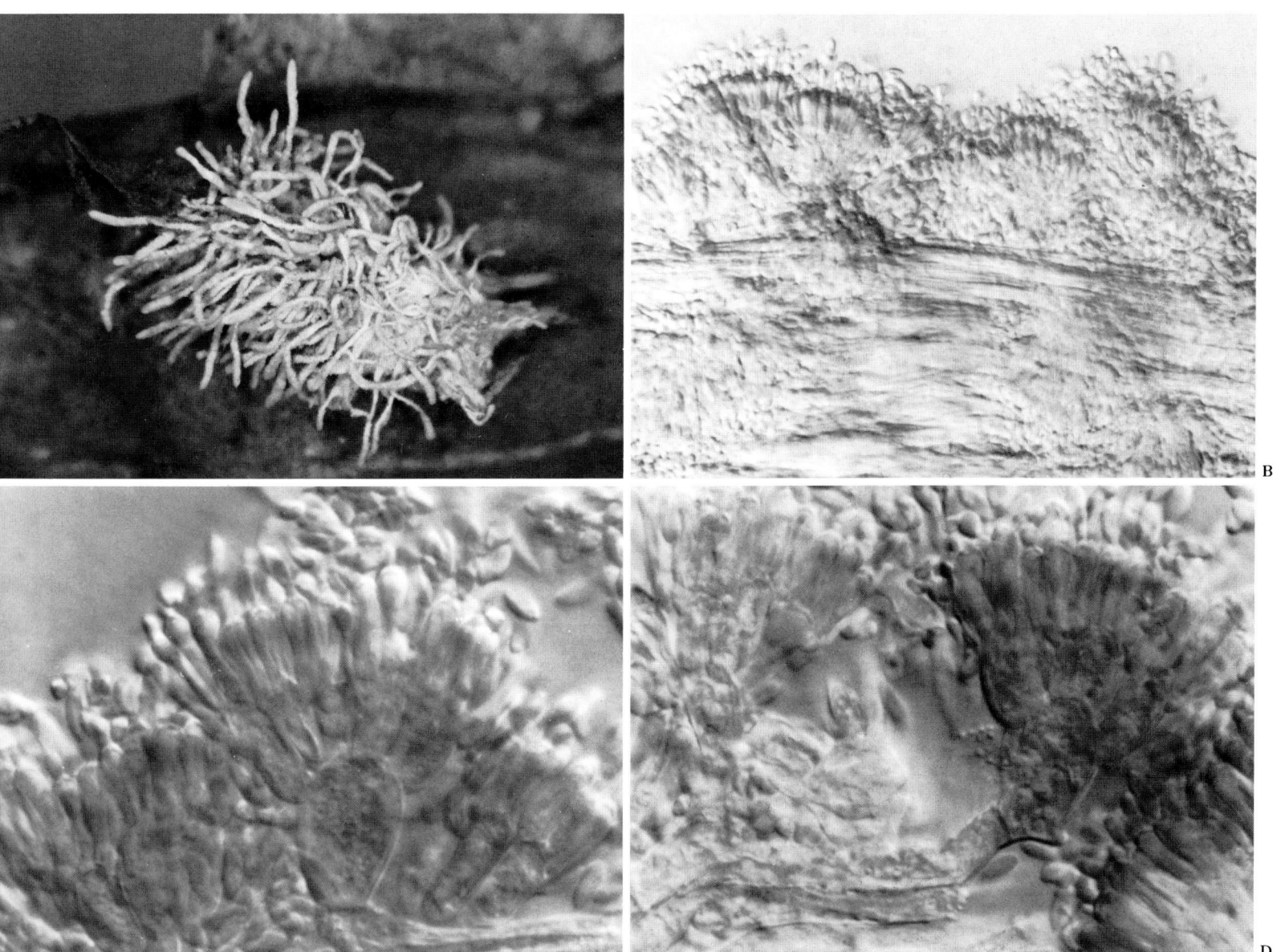

PLATE 72. *Gibellula pulchra*
A Synnemata on a large tropical spider, B part of synnema showing conidiophores (×200), C-D conidiophores showing vesicle, phialides and conidia (×750 and ×1875 resp.).

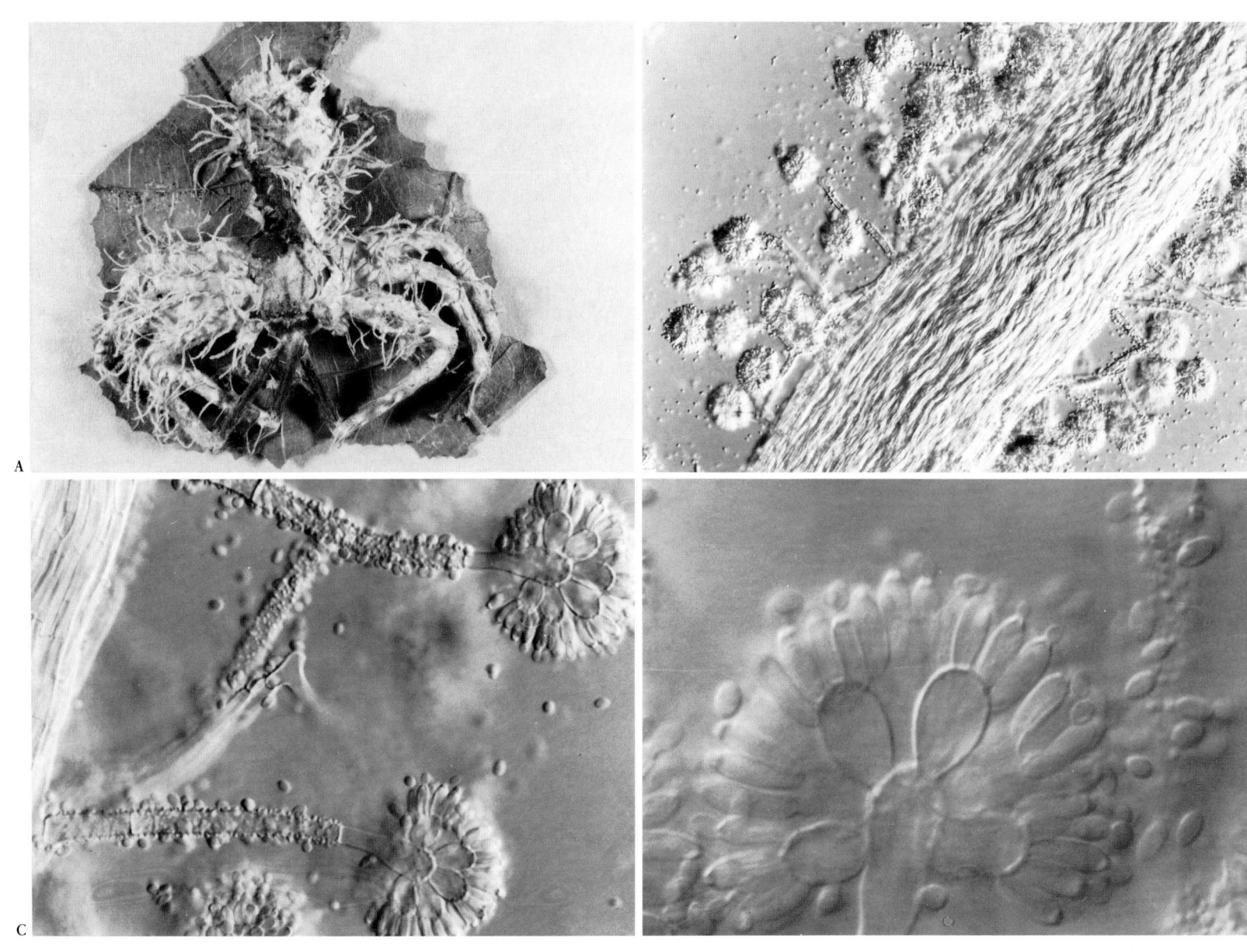

PLATE 73. *Granulomanus* state
A Conidiophore formation on a small salticid spider, B conidiophore development on legs of host (× 200), C polyblastic conidiogenous cells (× 750), D conidiogenous cells and rod-like conidium (× 1875).

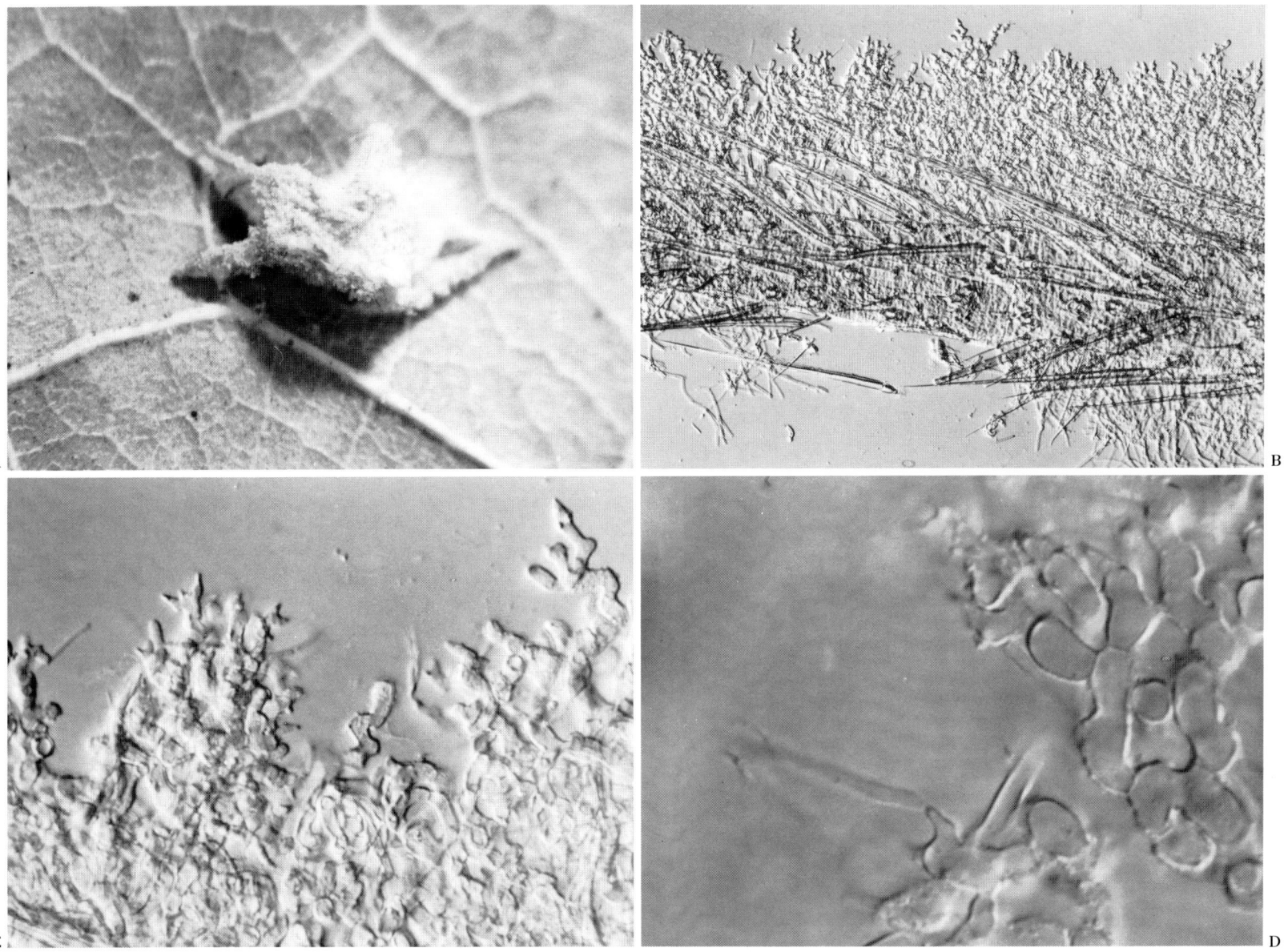

PLATE 74. *Hirsutella citriformis*
A Synnemata on brown planthopper of rice *(Nilaparvata* sp.), B branched synnema (× 75), C part of synnema showing phialides and conidia (× 750), D conidia held together in a mucus (× 1875).

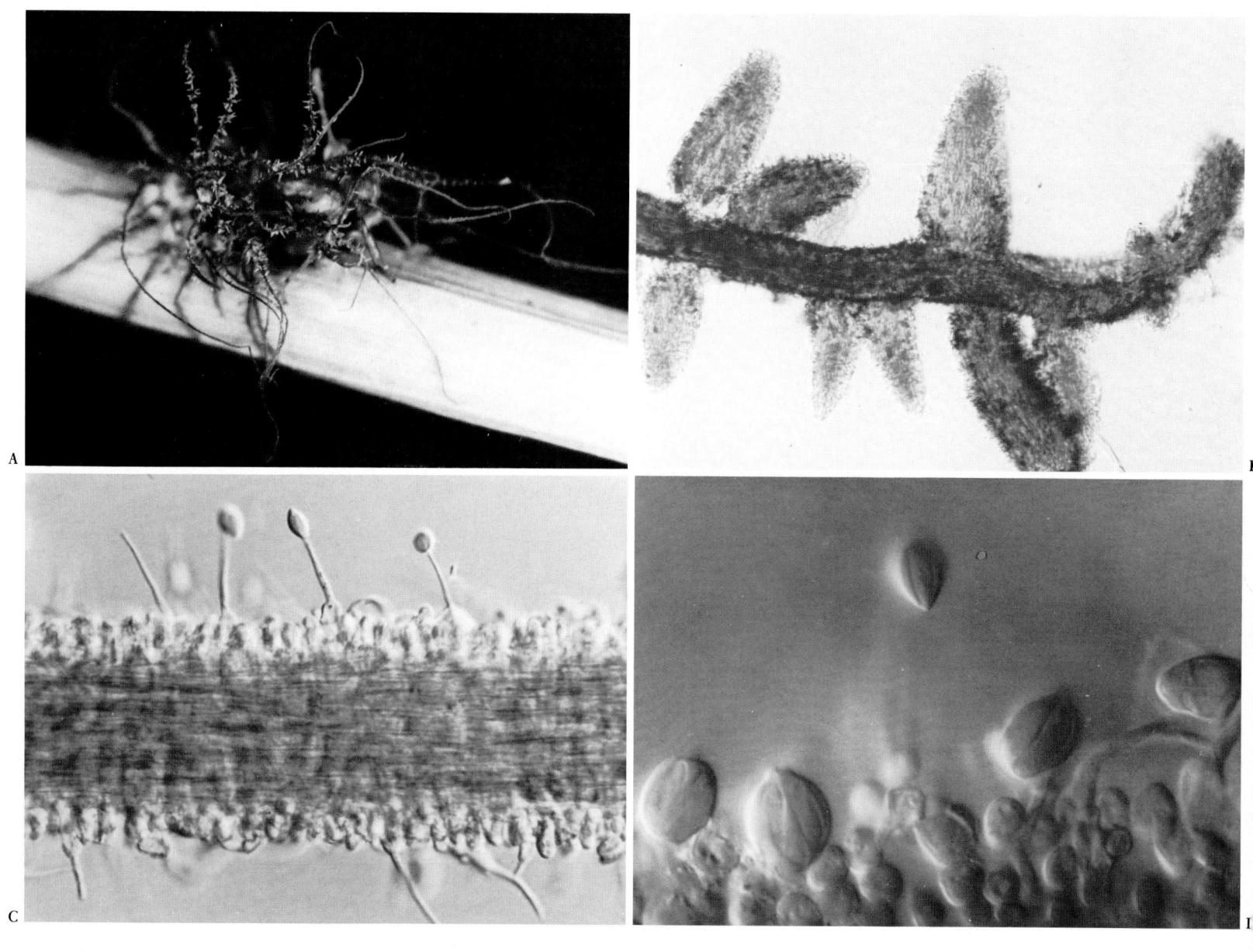

PLATE 75. *Hirsutella entomophila*
A Synnemata on adult coleopterans, B apical part of synnema (×75),
C-D part of synnema showing solitary awl-shaped phialides and
conidia (×750).

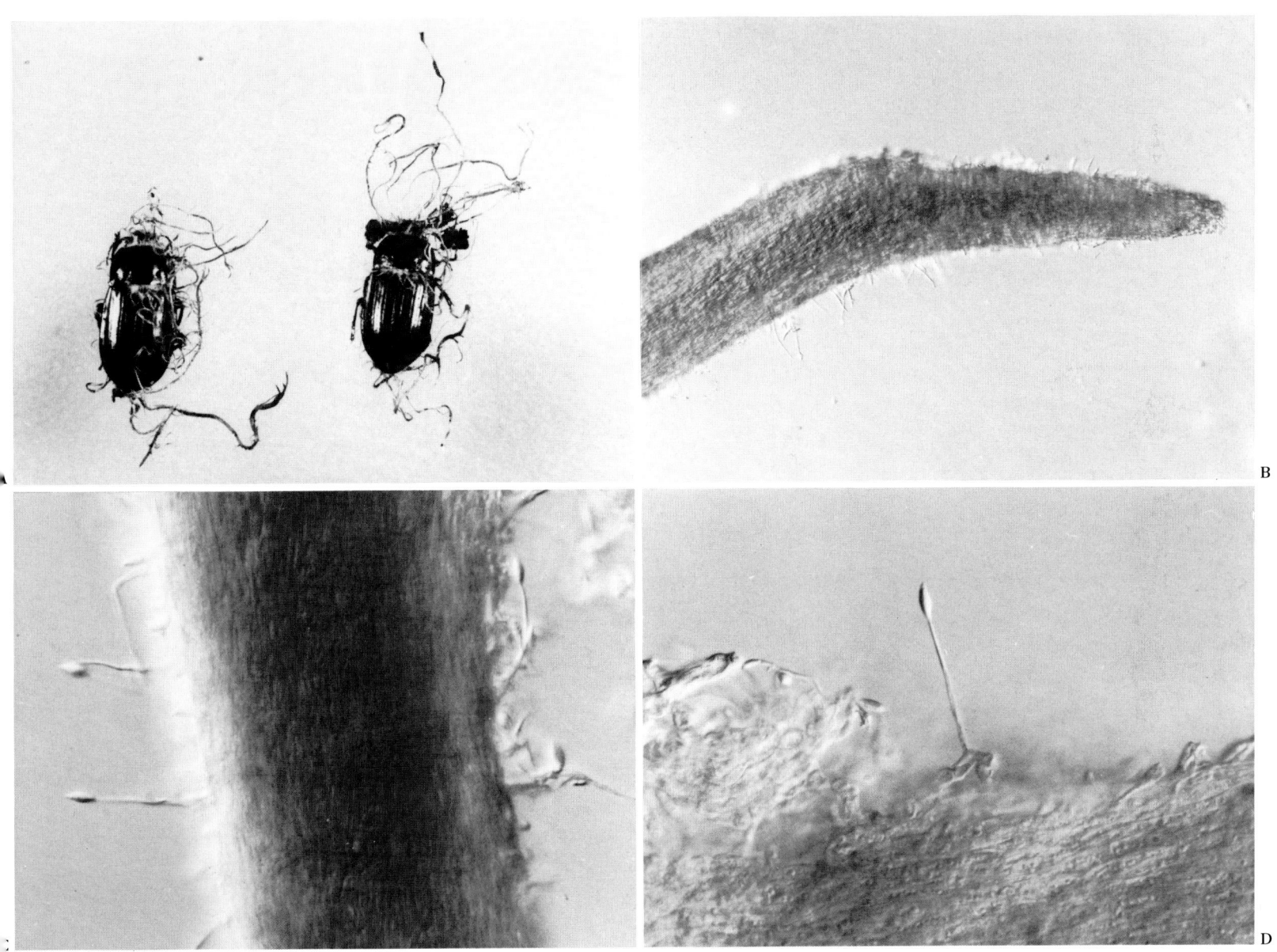

PLATE 76. *Hirsutella jonesii*
A Sclerotia and conidiophore production on a leafhopper *(Nephotettix* sp.), B sclerotia (×200), C-D part of synnema with long phialides (×200 and ×750 resp.), E-F apical part of synnema bearing a ball of conidia covered in mucus (×750 and ×1875 resp.).

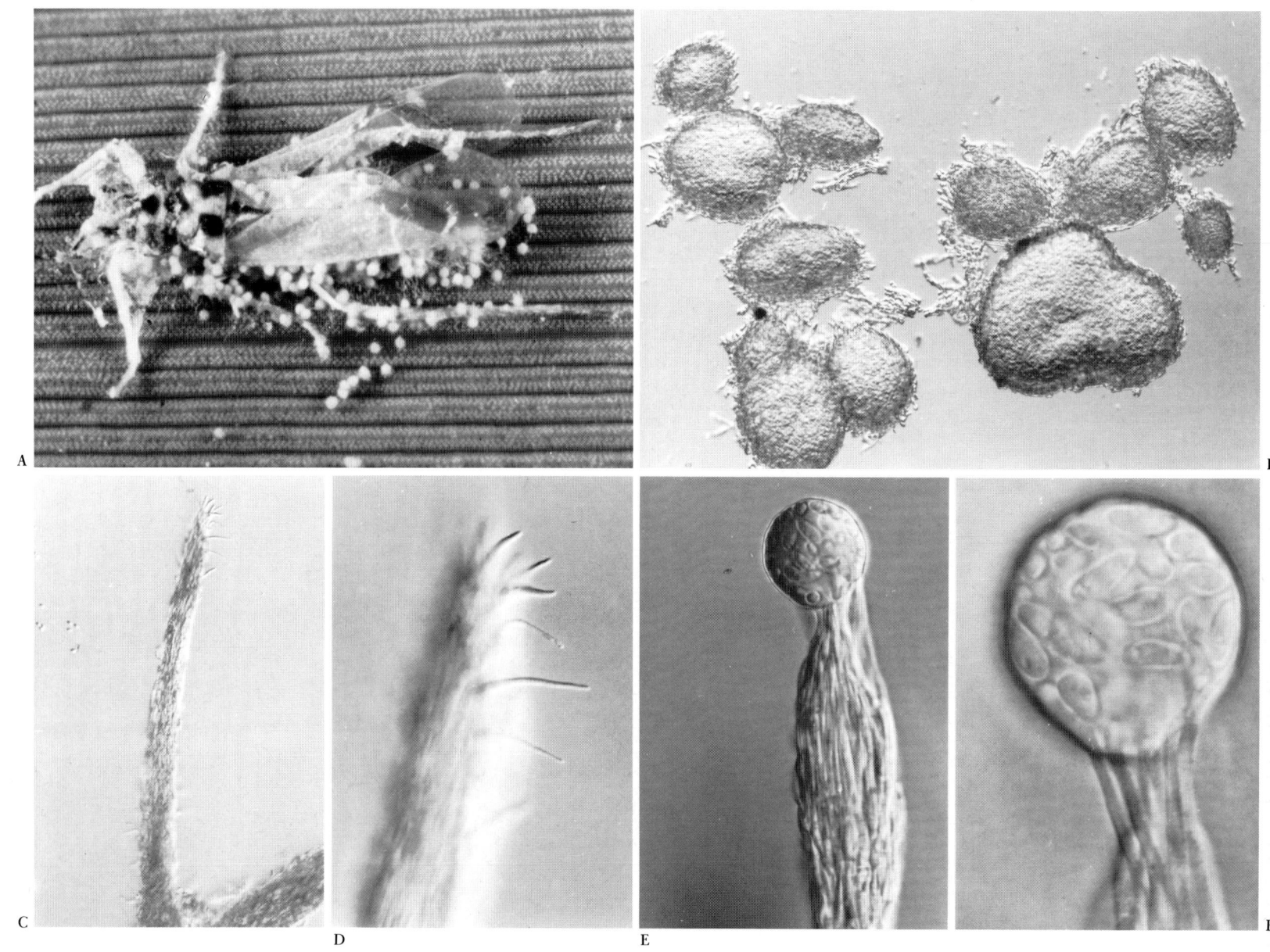

PLATE 77. *Hirsutella sausserei*
A Synnemata on *Polistes* wasps, B apical part of synnema (× 200),
C–D part of synnema showing phialides with long necks and inflated
base (× 750), E conidia (× 750).

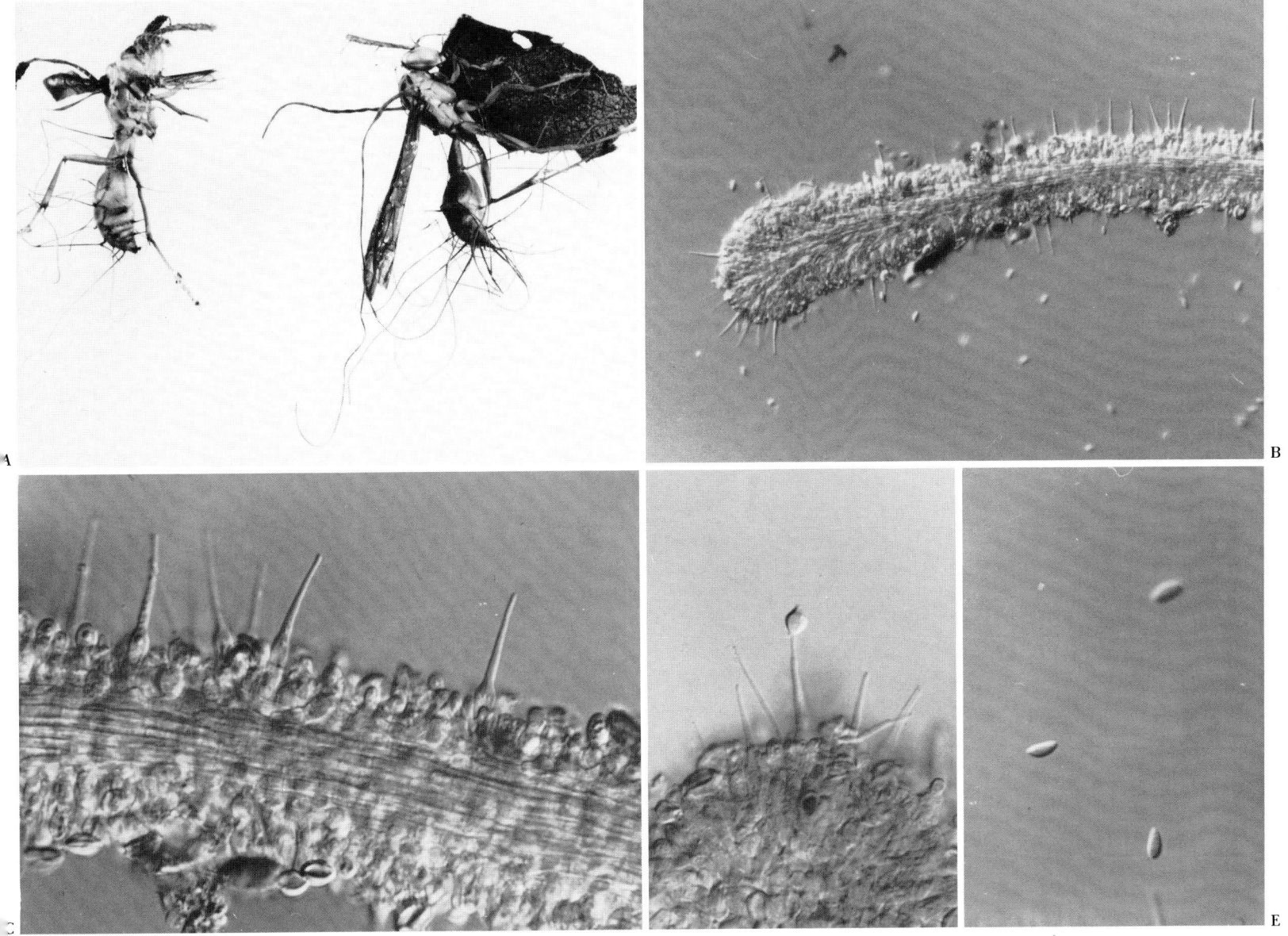

PLATE 78. *Hirsutella thompsonii*

A Fertile hyphae arising from eryophid mite, B host showing development of mycelium within the body (×750), C-D hyphae bearing solitary phialides and globose rough-walled conidia (×750 and ×1875 resp.), E culture of *H. thompsonii* on malt extract agar isolated from an *Abarus* mite, showing production of mononematous *Hirsutella* conidiogenous structures and distinct synnemata production of the *Akanthomyces* synanamorph.

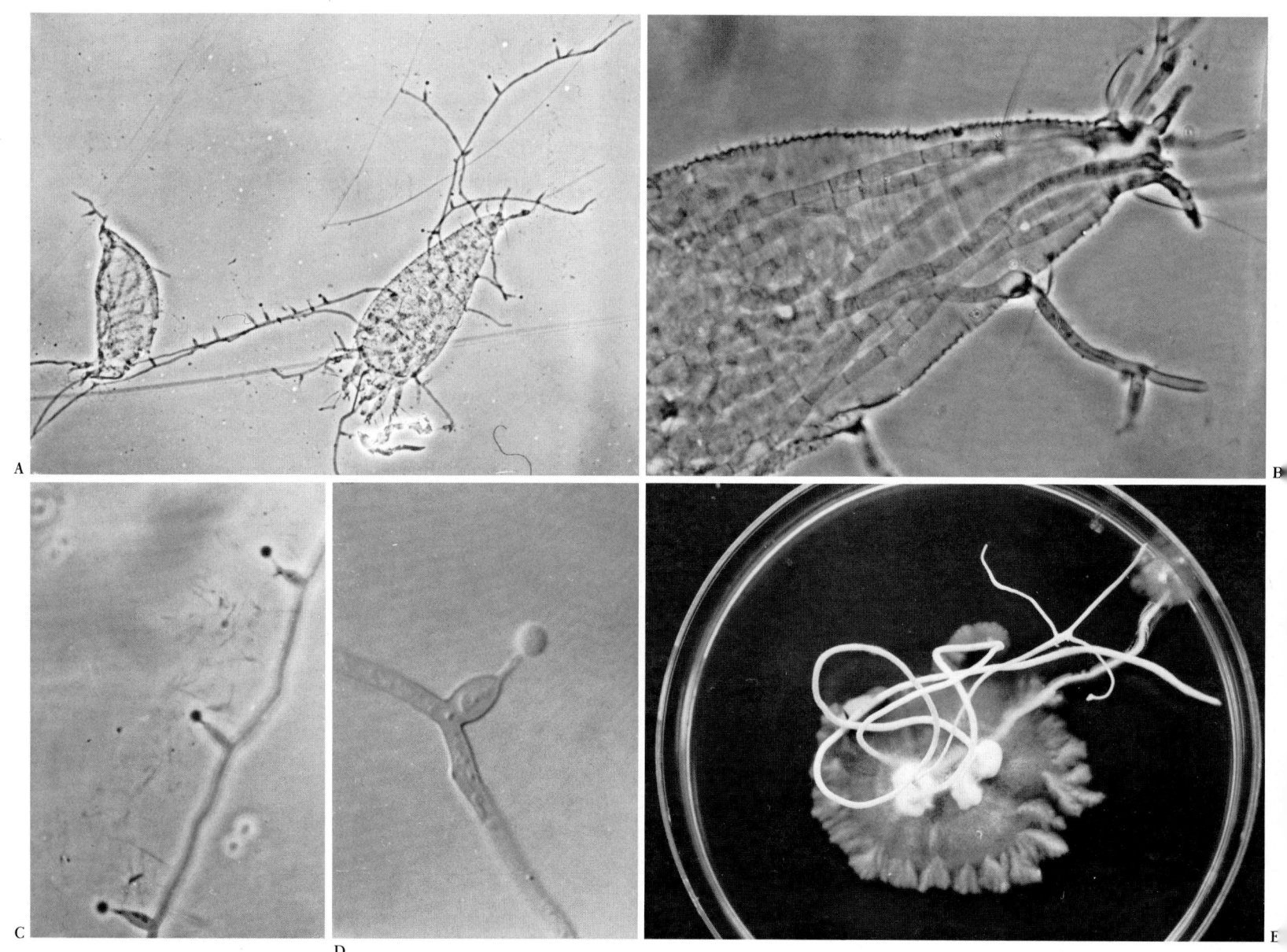

PLATE 79. *Hirsutella versicolor*
A Conidiophore development on mango leafhoppers *(Idiocerus* sp.), B-E phialidic conidiogenous cells with one or more necks and production of thin fusiform or globose conidia (×750).

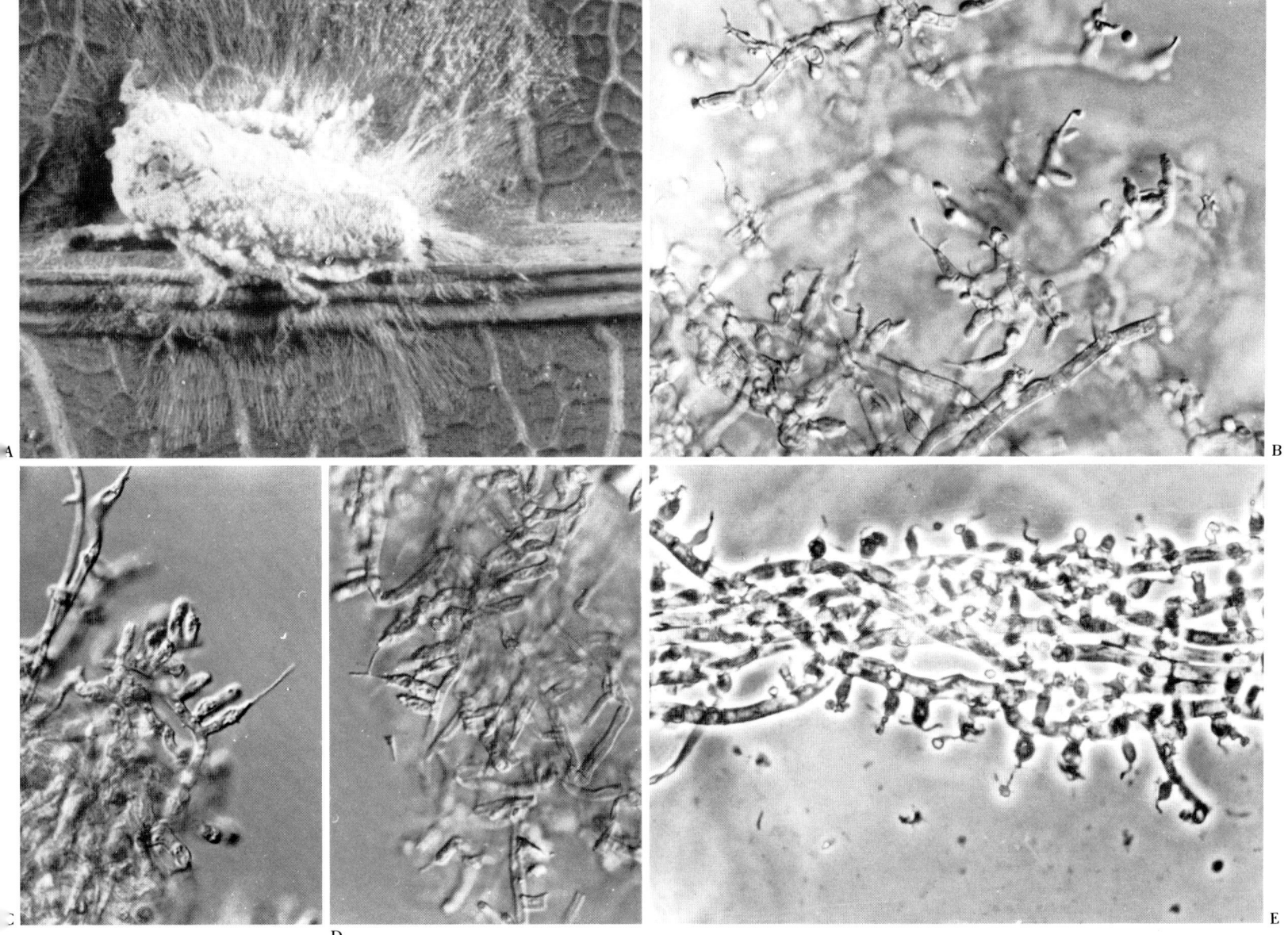

PLATE 80. *Hymenostilbe dipterigena*
A Synnemata of *H. dipterigena* and clavae of *Cordyceps dipterigena* on fruit flies (Tephritidae), B-D parts of synnema showing hymenium-like layer of polyblastic conidiogenous cells and conidia (B-C × 750, D × 1875).

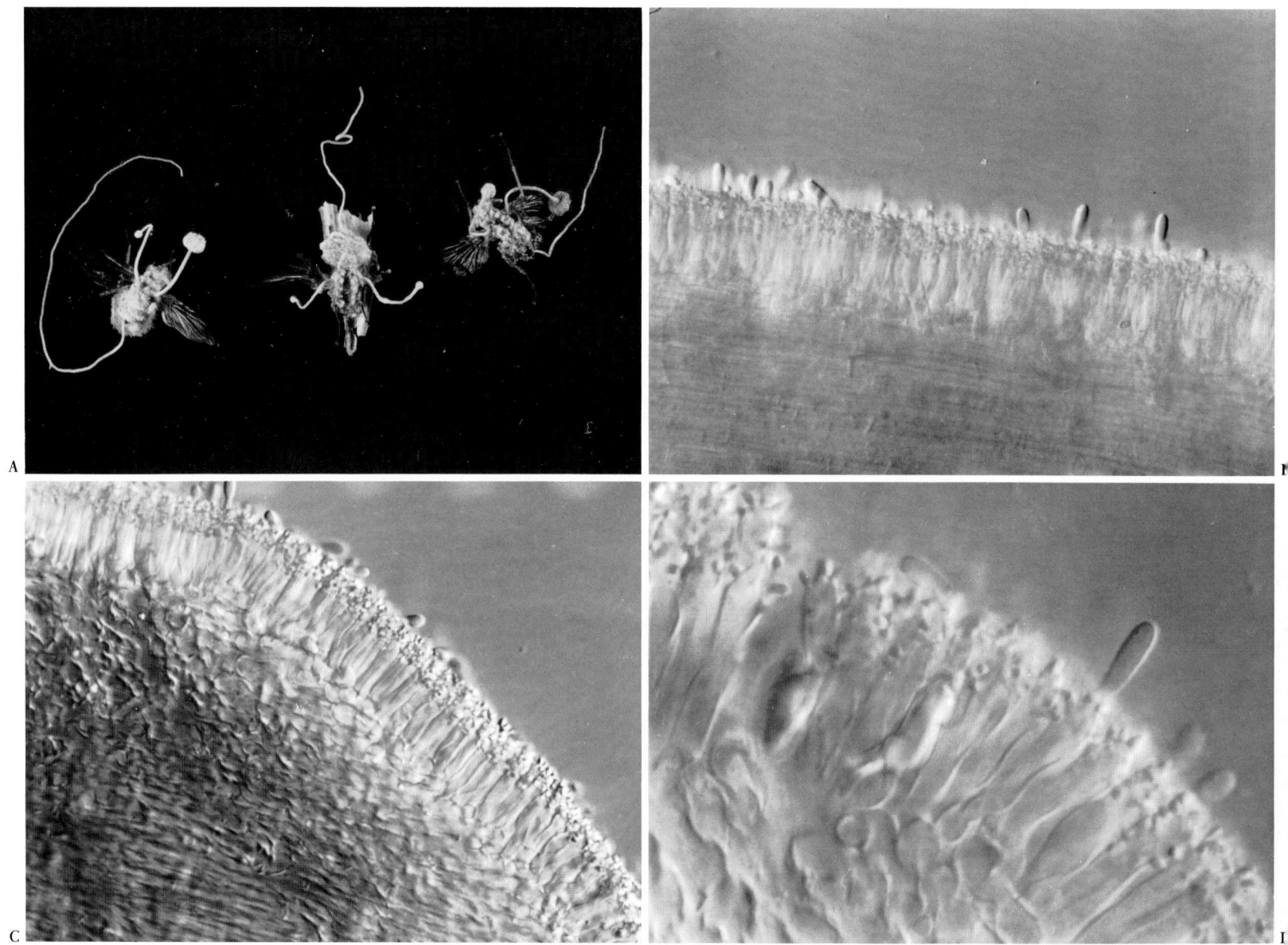

PLATE 81. *Hymenostilbe formicarum*
A Synnema of *H. formicarum* and clava of *Cordyceps lloydii* on *Camponotus* ant, B-C part of synnema showing hymenium-like layer of polyblastic conidiogenous cells (×750 and ×1875 resp.), D conidia (×750).

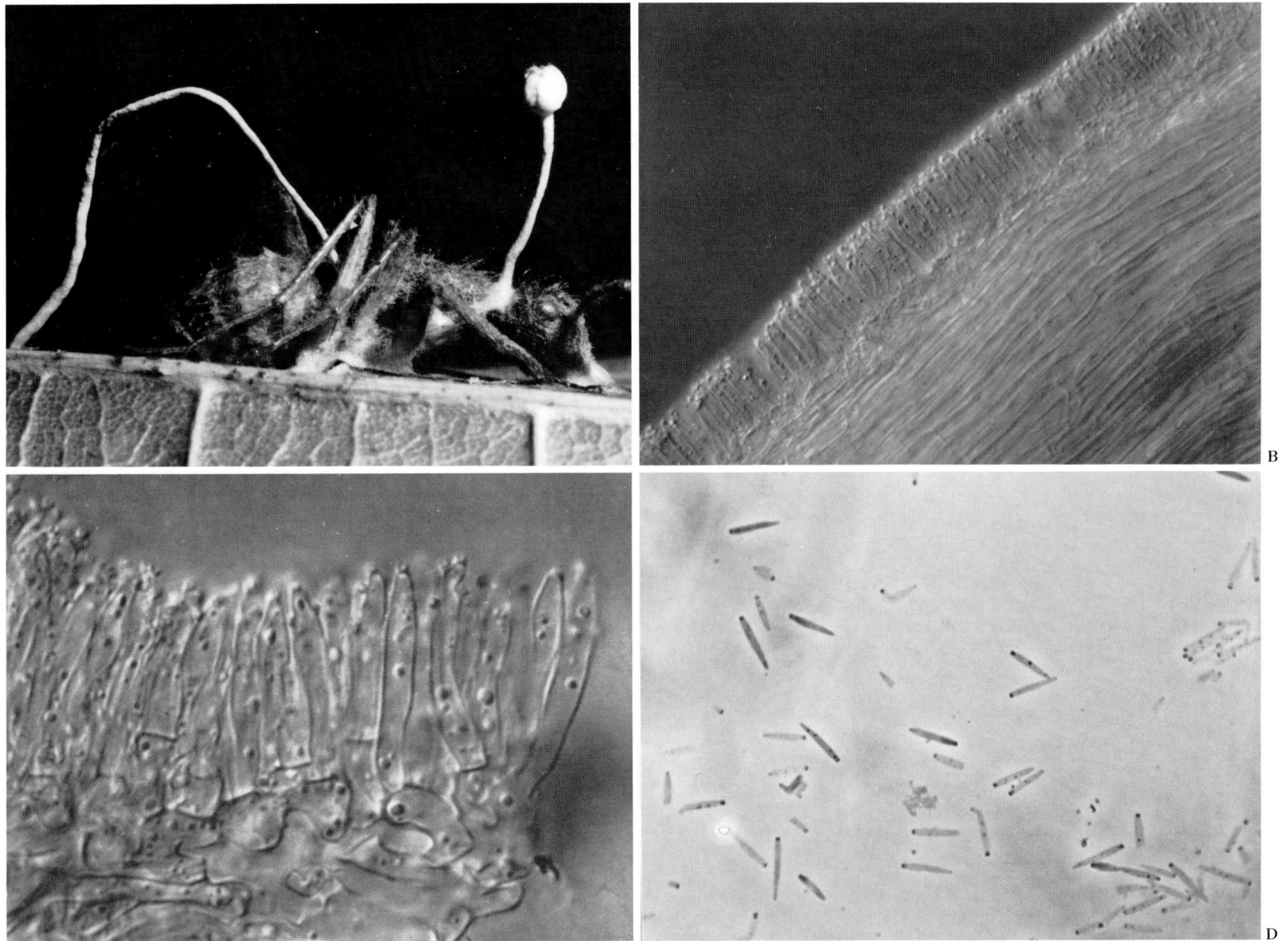

PLATE 82. *Hymenostilbe muscaria*
A Synnemata on fly, B part of synnema (×750), C conidia (×750), D polyblastic conidiogenous cells (×1875).

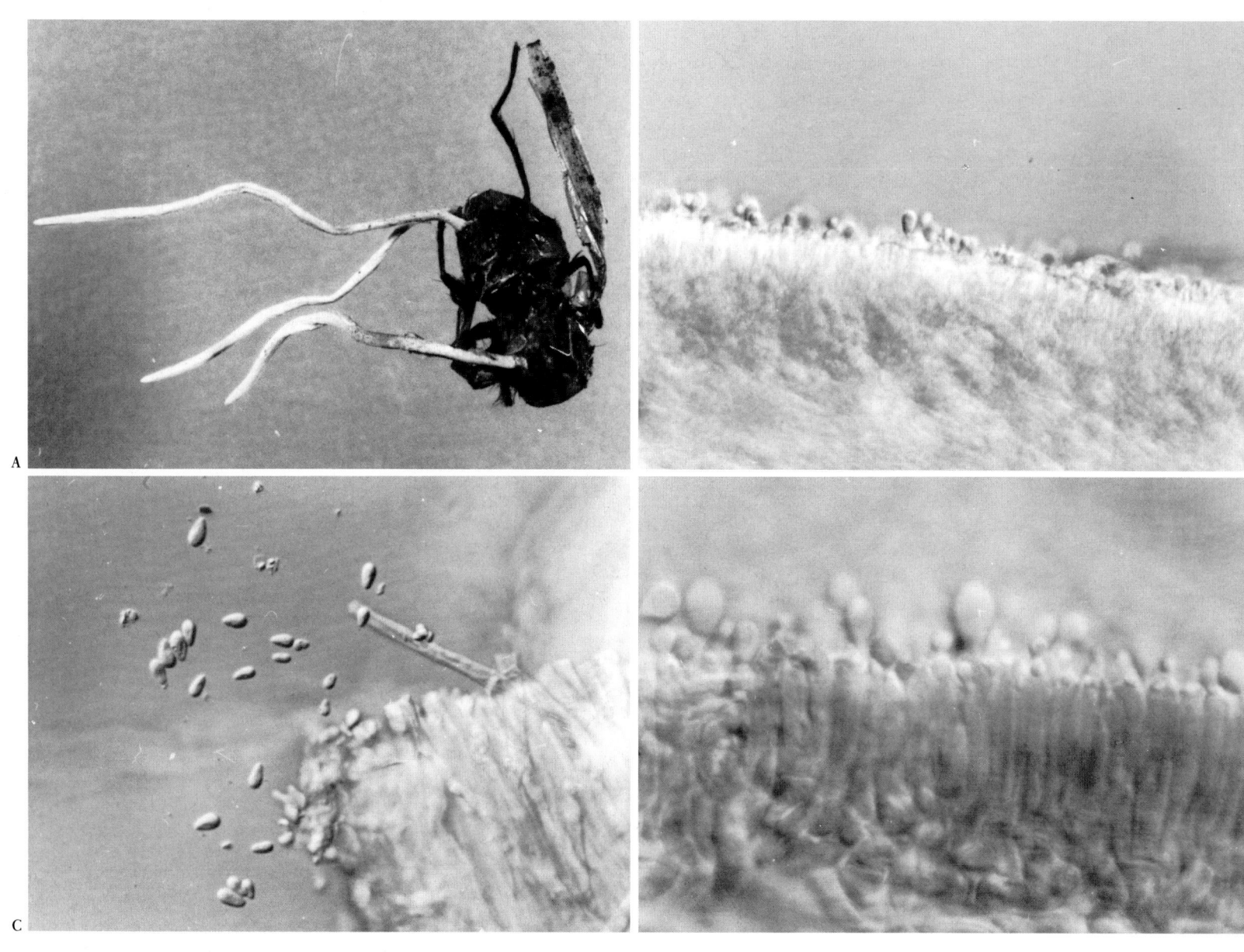

PLATE 83. *Hymenostilbe* species
A Synnemata on Gryllidae, B section through synnema showing globose broader apical region (× 200), C-D polyblastic conidiogenous cells and conidia (× 750 and × 1875 resp.).

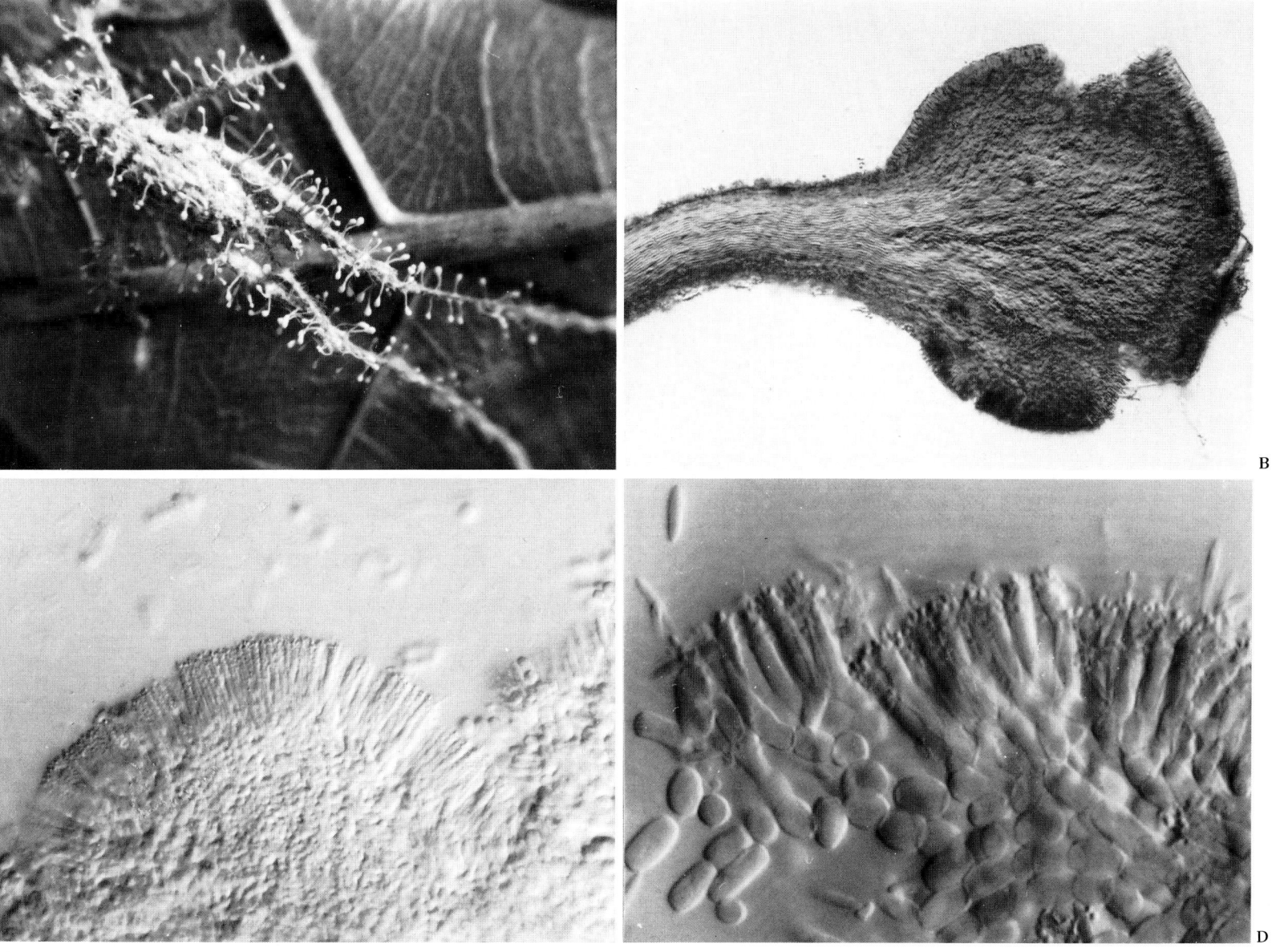

PLATE 84. *Metarhizium album*
A Colony on a rice leafhopper *(Cofana* sp.), B conidiophores and phialides (× 1875), C sclerotium production when grown on artificial media, D hyphal bodies within host (× 750), E conidia (× 1875).

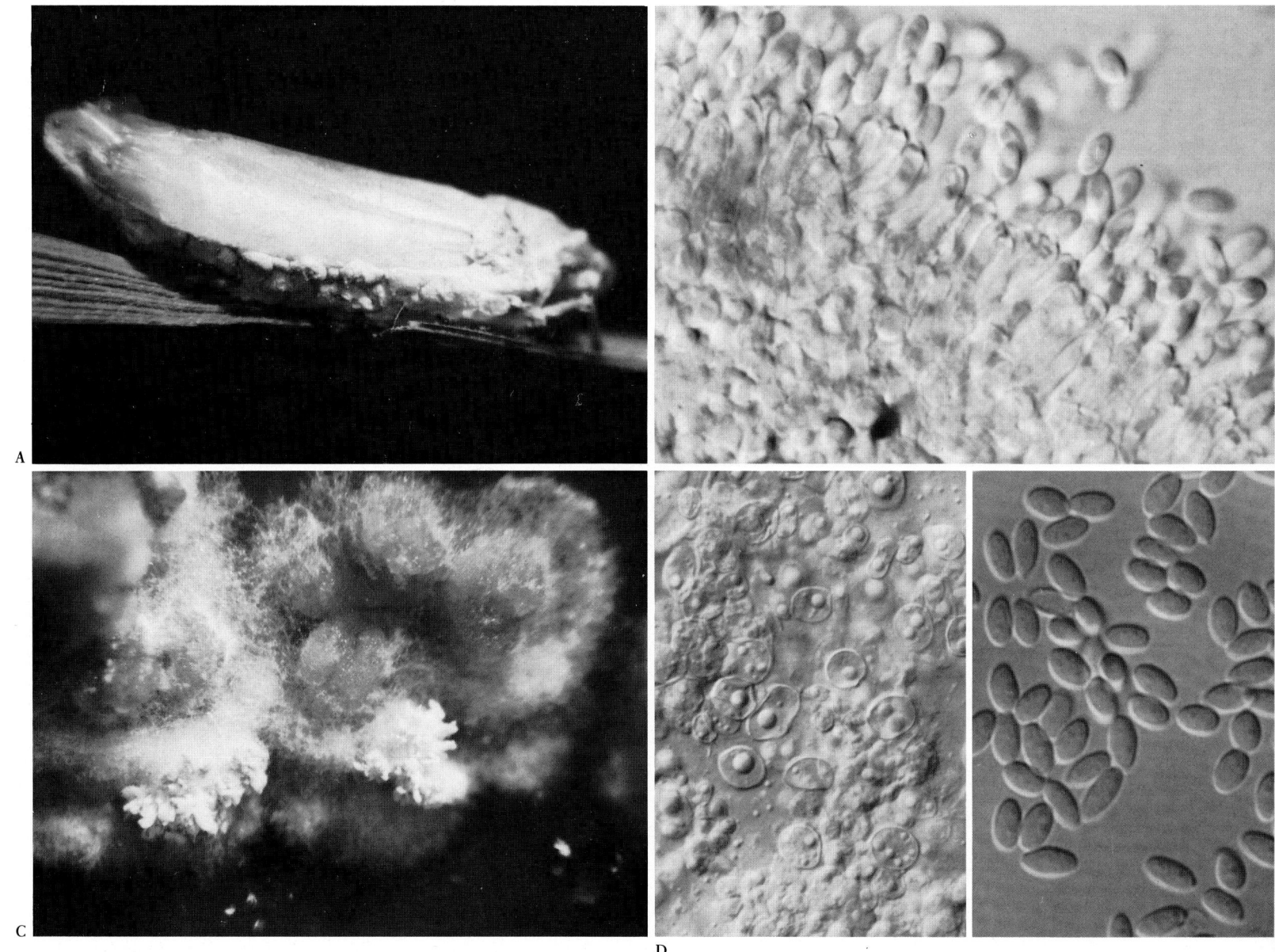

PLATE 85. *Metarhizium anisopliae*
Mononematous conidiophores on A a Blattidae adult and B on *Oryctes* beetle and synnematous conidiophores on C soil-borne coleopteran larva and D subterranean gryllid.

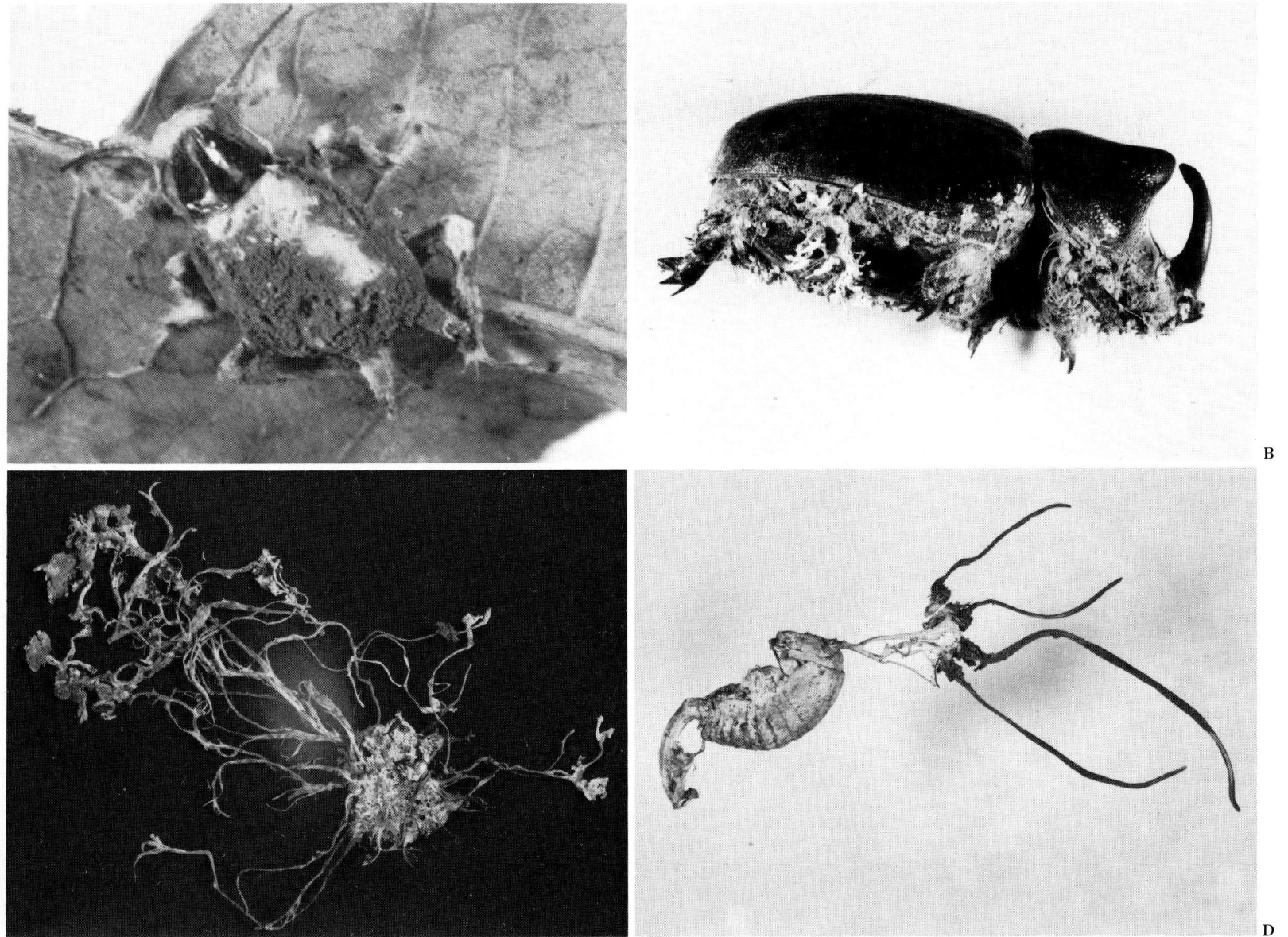

PLATE 86. *Metarhizium anisopliae* var. *anisopliae*
A Typical palisade production of conidia, B-D phialides and conidia (× 1875).

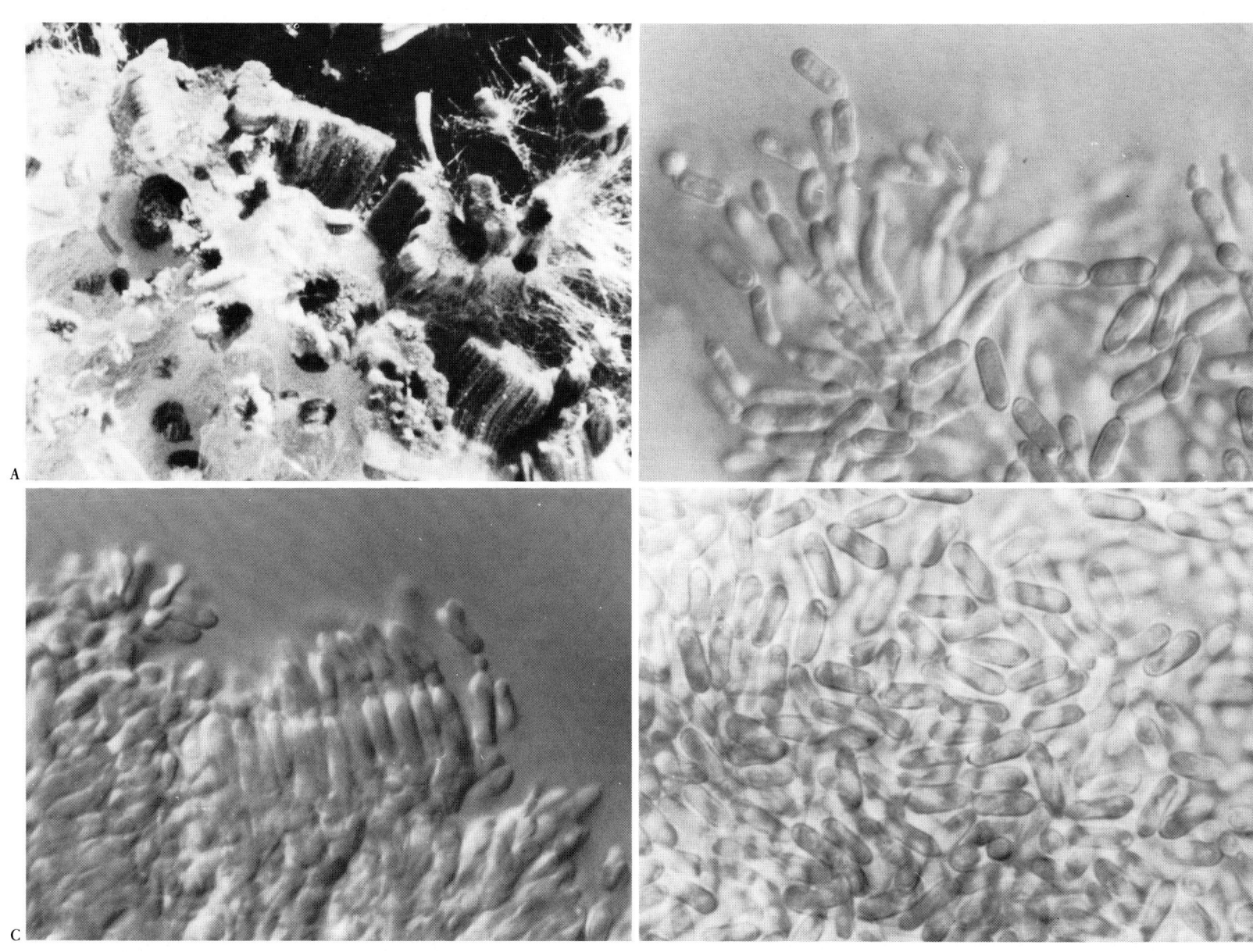

PLATE 87. *Metarhizium anisopliae* var. majus Conidiophores on A *Oryctes* beetle larvae and B *Scapanes australis* adults, C conidiophores and phialides (× 750), D conidia (× 1875).

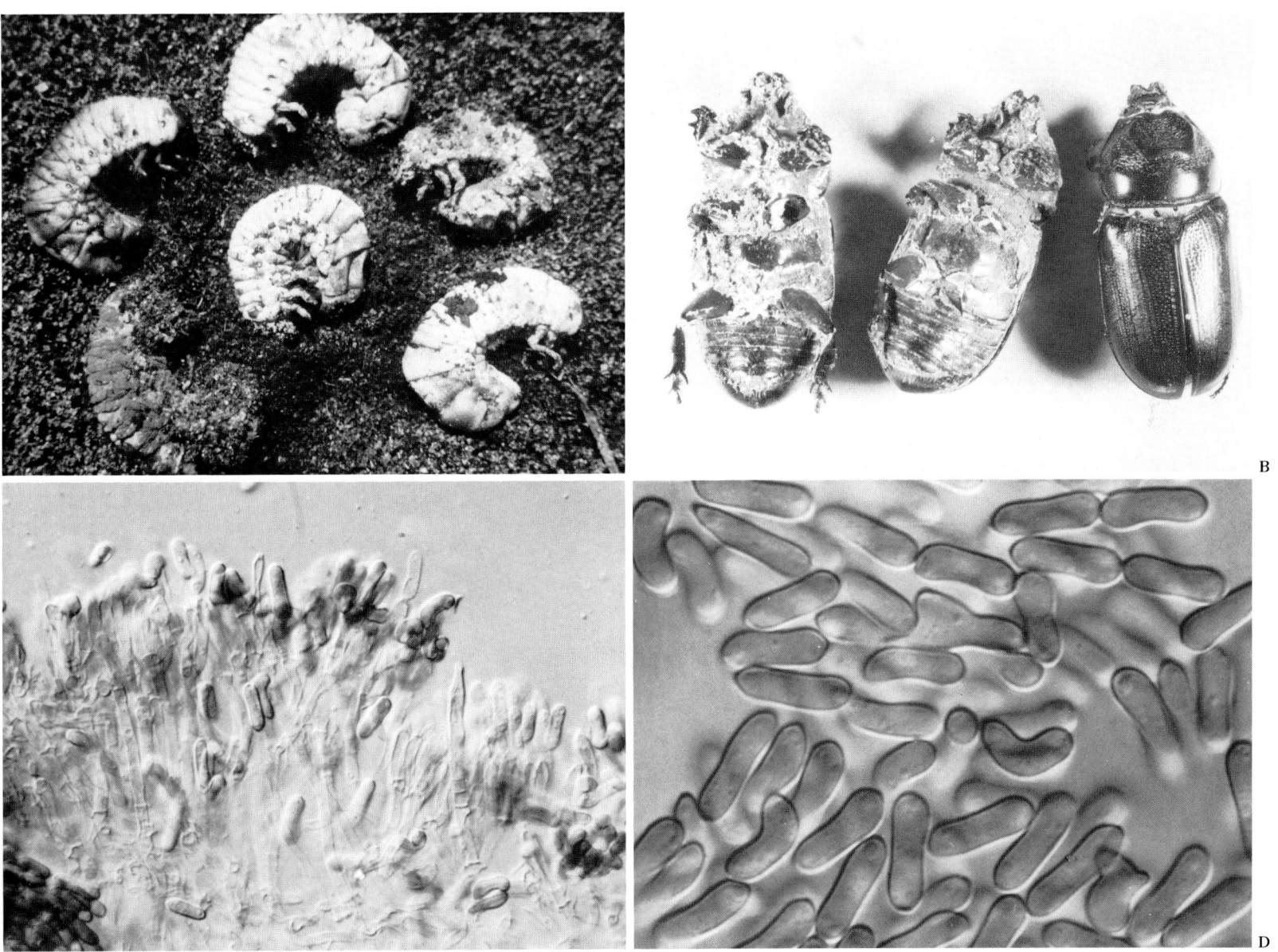

PLATE 88. *Metarhizium flavoviride*
A Conidiophores on brown planthopper of rice *(Nilaparvata* sp.), B conidiophores (×750), C phialides (×1875), D conidia (×1875).

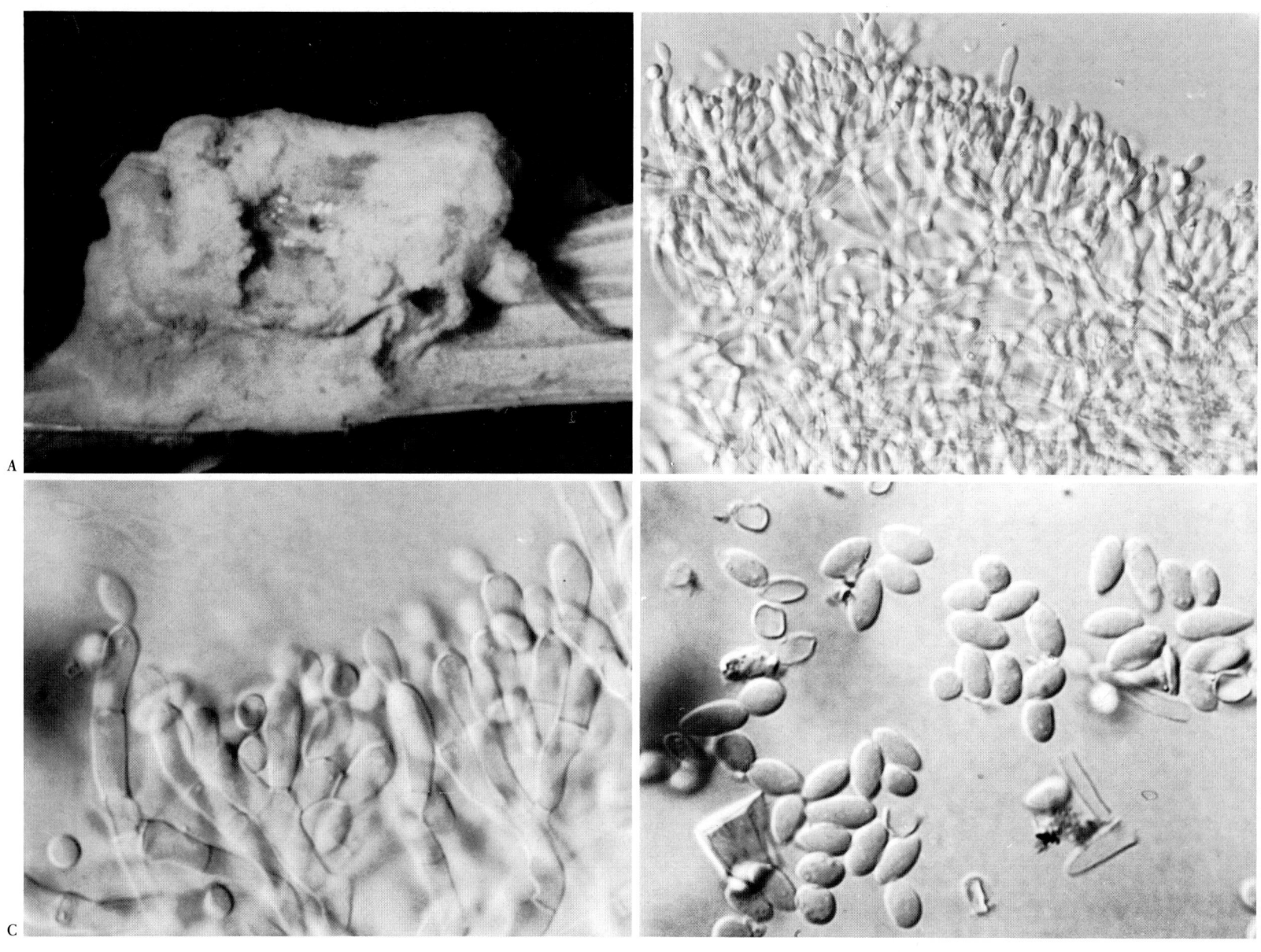

PLATE 89. *Nomuraea atypicola*
A On a large spider *(Mygalidae* sp.), B-C conidiophores showing verticillate whorls of branches and phialides with indistinct necks (×750 and ×1875 resp.), D conidia (×1875).

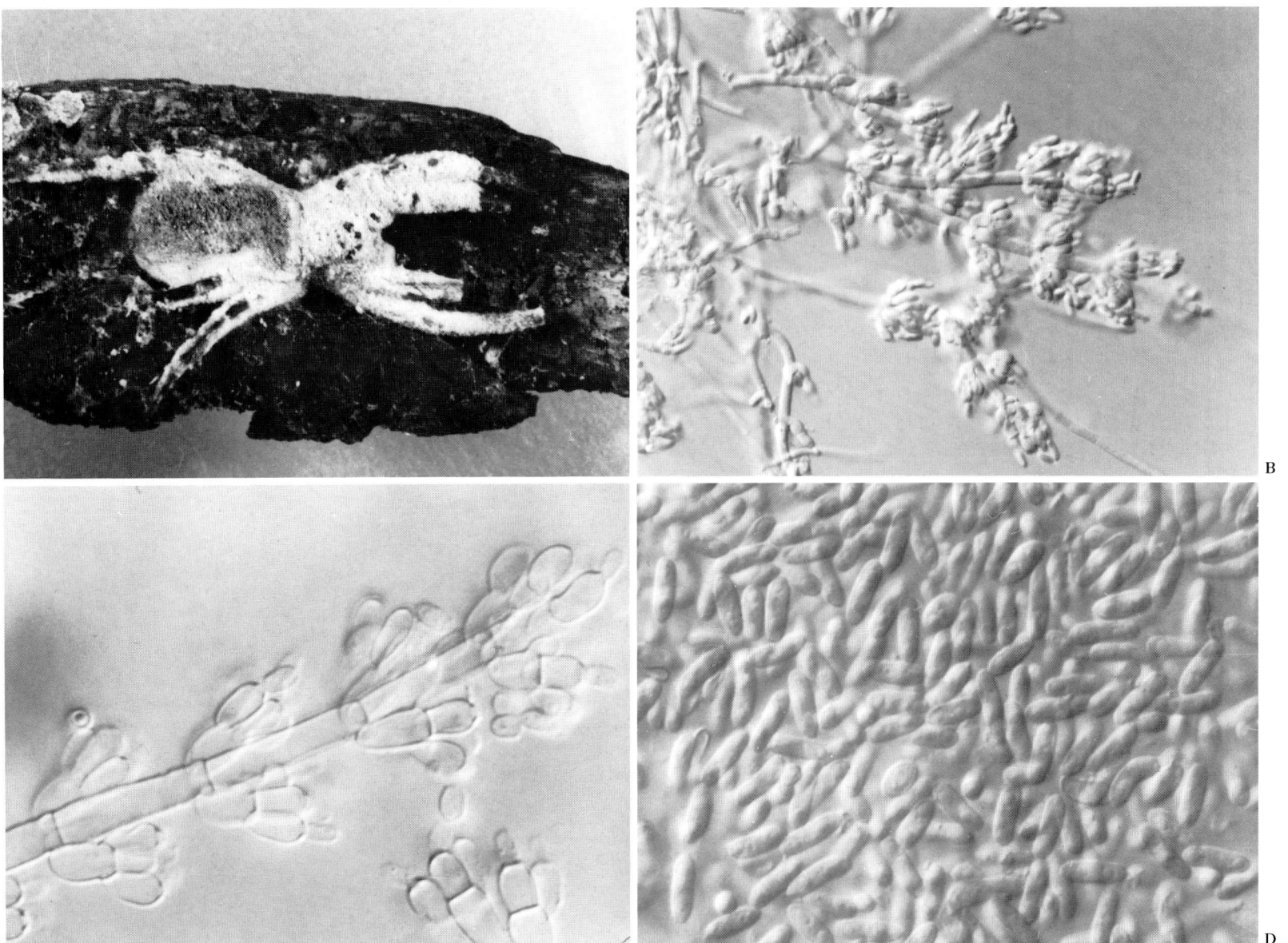

PLATE 90. *Nomuraea rileyi*
A On a soyabean looper *(Spodoptera exigua)*, B conidiophores (× 750), C-D conidiophores showing metulae and phialides (× 1875), (× 1875), E conidia (× 1875).

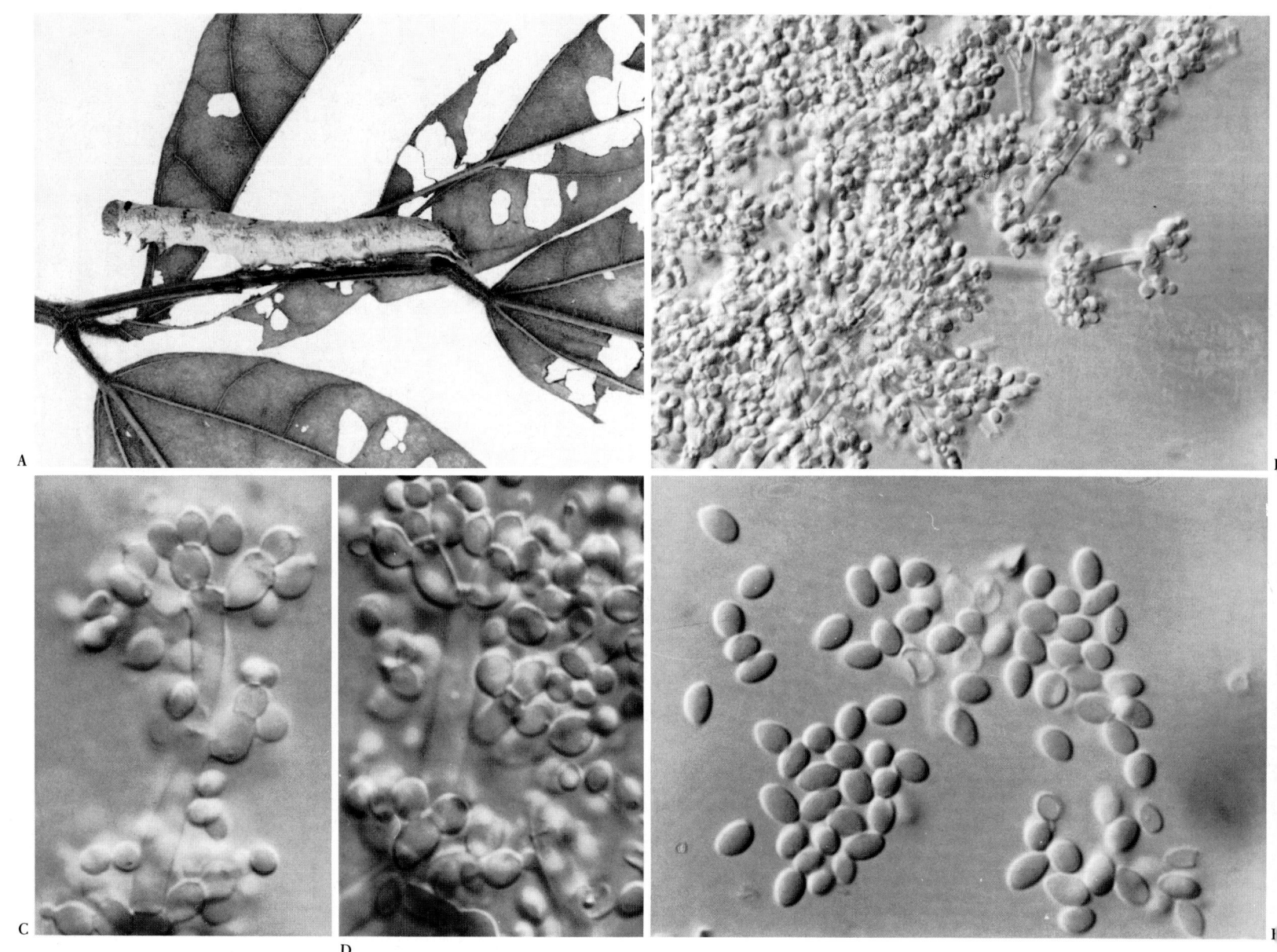

PLATE 91. *Paecilomyces amoeneroseus*
A Synnemata on lepidopteran pupa, B conidiophores (× 750), C conidiophores showing metulae and phialides (× 1875), D phialides and conidia (× 1875).

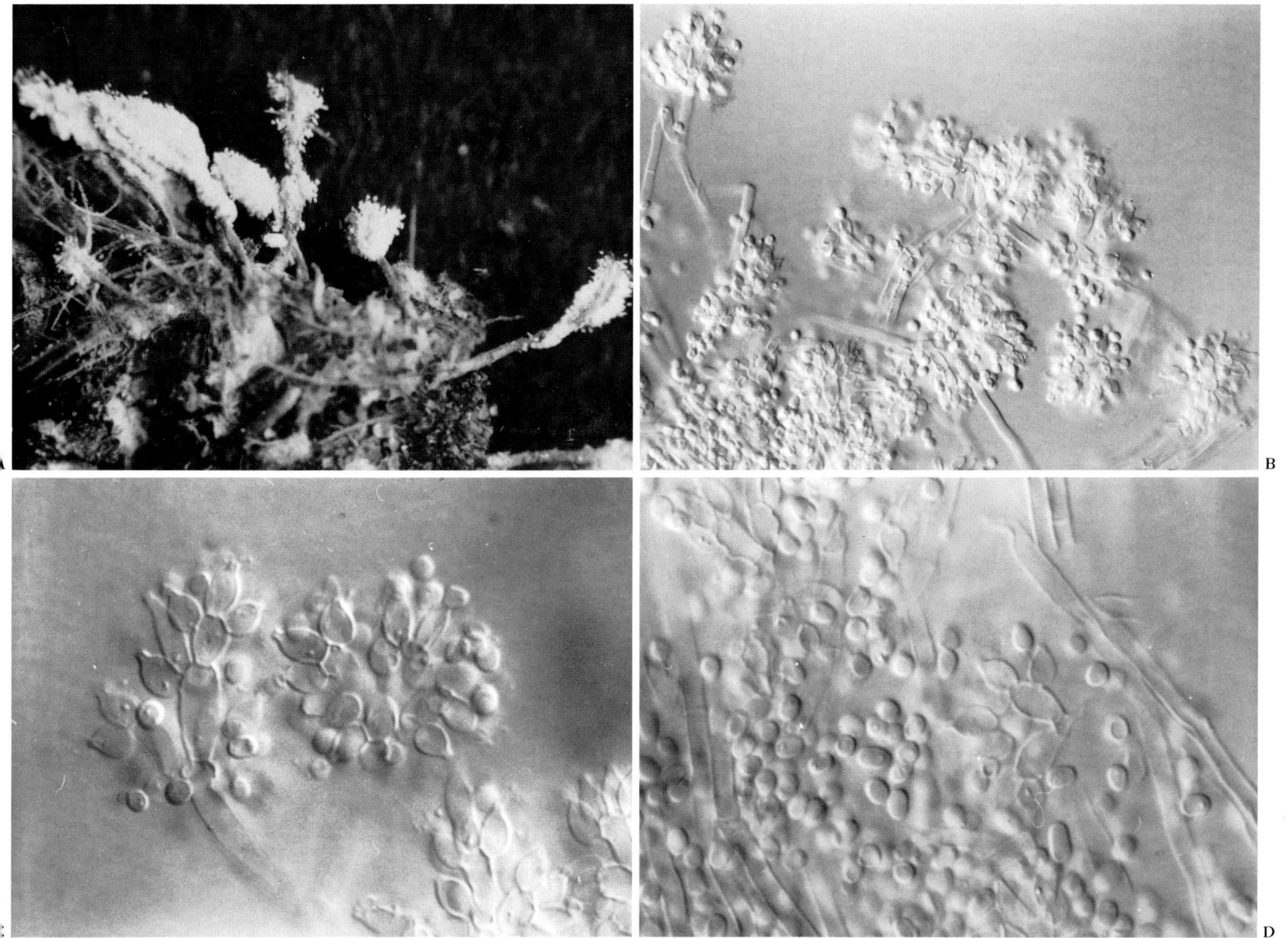

PLATE 92. *Paecilomyces cicadae*
A Synnemata on pupa of cicada (*Cicadidae* sp.), B conidiophores
(× 750), D-E phialides and conidia (× 1875).

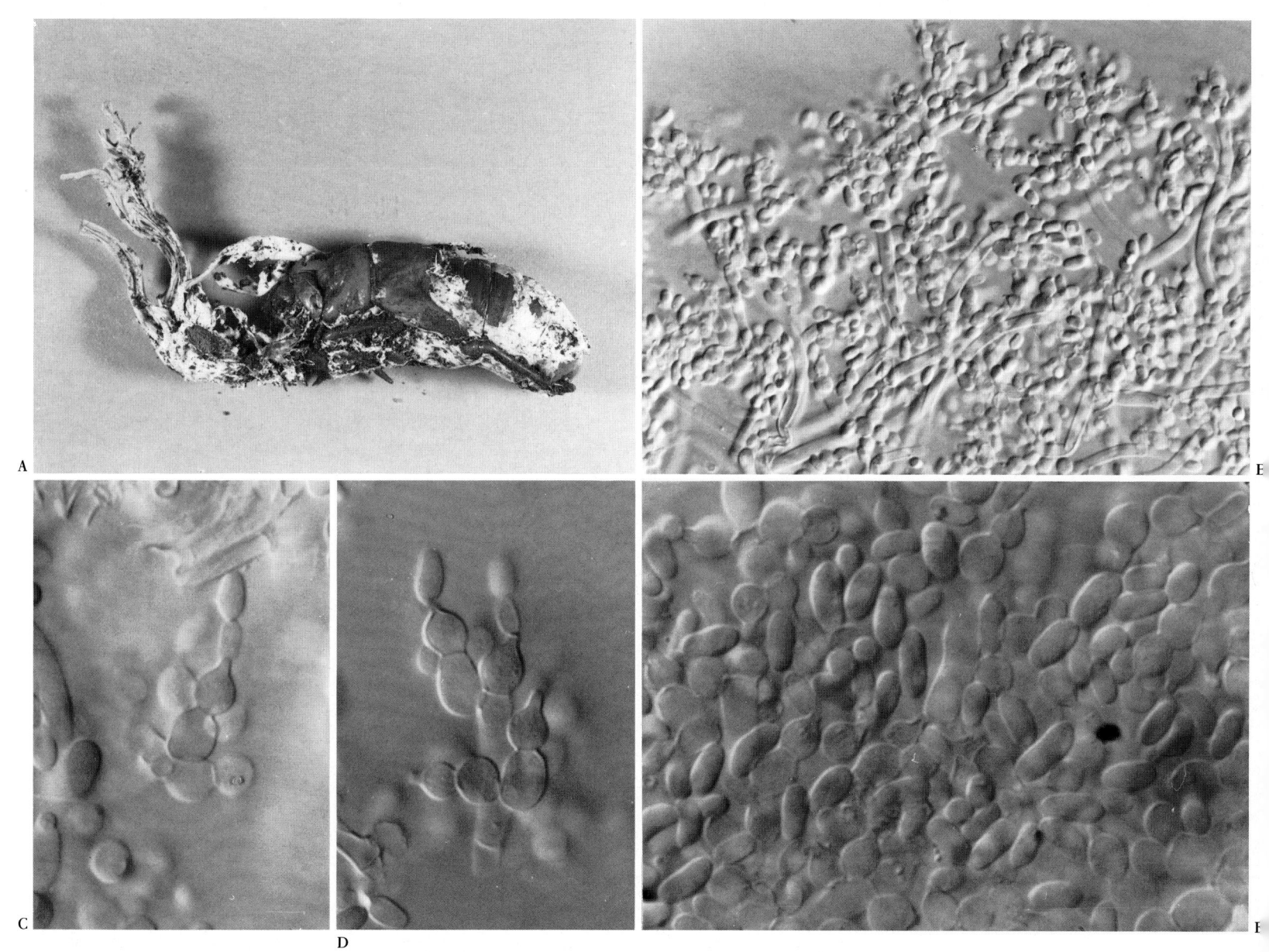

PLATE 93. *Paecilomyces farinosus*
A–B Synnemata on lepidopteran pupae, C conidiophores (× 750), D conidiophores showing metulae and phialides (× 1875), E conidia (× 1875).

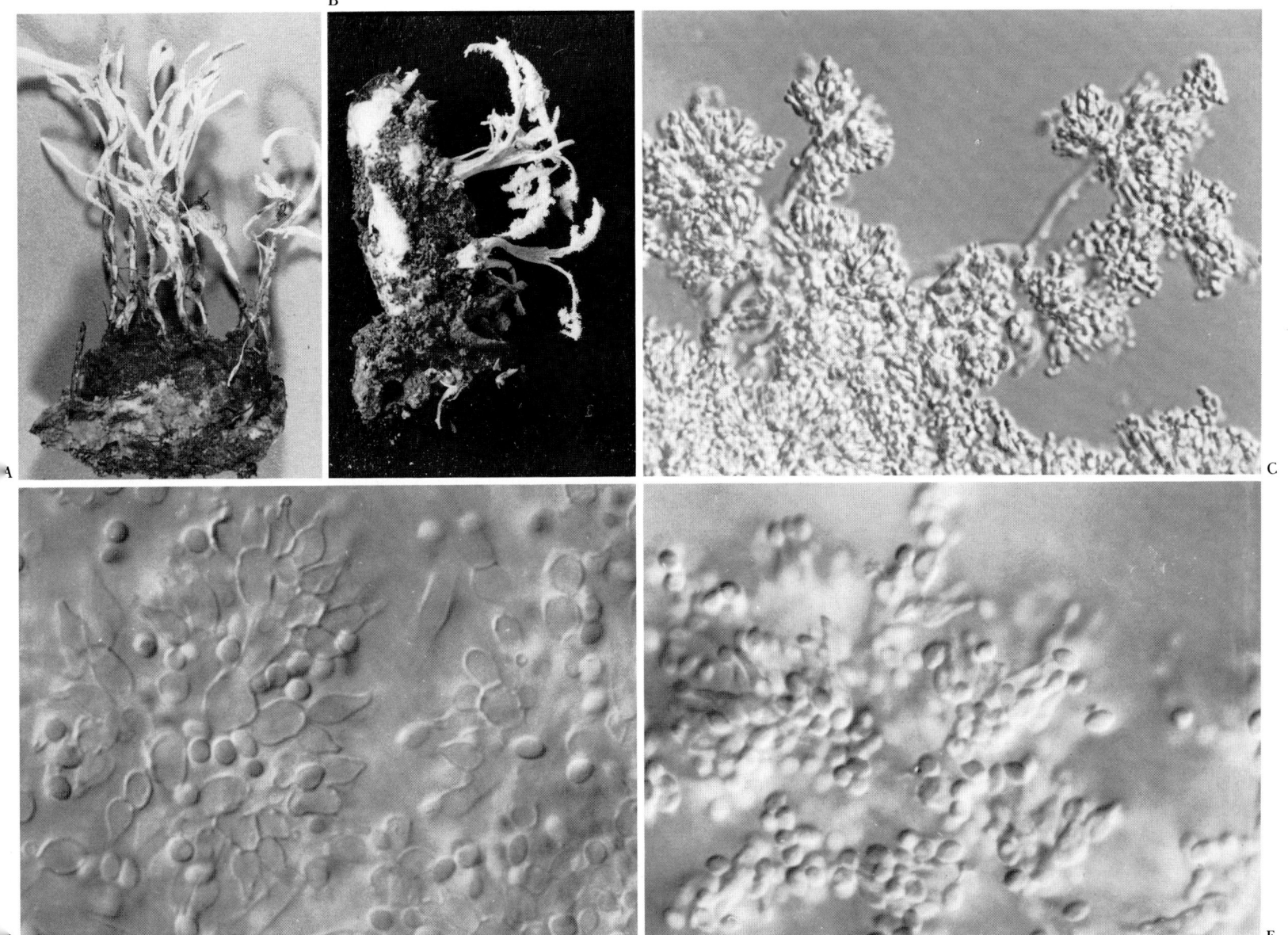

PLATE 94. *Paecilomyces lilacinus*
A Synnemata on Cydnidae *(Aethus* sp.) B synnema (× 200), C conidiphores (× 750), D conidia (× 1875).

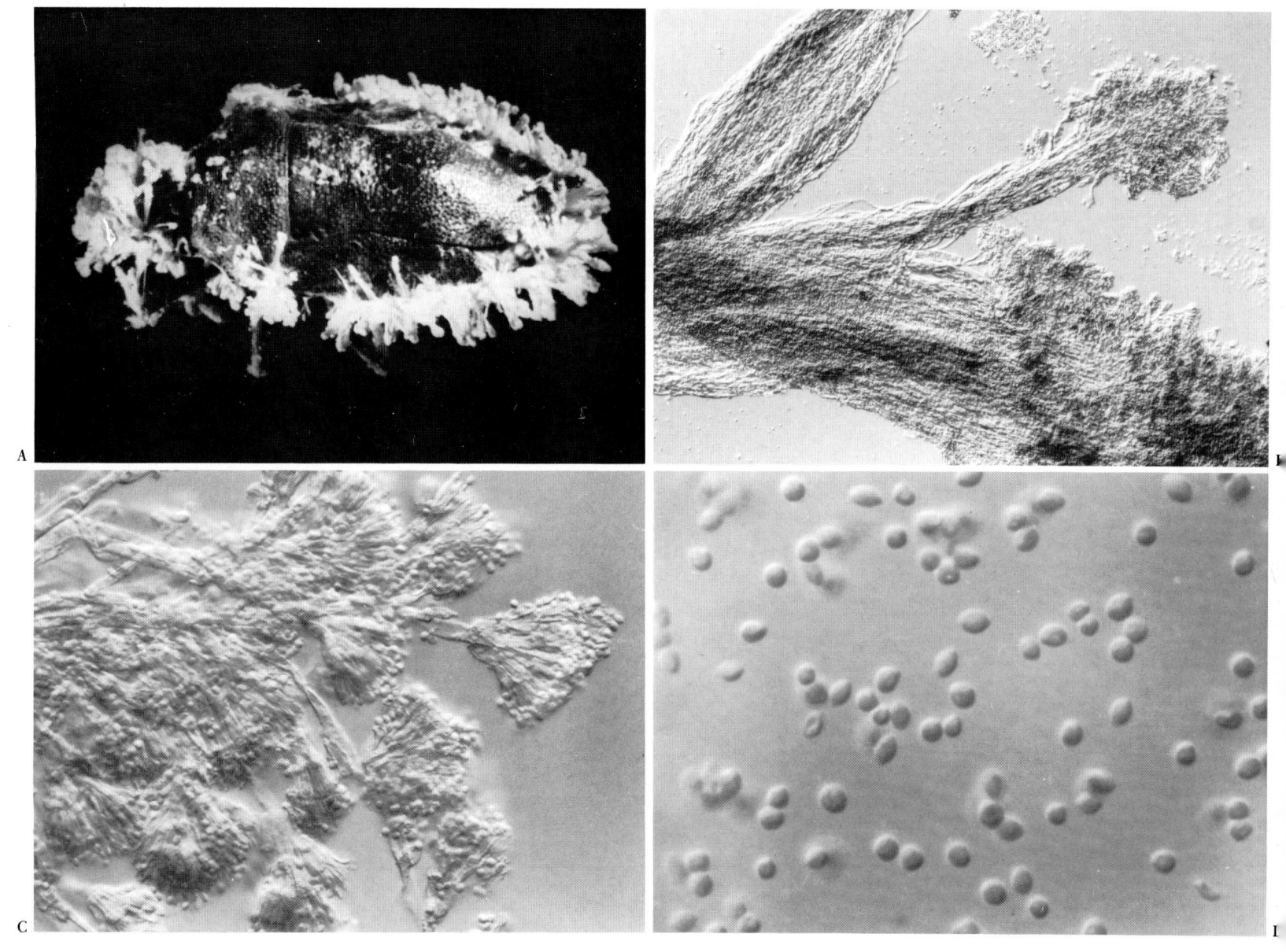

PLATE 95. *Paecilomyces tenuipes*
A Synnemata on lepidopteran pupa, B part of synnema with conidiophores (× 750), C conidiophore showing metulae and phialides (× 1875), D conidia (× 1875).

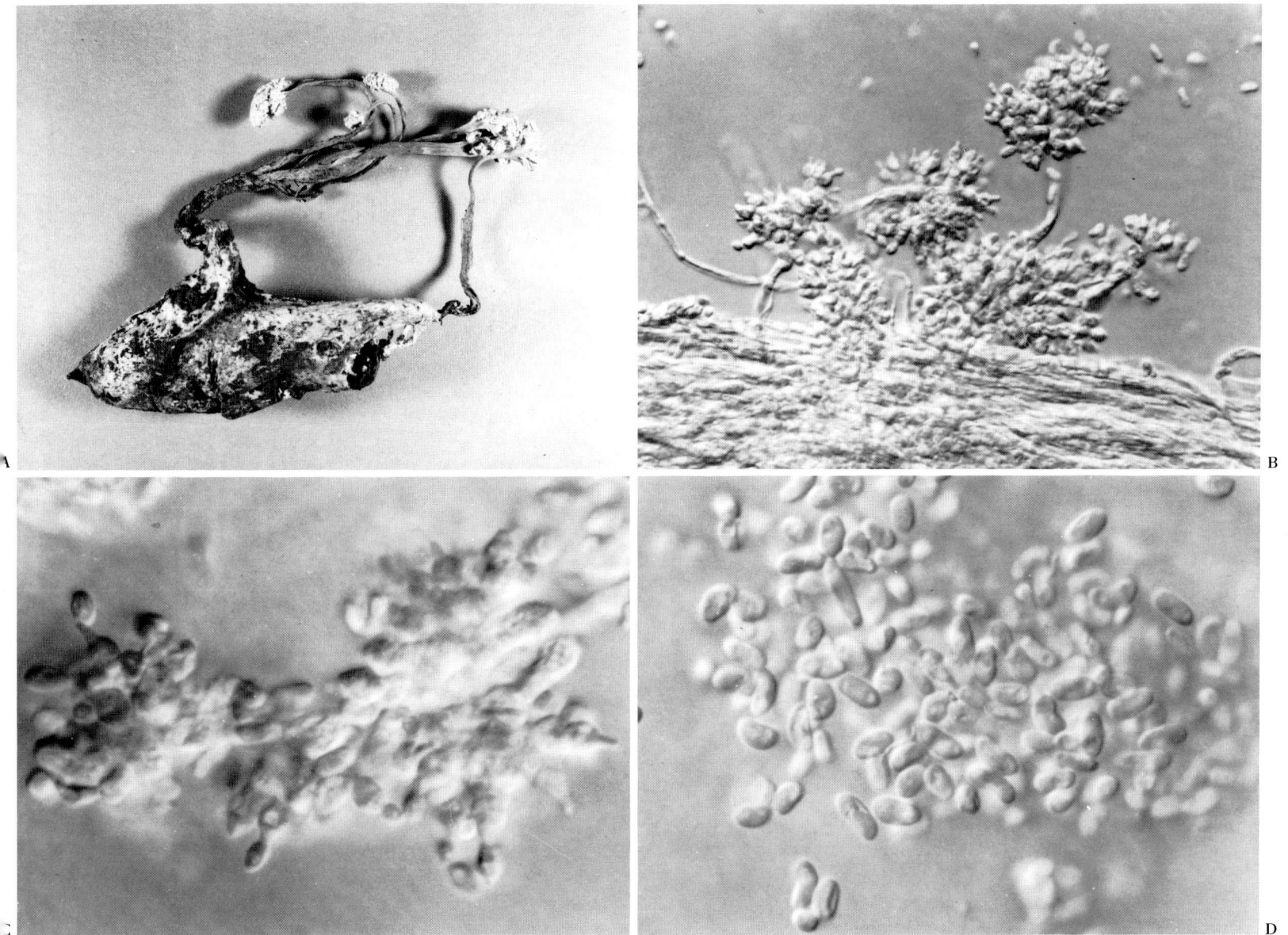

PLATE 96. *Paraisaria dubia*
A Synnemata on swift moth larvae *(Hepialis* sp.), B part of synnema with conidiophores (×200), C-D conidiophores with branches and phialides (×750), E phialides with typical proliferating necks (×1875).

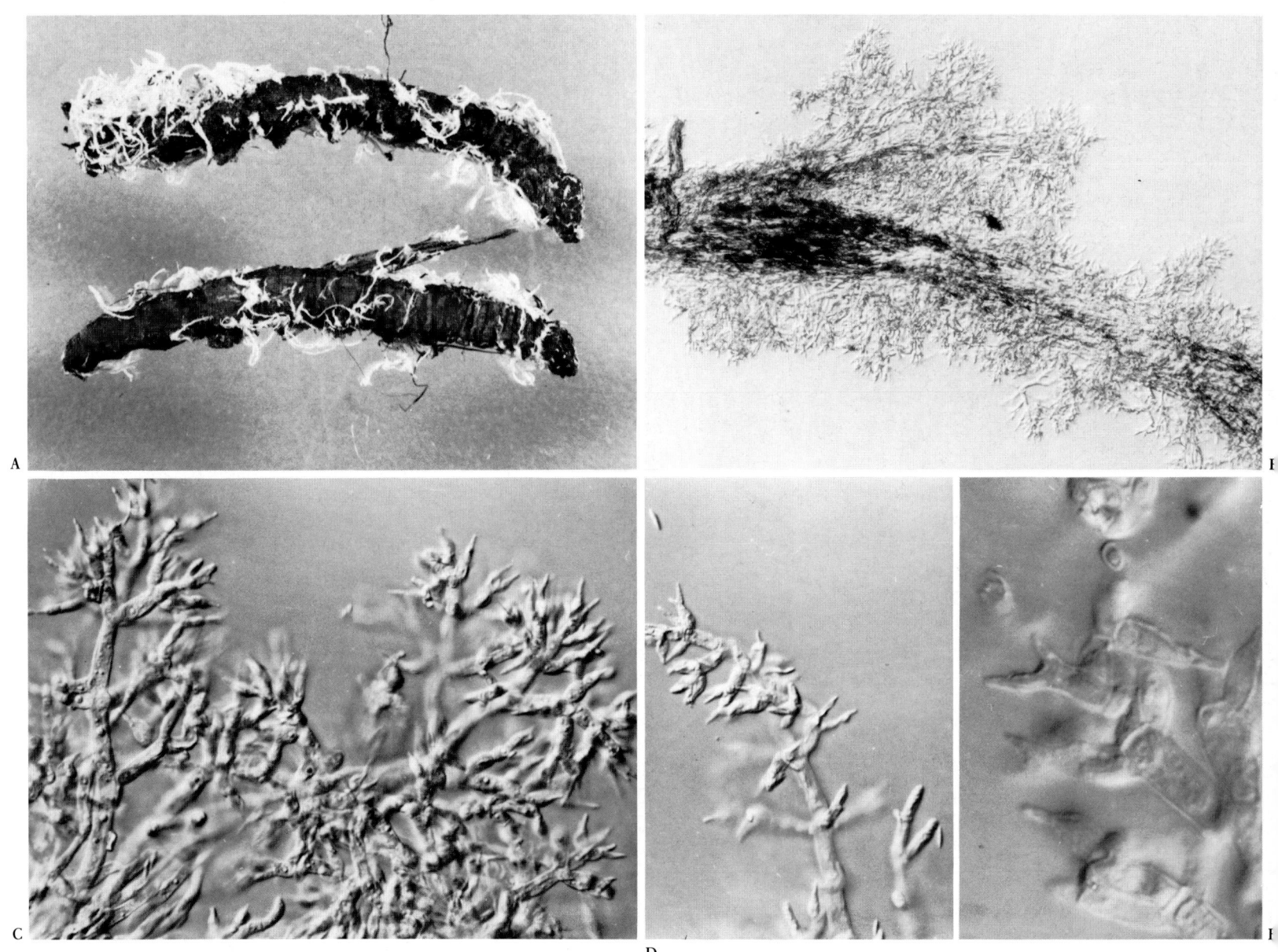

PLATE 97. *Pleurodesmospora coccorum*
A Conidiophores on coccid (Diaspidae), B fertile hyphae (× 750), C–D conidiogenous cells showing simple phialides and conidia (× 1875).

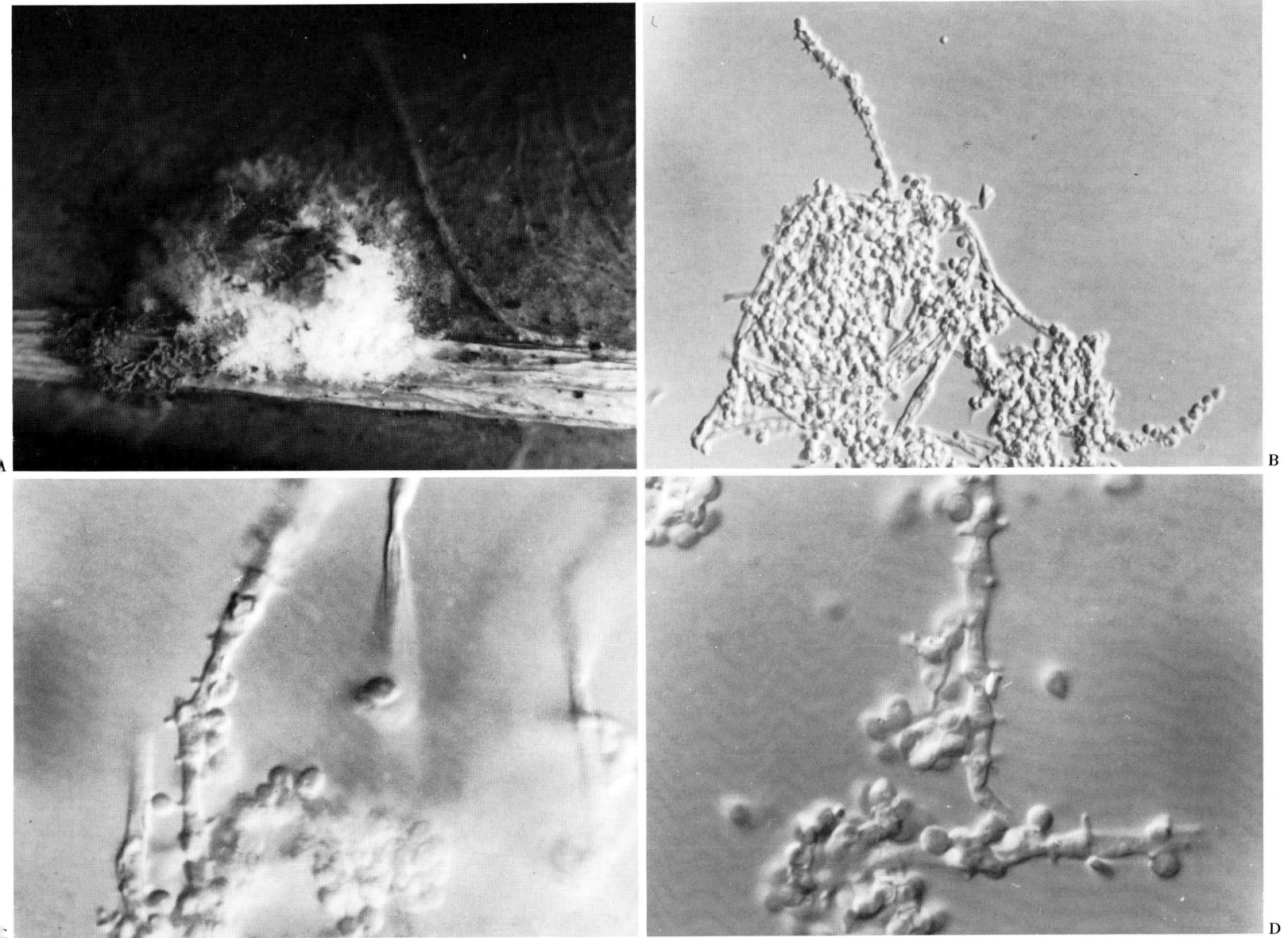

PLATE 98. *Polycephalomyces ramosus*
A Synnemata on troglobiotic (cave-inhabiting) dipterans, B synnema (× 200), C–D part of synnema showing layers of phialides (× 750), E phialides and conidia (× 1875).

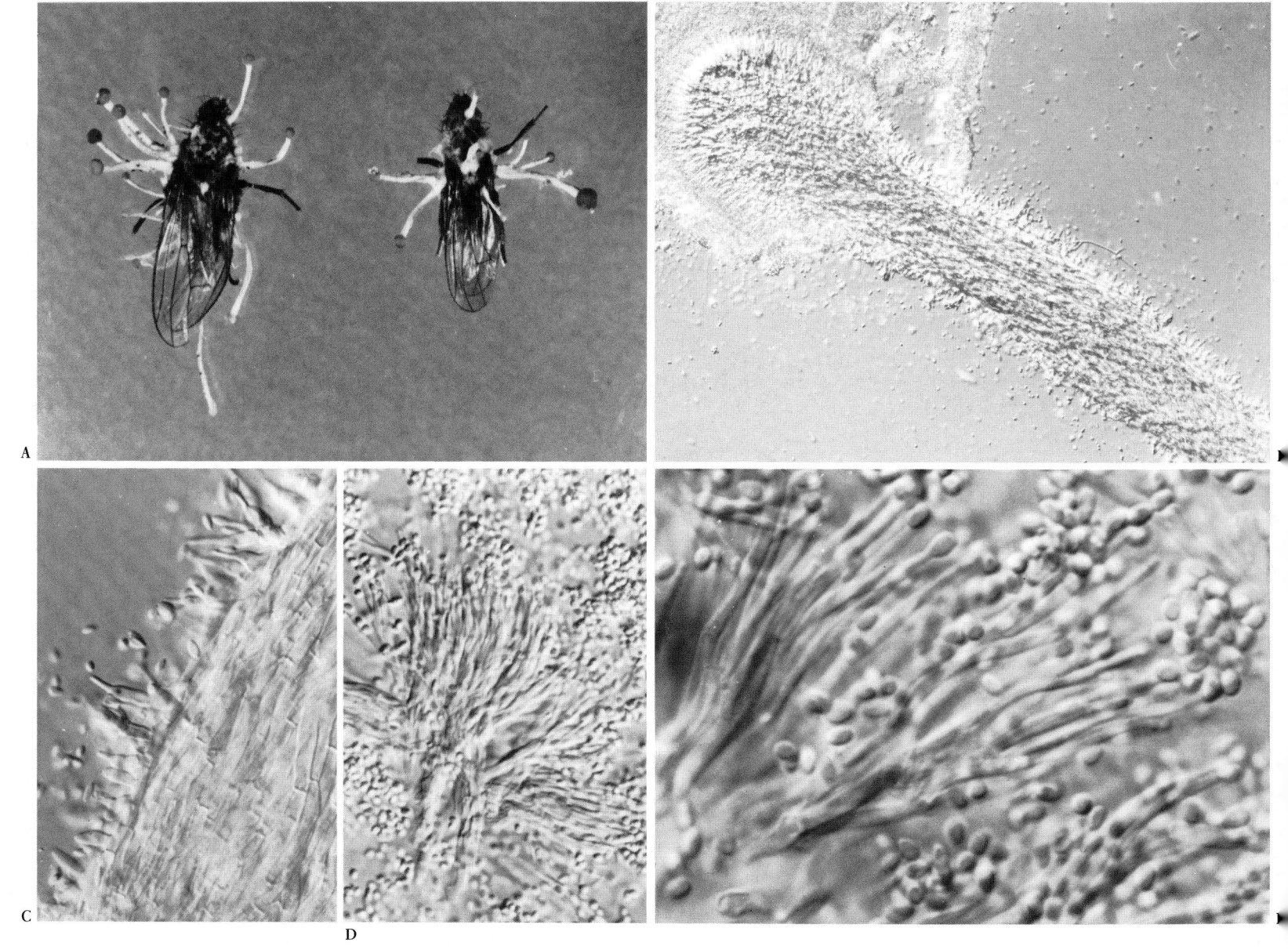

PLATE 99. *Pseudogibellula formicarum*
A Synnemata on ponerine ant (*Paltothyreus tarsatus*), B-C synnema (C × 200), D conidiophore (× 750), E apical part of conidiophore showing metulae and conidiogenous cells (× 1875), F conidia (× 1875).

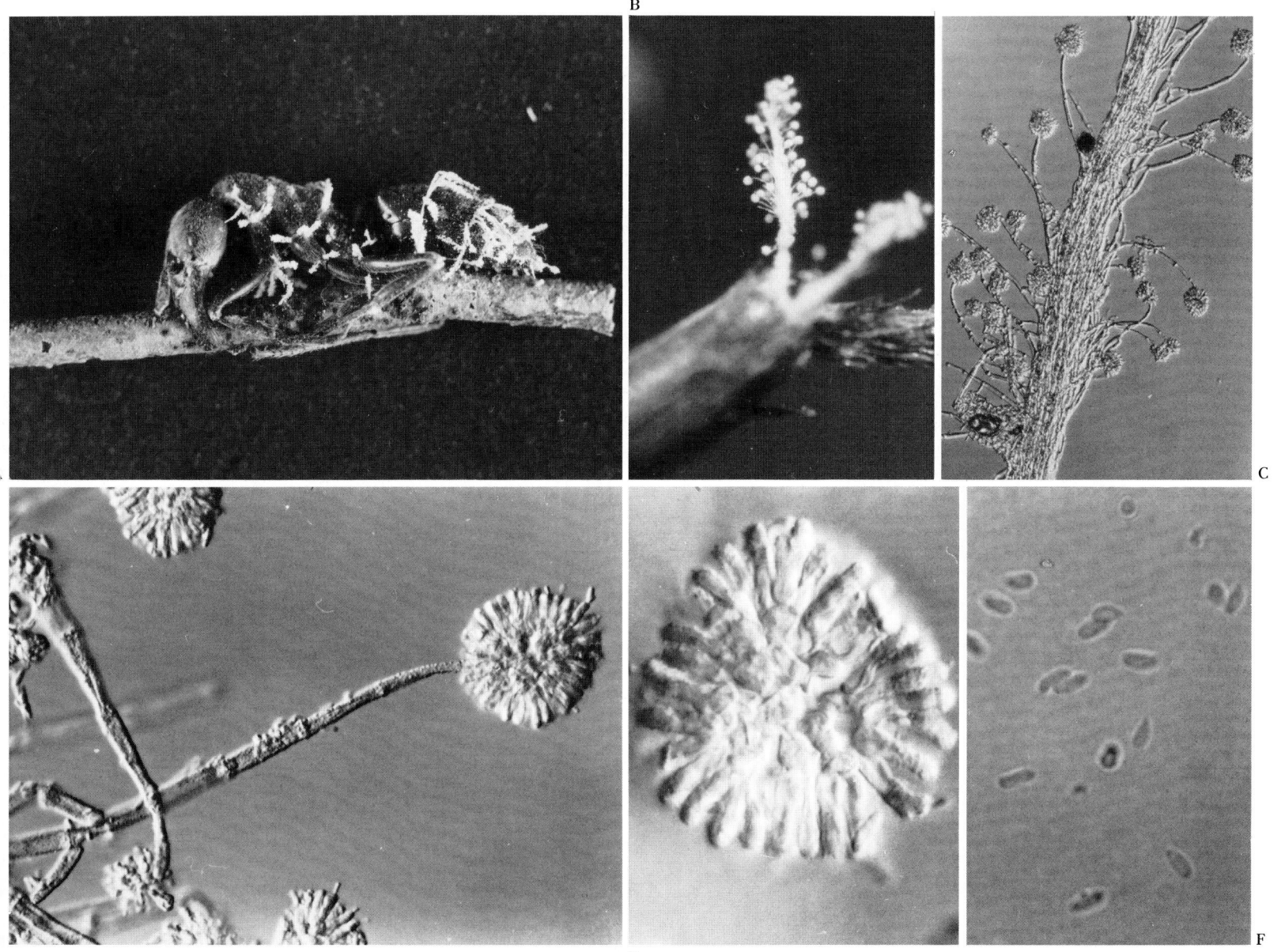

PLATE 100 *Sporothrix isarioides*
A Synnemata on a homopteran larva, B synnema (× 200), C part of synnema (× 750), D sympodial conidiogenous cells (× 1875).

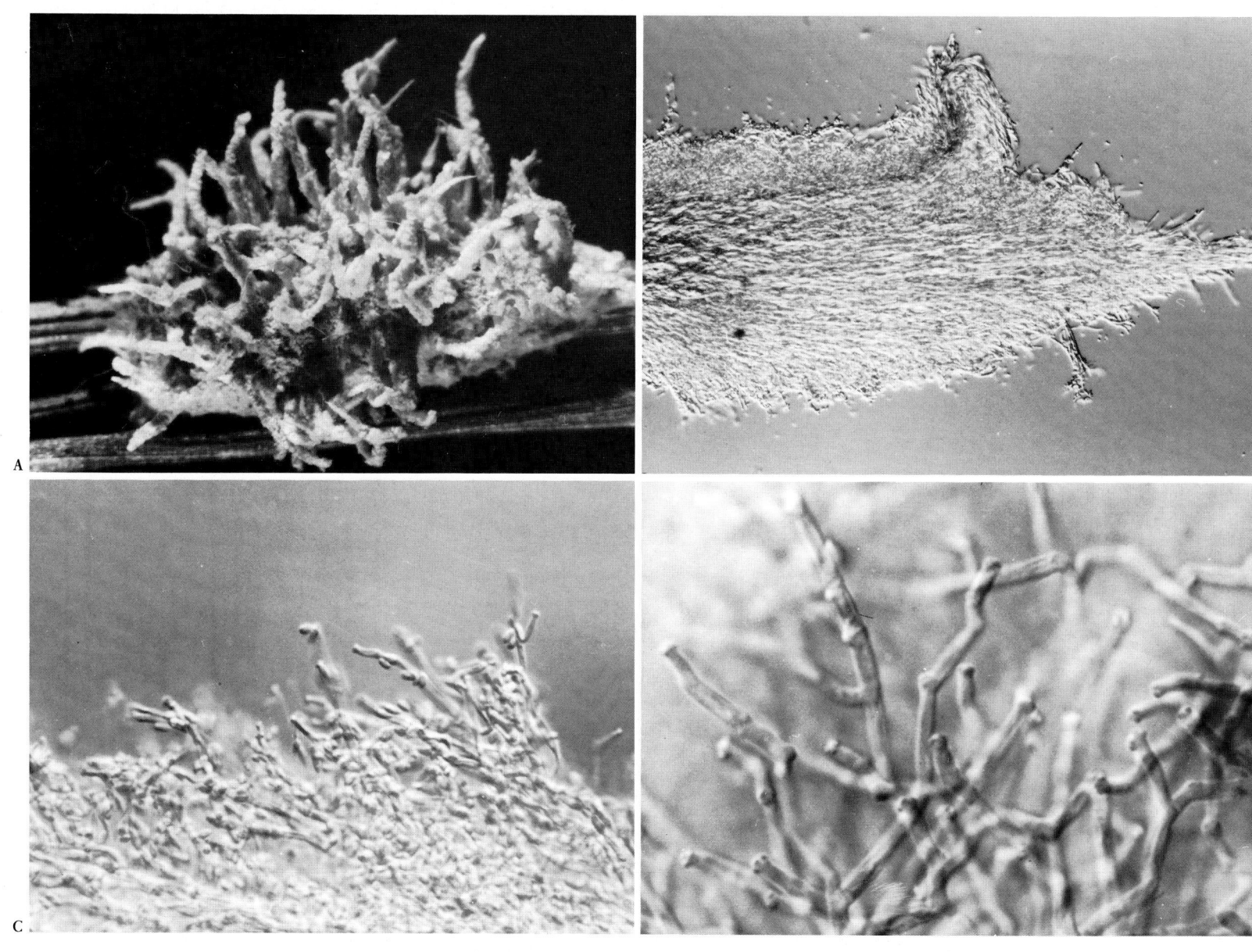

PLATE 101. *Sporothrix insectorum*
A Synnemata on ponerine ant (*Paltothyreus tarsatus*), B synnema
(×200), C part of synnema showing sympodial conidiogenous cells
(×750), D conidia (×1875).

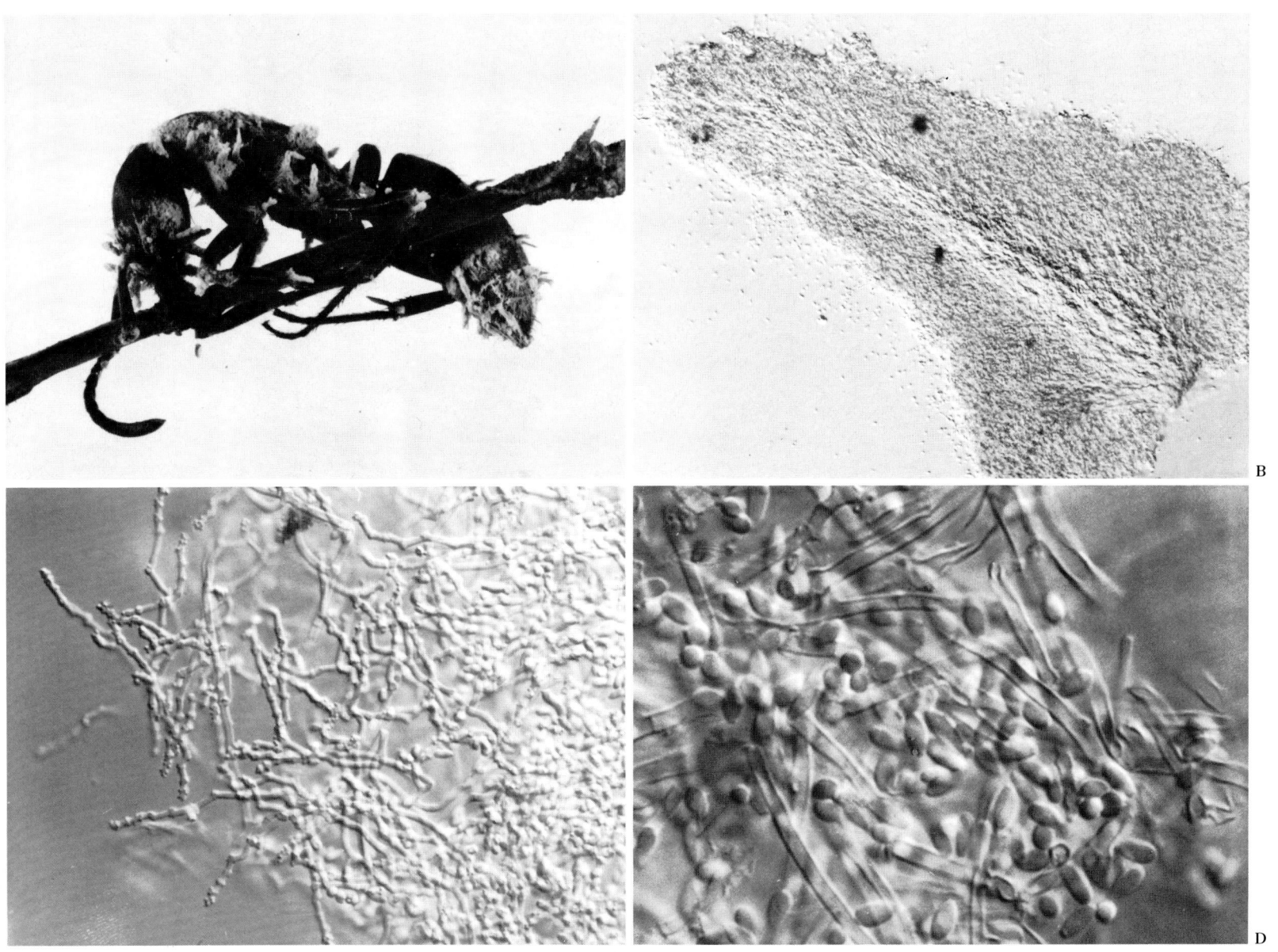

PLATE 102. *Stilbella buquetii* var. buquetii
A Synnemata on tropical beetle, B synnema (× 75), C apical part of synnema showing layer with cylindrical phialides (× 750), D conidia (× 1875).

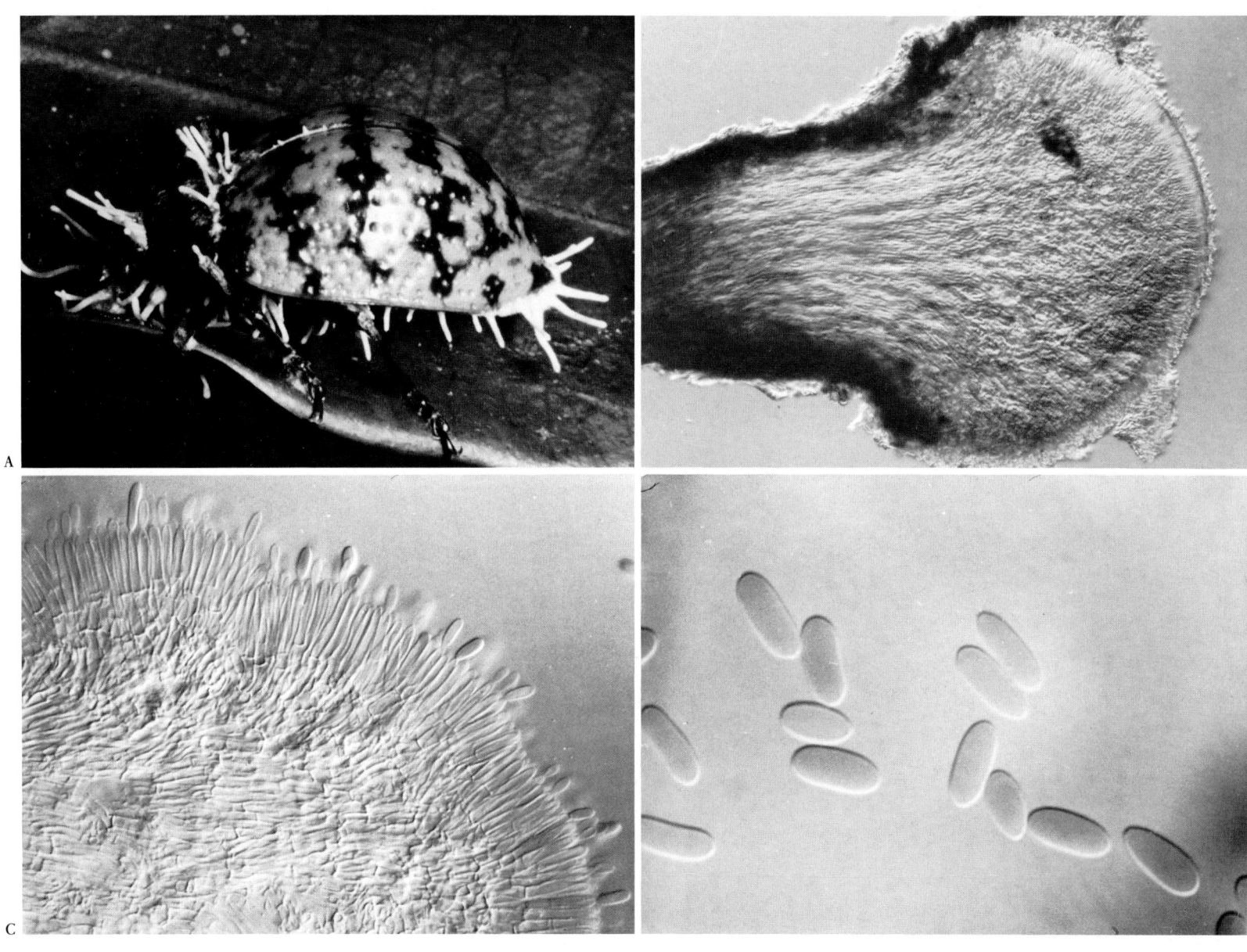

PLATE 103. *Stilbella buquetii* var. *formicarum*
A Synnema on ponerine ant (*Paltothyreus tarsatus*), B synnema
(×75), C-D apical part of synnema showing dense layer of phialides
(×750 and ×1875 resp.), E conidia (×1875).

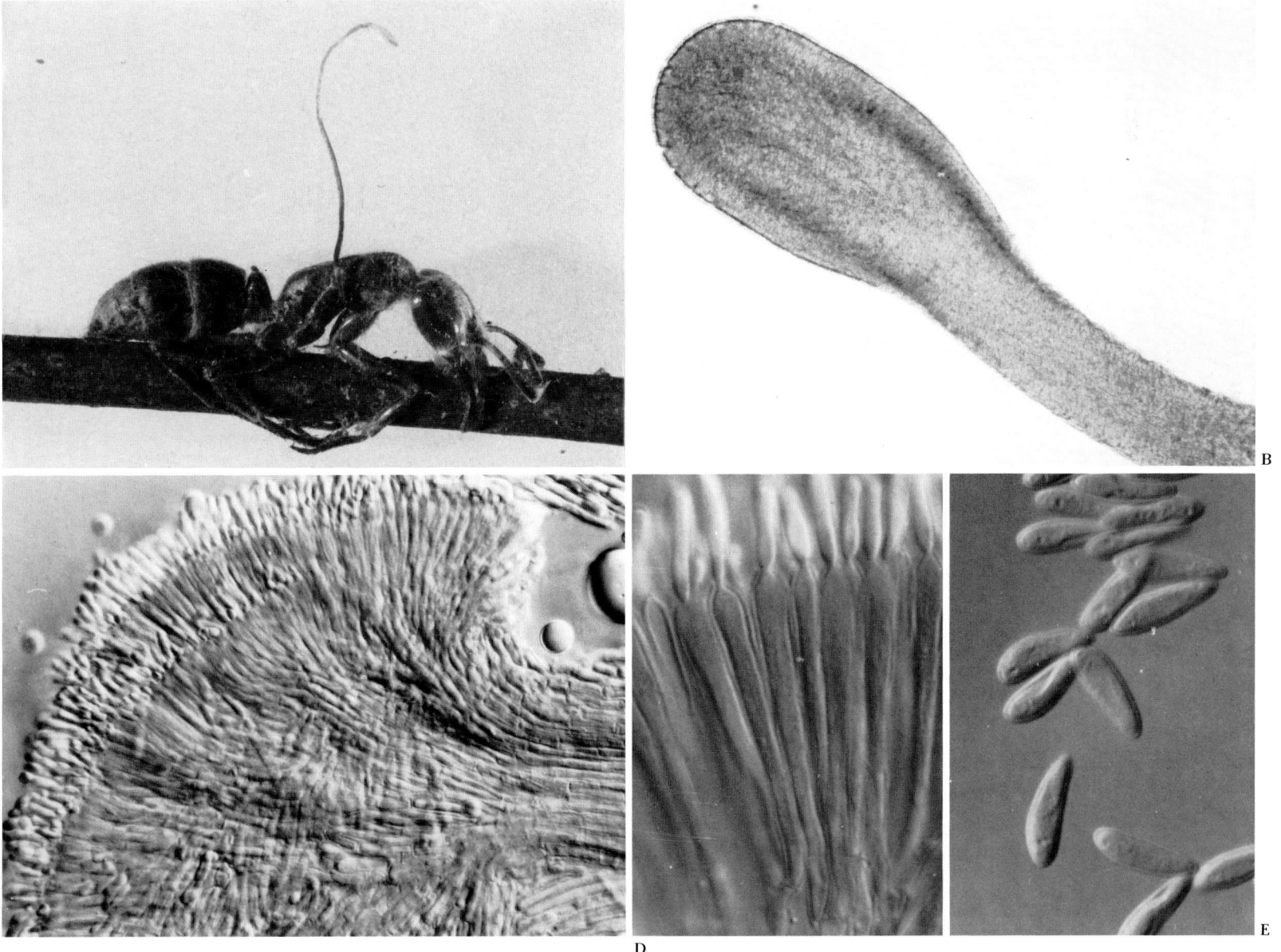

PLATE 104. *Tetracrium coccicolum*
A Sporodochia on citrus scales (*Lepidosaphes* sp.), B part of sporodochium (× 200), C-E conidia (C × 200, D-E × 750).

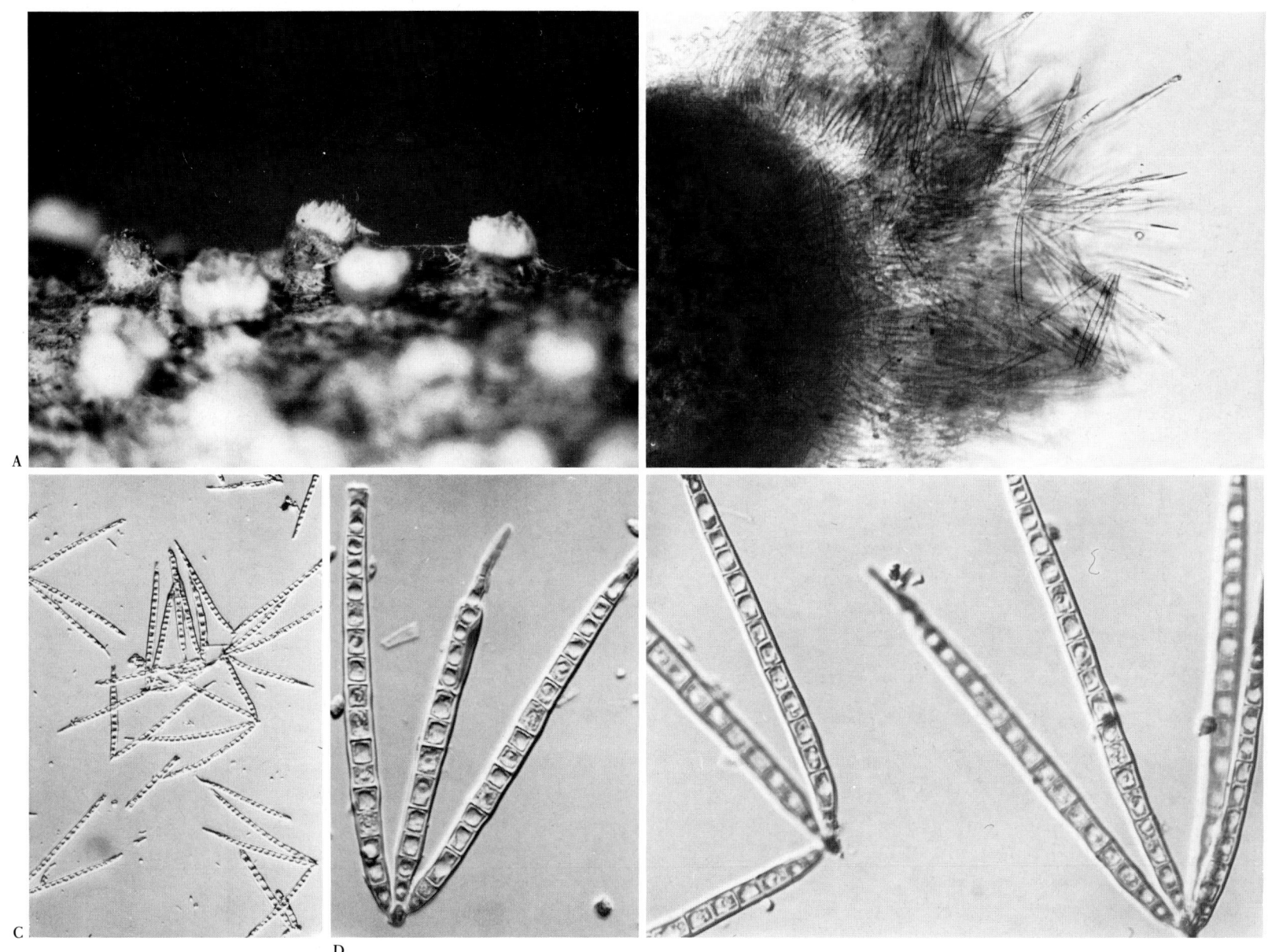

PLATE 105. *Tilachlidiopsis nigra*
A-B Synnemata on stroma of *Cordyceps entomorhiza* on soil borne beetle, C synnema (× 75), D apical part of synnema showing layer of phialides (× 1875), E conidia (× 1875).

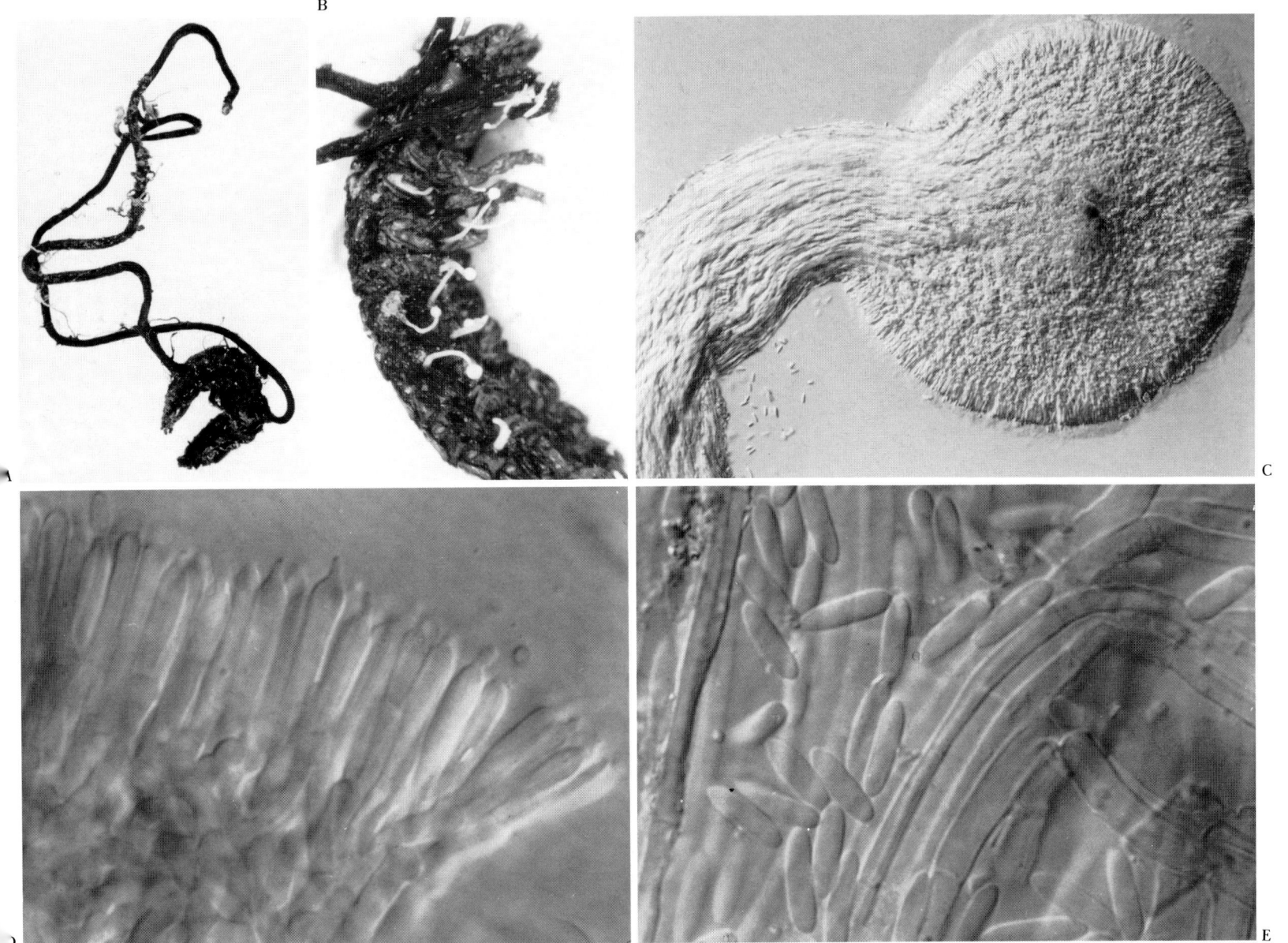

PLATE 106. *Tilachlidium liberianum*
A Synnemata on ponerine ant (*Paltothyreus tarsatus*), B synnema
(×200), C-D part of synnema showing awl-shaped phialides (×750),
E conidia (×1875).

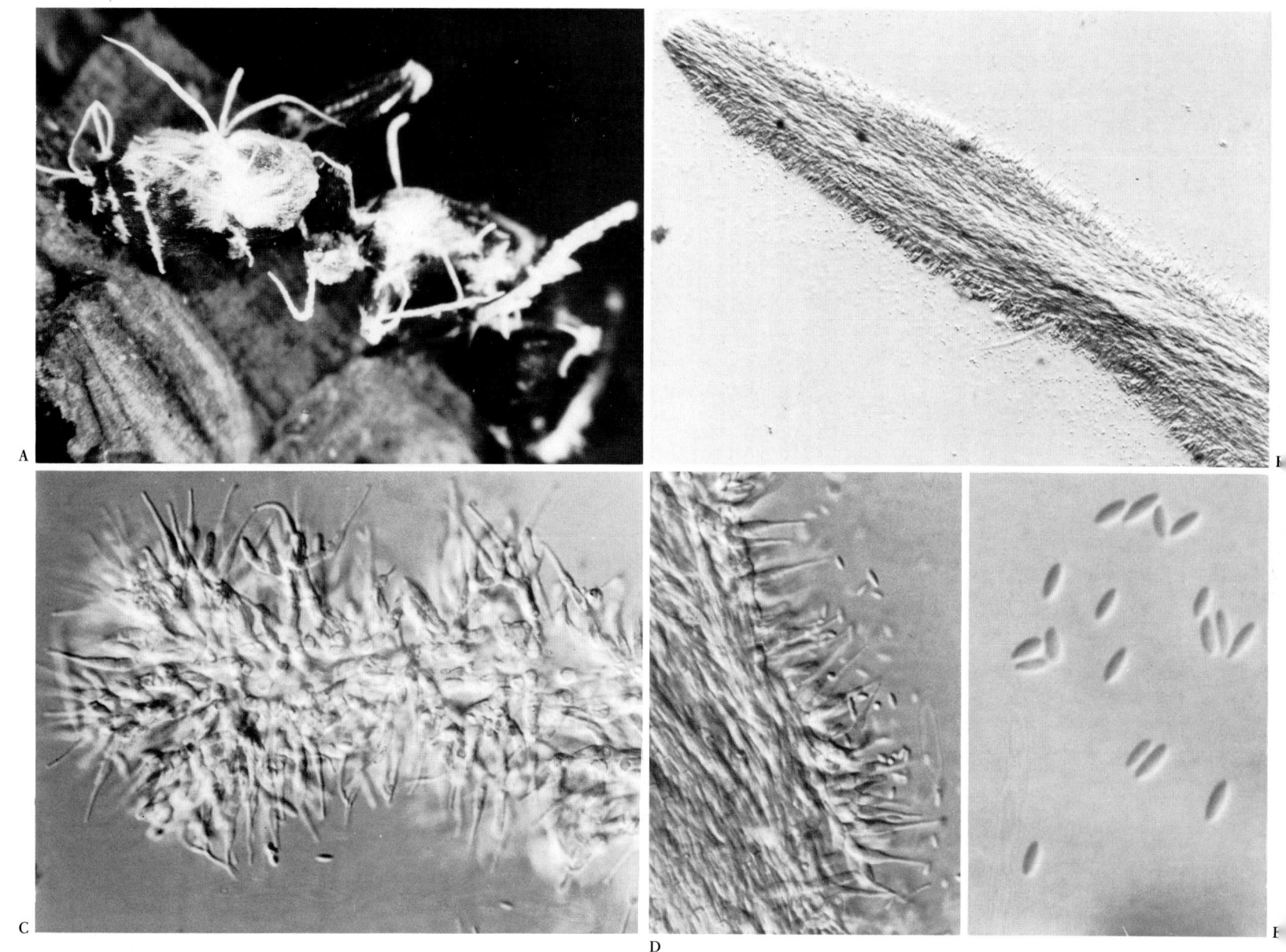

PLATE 107. *Tolypocladium cylindrosporum*
A Colony on mosquito larva (*Aedes aegypti*), B–C early infection and penetration, D conidiogenous cell with reduced phialidic neck on host (× 2500), E phialides formed on conidiophores when grown in culture (× 1875), F conidia (× 1875).

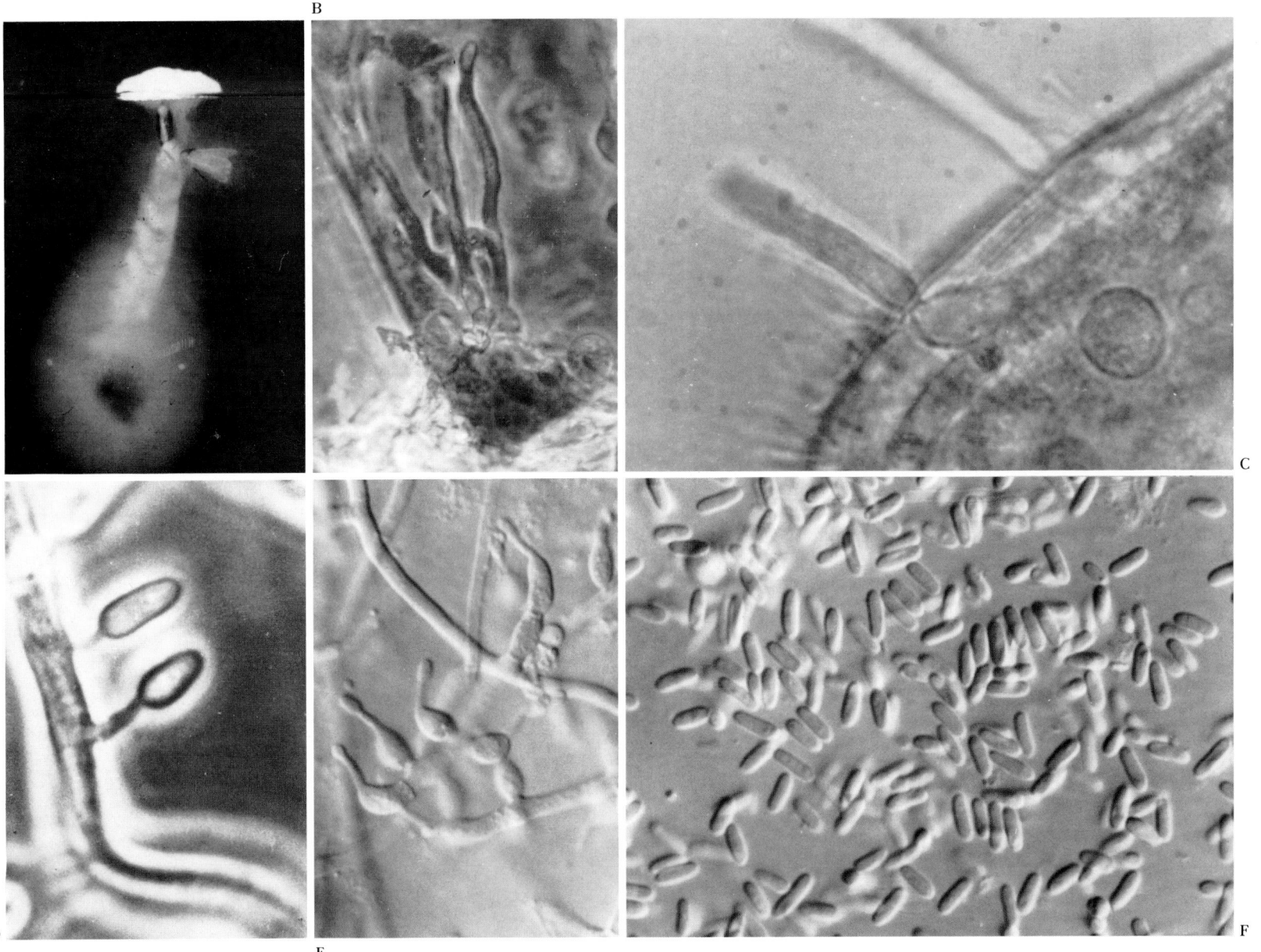

PLATE 108. *Verticillium lecanii*
A Colony on coffee green scale (*Coccus* sp.), B-D conidiophores showing verticillate whorls of awl-shaped phialides (× 750), E conidia (× 1875).

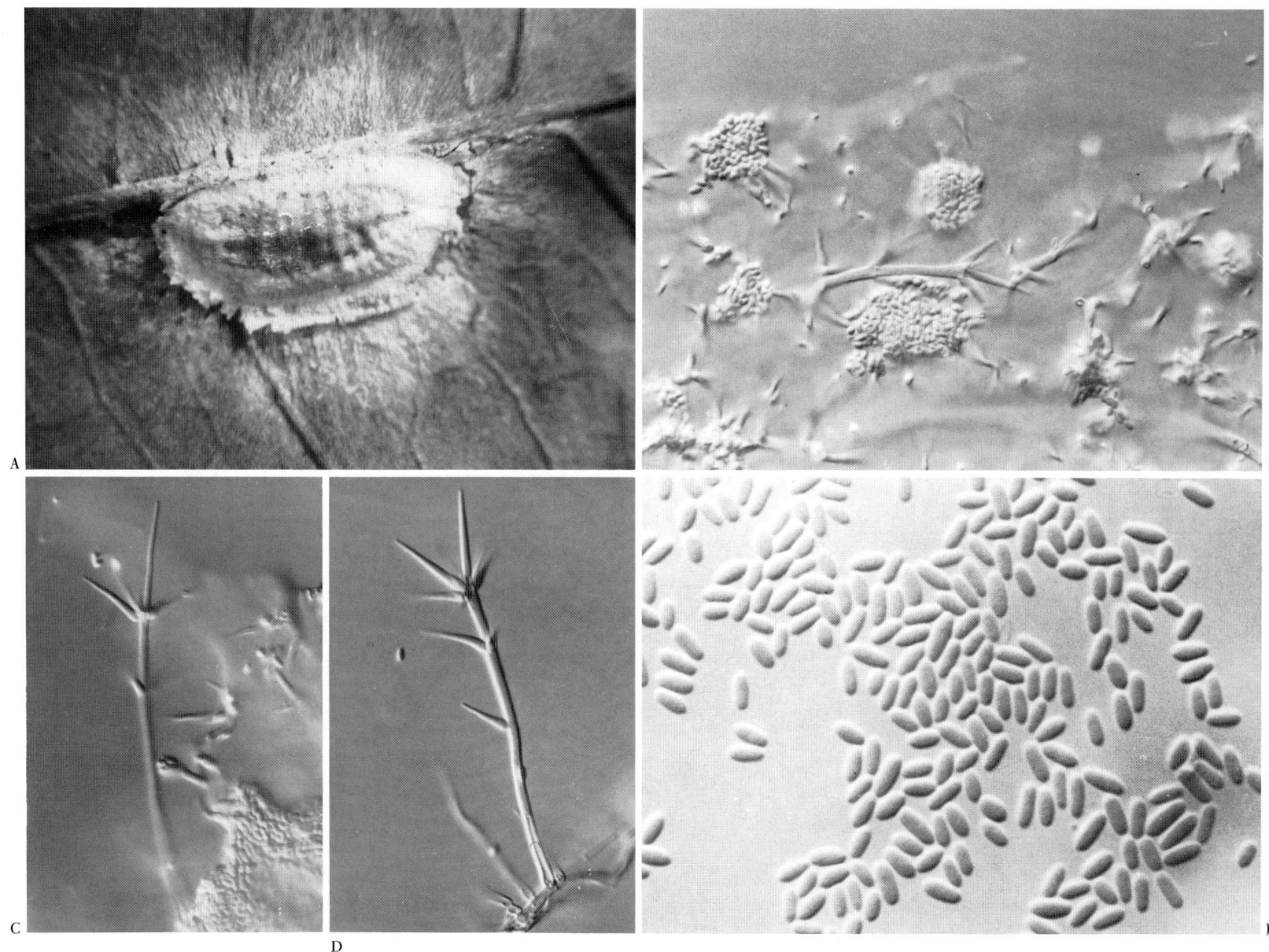

PLATE 109. *Verticillium lecanii*
A Conidiophores on aphid, B typical heads of conidia (×750), C conidiophores with whorls of awl-shaped phialides (×750), D conidia (×750).

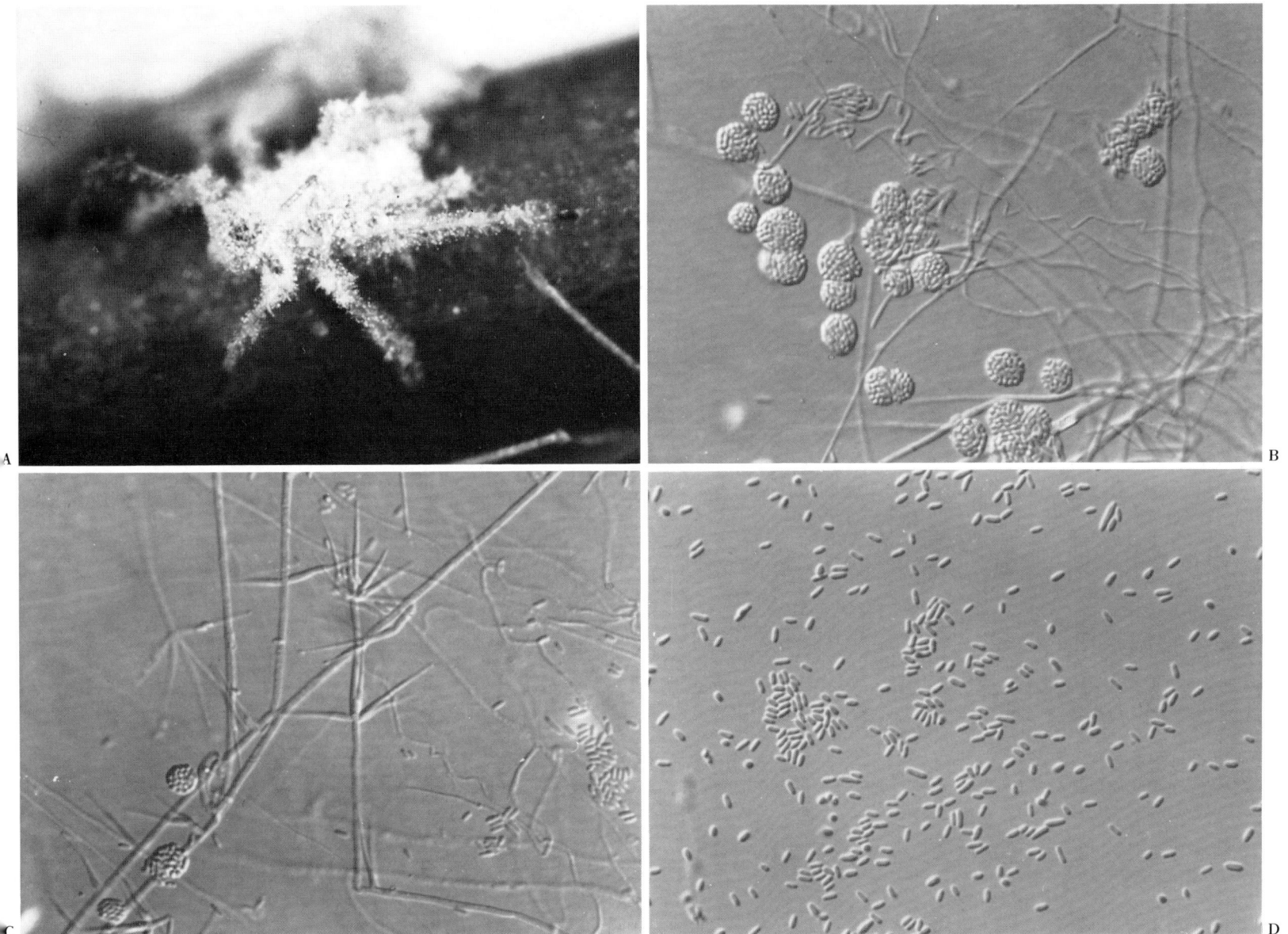

Chapter 4

Fungal pathogenesis

Introduction

Entomopathogenic fungi, like most fungal pathogens of plants and vertebrates, infect their host through the external cuticle. This mode of infection is unique and characteristic of the fungi since all the other entomopathogenic microorganisms, including bacteria, viruses and microsporidia, penetrate the host via the mid-gut. Three phases have been recognized in the development of an insect mycosis: *a* adhesion and germination of the spore on the host cuticle; *b* penetration of the insect integument by a germ tube; *c* development of the fungus inside the insect body, generally resulting in death of the infected host (fig. 4-1). The morphology of the infection process has been comprehensively studied but the physiological and biochemical mechanisms determining the susceptibility or resistance of an arthropod to a potentially active fungus are much less understood in invertebrate pathology than in plant or human pathology. However, within the last ten years more emphasis has been put by insect pathologists on studies of host-pathogen relationships, since the lack of knowledge in this field has probably contributed to the many failures in the use of fungi as biological control agents of insect pests. Interest in the study of invertebrate diseases has increased recently after the discovery in invertebrates of molecules with antibody and complement functions (Laulan et al., 1986). Research on the humoral and cellular defence reactions associated with arthropod resistance against mycopathogens can be expected, therefore, to provide insight into vertebrate defence systems against fungal diseases.

This chapter will review the current status of arthropod-fungal pathogenesis. Because of the limited number of examples studied so far (less than ten fungal species), the conclusions reported in this chapter must be regarded as provisional. In order to generate new research lines in insect mycology, comparisons should be made with the numerous data bases established for vertebrate and plant pathology.

Attachment of the spore to the cuticle

Contact between a fungal spore and its insect host is the prerequisite for the establishment of a mycosis. In the case of most entomopathogenic fungi which produce airborne spores, such contact occurs at random; the chances of success being highly dependent on climatic conditions, on the amount of fungal inoculum and on host density. In contrast, the search for the host by motile spores such as zoospores of *Aphanomyces* or the zygotes of *Coelomomyces* is usually an active chemotactic process; the fungus being chemically attracted by metabolites released by the host (Cerenius & Söderhäll, 1984; Travland, 1979a).

The epicuticle or outermost layer of the host integument is the site for the initial fungus-host interaction. This is a highly complex structure appearing as a series of several electron dense and translucent layers (Filshie, 1970; 1982). The chemical composition of the epicuticle, with the exception of the lipid fraction, is still in-

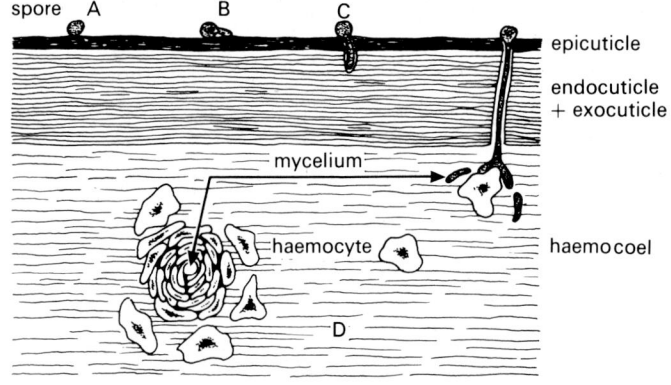

FIGURE 4-1. Establishment of a mycosis in an invertebrate. A Adhesion of the spore to the cuticle, B germination, C penetration of the cuticle by the germ tube, D colonization of the haemocoel and host defence reaction.

sufficiently understood (Blomquist & Jackson, 1979; Blomquist, 1984; Escoubas et al., 1986). Although lipids, lipoproteins, polyphenols and proteins (particularly enzymes) have been recognized as the main constituents of the epicuticle, the location of the different polymers and their spatial and chemical arrangements in this cuticular structure are still unknown (Locke, 1984). Moreover, this layer has often been considered as completely impermeable, but recent transmission electron microscope studies have shown that very fine channels continuously traverse the layer, supplying to the epicuticular level not only waxes but also sugars and proteins which could be of primary importance in the signal-receptor interactions at the spore-cuticle interface (Locke, 1984). The importance of such components in the recognition and adhesion processes has been emphasized in plant pathogenic fungi (Ouchi, 1984). Non-structural sugars and nitrogenous compounds originating from plant leachates or insect excretions may also contaminate the surface of the cuticle and could therefore influence the attachment process.

The composition of the outer layer of the spore wall of most entomopathogenic fungi has also been poorly studied. The external wall structure will, of course, depend on the type of spore under consideration. Slimy phialoconidia, such as those of *Verticillium lecanii* and *Hirsutella thompsonii* or certain ballistospores of the Entomophthorales (e.g. *Entomophthora muscae*), are always covered by an amorphous mucus which can facilitate the adhesion of the spore to the cuticle (Samson et al., 1980; Eilenberg et al., 1986) (fig. 4-2 A,B). The nature of this mucilaginous matrix is unknown in the entomopathogenic fungi and requires further study. Similar matrices found in phytopathogenic fungi are of primary importance in the adhesion process. Composed mainly of glycoproteins, these matrices protect the spores both from dessication and from potentially toxic polyphenols present on the host. Certain glycoproteins may also have an enzymatic function, possibly being involved in the dissolution of the cuticle and the subsequent uptake of nutrients necessary for germination (Nicholson & Moraes, 1980; Bergstrom & Nicholson, 1981). Contrary to 'wet' spores, dry blastic conidia, like those of *Nomuraea rileyi*, do not possess an amorphous outer layer and are covered by interwoven fascicles of rodlets (C). The ballistospores of *Conidiobolus obscurus* represent an intermediate form between the two types of spores mentioned above (Latgé et al., 1986b). The spores of this fungus are covered by a gelatinous mucus. When the forcibly discharged spores land on the insect, the mucus spreads out and dries on the cuticle, uncovering a layer of regularly arranged rodlets on the surface of the spore. The mucus, mainly composed of fibrillar glucans associated with amorphous proteins and loose rodlets, functions as an adhesive fixing the spore to the integument (D-G). When

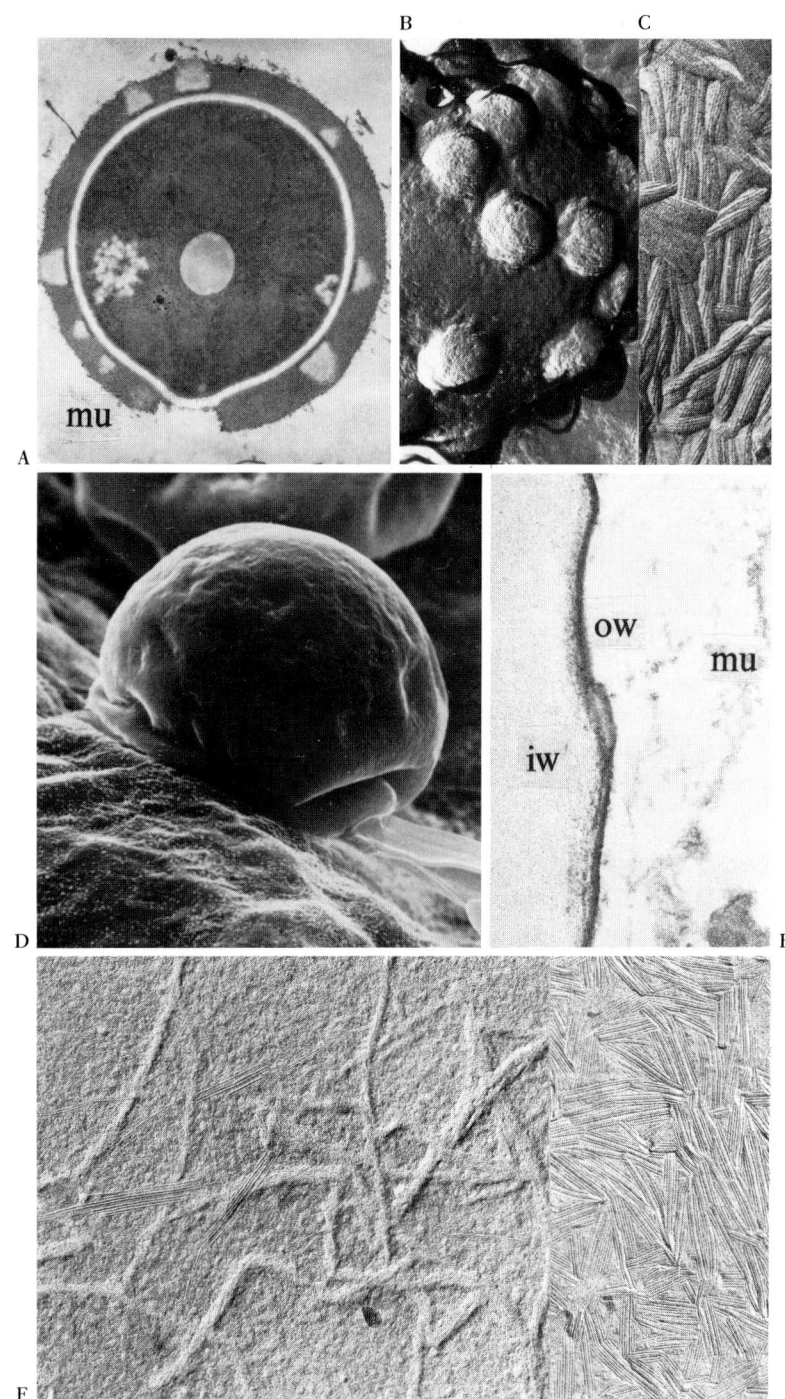

FIGURE 4-2. Outer layers of spores and conidia of entomopathogenic fungi. A Sections of *Hirsutella thompsonii* conidia showing the presence of mucus (mu) on the surface of the conidium ($\times$ 12.000), B carbon-platina replica of the amorphous outer layer of *H. thompsonii* ($\times$ 12.000), C rodlet layer on the surface of *Nomuraea rileyi* conidia ($\times$ 45.000), D spore of *Conidiobolus obscurus* adhering to the aphid cuticle ($\times$ 1000), E section of the wall of *C. obscurus* primary spore (iw: inner wall, ow: outer wall, mu: fibrillar mucus) ($\times$ 60.000), F and G replicas of the fibrillar mucus (F, $\times$ 60.000) and the outer surface of the wall (G, $\times$ 60.000).

present, the rodlet layer may play a key role in the attachment process. Rodlets have also been described in many non-entomopathogenic organisms and, their composition is considered to be predominantly lipoproteic (Cole & Pope, 1981). The presence of lipoproteins on the surface of the spore would undoubtedly facilitate its attachment to the hydrophobic lipophilic insect epicuticle. However, the chemical nature of rodlets is highly variable from one fungal species to another (Cole & Pope, 1981) and similar variations occur amongst entomopathogenic fungi (Boucias et al., 1988). For example, the rodlets of *N. rileyi* do not appear to be of a lipoproteic nature since their typical structural arrangement remains unchanged after boiling the spores in a chloroform-methanol mixture or in various detergent solutions (Boucias & Latgé, 1986). In contrast to *N. rileyi*, the rodlet layer of *C. obscurus* is removed easily by simple water washing (Latgé et al., 1986b).

In many cases, however, attachment of the spores to the insect results from an apparently passive mechanism: no exogenous adhesive material can be demonstrated at the contact point between the spores and their host integument (Zaccharuck, 1970a; Michel, 1981). Nevertheless, in aquatic genera, such as *Culicinomyces* and *Coelomomyces*, adhesion may be due to an active biosynthetic process consecutive to the recognition of the host by the fungus (Travland, 1979b; Sweeney et al., 1984). The deposition of an adhesive material usually occurs during encystment of the motile spores, analagous to the process observed in phytopathogenic Oomycetes (Nicholson, 1984).

The degree of adhesion depends on the fungal species under consideration. Most of the dry conidia are passively attached to the host and may thus be readily removed by rinsing the host with a detergent solution. However, the binding of the conidia of *N. rileyi* to the insect cuticle is very strong. Attempts to remove spores attached to cuticle substrates by various treatments, including boiling in 1% SDS, chloroform: methanol (2:1), mild alkali, acid and enzyme treatments have been unsuccessful, even when spores were adhering to non-host cuticle such as the pea aphid. Paradoxically, the ballistospores of the aphid pathogen *C. obscurus* are easily detached from the aphid cuticle (Boucias et al., 1988). Spore adhesion has also been frequently correlated with the aggressiveness or host specificity of a fungal species. Zebold et al. (1979) and Fargues (1981) were able to correlate host specificity of *Coelomomyces psorophorae* and *Metarhizium anisopliae* with the ability of spores to attach to the cuticle of mosquitoes and scarabid larvae, respectively. Similarly, hypovirulence of certain *M. anisopliae* strains pathogenic to mosquito larvae resulted from a defect in attachment of the floating conidia to the larval siphon (Al-Aidroos & Roberts, 1978). However, adhesion is not always related to aggressiveness: for example, no difference has been found in the adherence of conidia of aggressive and non-aggressive strains of *C. obscurus* to the pea aphid cuticle (Latgé et al., 1982) or of *N. rileyi* to the cuticle of *Anticarsia gemmatalis* (Boucias et al., 1988).

These results suggest that in some fungi adhesion is a non-specific phenomenon whilst in others it is a specific process involving the mutual recognition of the fungus and the insect. Such factors could be investigated by comparing the adhesion process on heterologous (non-susceptible) insects or artificial surfaces with that on the cuticle of the specific insect host. At the present time, the energetics and kinetics of macromolecular attachment, the nature of the exchange reactions and the dependence of these features on the structure and chemistry of the fungal and insect surfaces, as well as on the environmental conditions, are almost totally unknown in entomopathology. The involvement of electrostatic forces was the first hypothesis advanced to explain fungal adhesion (Fargues, 1984). At the initial contact, adsorption will involve charged groups on both the spore and host surfaces (Pendland & Boucias, 1984 a), implicating hydrogen or van der Waals bonds. The presence of electrical discontinuity points on a cuticular surface has been correlated already with the preferential sites of attachment and infection (Cherbit & Delmas, 1979). Electrostatic binding of *N. rileyi* to DEAE resins or chitosan flakes can be eliminated if conidia are pretreated with the polycation, polylysine. However, polylysine does not inhibit the binding of *N. rileyi* conidia to the insect cuticle (Boucias & Latgé, 1986). Based on current concepts of the recognition and adhesion processes in various biological systems (Monsigny, 1984), attachment of the spores should be dependent upon molecular interactions between the two organisms. The most likely type of molecules functioning as specific surface receptors are glycoproteins. Recent results suggest that such chemical interactions could occur between the insect host and the spore. Haemagglutinins are present on the conidial surface of *Beauveria bassiana* and may be involved in the infection process; this haemagglutinating activity being inhibited by glucose and N-acetyl-glucosamine (Grula et al., 1984). Glucose and N-acetylglucosamine binding proteins have also been demonstrated in the mucus of *C. obscurus* ballistospores (Latgé et al., 1984 b). Haemagglutinins interacting with galactosidic sugars have also been detected in a water-sonicated extract of *M. anisopliae* conidia, but the blockage of this lectin by its corresponding sugar does not inhibit conidia from binding to the insect cuticle (Boucias et al., 1988). Carbohydrates have also been implicated in host recognition by the mosquito pathogens *L. giganteum* and *C. psorophorae* (Kerwin & Washino, 1986 b).

Germination of the spore

Once the spore has attached to the insect, it must germinate to produce a germ tube which will then penetrate the host cuticle. In addition to serving as the penetrant hypha, the germination structures also play a role in strengthening the adhesion of the fungus to the insect cuticle. Germinating spores of several entomopathogenic species, such as *M. anisopliae*, *B. bassiana*, *C. psorophorae* and *Neozygites fresenii*, produce an appressorial cell at the germ tube-epicuticle interface (Zaccharuck, 1970 a; Brobyn & Wilding, 1977; Travland, 1979b; Michel, 1981). The appressorium has been reported to be coated with an amorphous mucilaginous material responsible for the attachment of this structure to the epicuticular surface (Zaccharuck, 1981). Similar material may also be deposited around the germ tubes of species which do not form appressoria, such as *N. rileyi* and *C. obscurus* (Pendland & Boucias, 1984a; Brey et al., 1986) (fig. 4-3). Although the germination phase has long been recognized as a critical step in the establishment of infection and in the determination of the pathogenicity of a fungal strain, the key factors controlling the germination of entomopathogenic spores are poorly understood: the main reason being that the majority of germination tests are carried out in vitro and are rarely correlated with in vivo observations. Components extracted from an insect cuticle should be tested: *a* at realistic doses similar to the concentrations of the same components found in the cuticle; *b* in environmental conditions (pH, temperature, light, humidity) as close as possible to those encountered on the insect cuticle. Failure to simulate natural conditions could lead to erroneous conclusions. Another possible source of error is the fact that germination studies consider only the total percentage germination. Additional criteria such as the duration of the germination time and the mode of germination, are of primary importance in determining the success of the infection process. For example, Pekrul & Grula (1979) reported similar total percentage germination between aggressive and non-aggressive strains of *B. bassiana*; however, highly pathogenic strains germinated quicker and penetrated the epicuticle directly whilst strains of low pathogenicity took longer to germinate and grew extensively over the cuticle surface with only limited penetration. Similar observations have been made with aggressive and non-aggressive strains of *Paecilomyces fumoso-roseus* and *V. lecanii* directed towards *Pieris brassicae* and *Macrosiphoniella sanborni* respectively (Delmas, unpubl; Jackson et al., 1985). However, with *N. rileyi* contrary results were obtained; the aggressive strains germinating slower and at a lower frequency than the non-aggressive isolates (Boucias & Pendland, 1984). In the Entomophthorales, conidia adhering to an insect host may either produce an infective germ tube, or a non-infectious secondary spore or simply die. In such cases, the type of structure formed is more critical in the establishment of an infection than the total percentage germination (Latgé et al., 1987b).

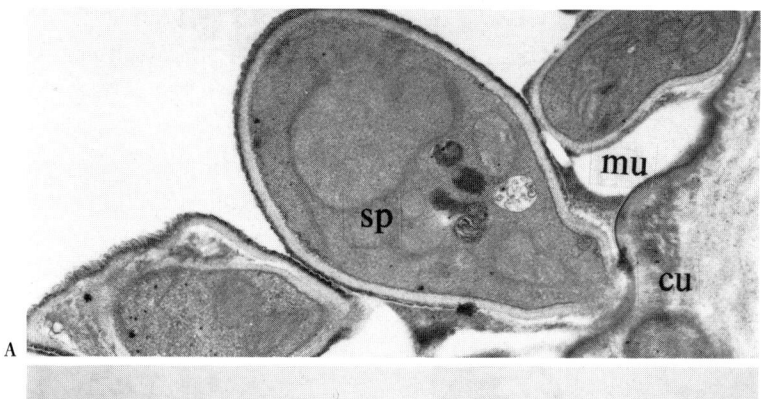
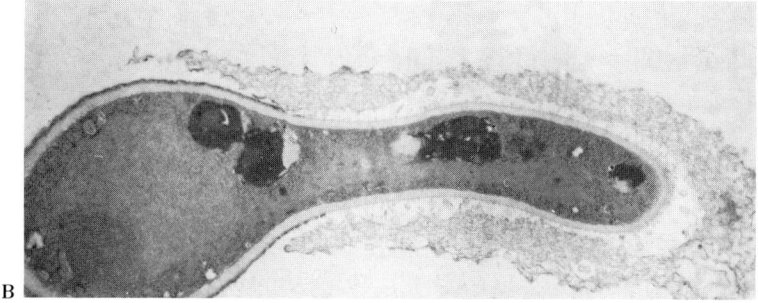

FIGURE 4-3. Mucilaginous material (mu) deposited around germ tubes of *Nomuraea rileyi* spores (sp) adhering to the insect cuticle (cu) (× 10.000). (Figures by courtesy of J. Pendland & D. Boucias.)

On the basis of in vitro data, spore germination has mainly been regarded as being dependent on macroclimatic factors, especially temperature and humidity. In vitro, relative humidities superior to 90% are always necessary to induce germ tube formation. In certain Deuteromycetes, the presence of a film of free water is necessary to obtain maximum infection levels (Hall, 1981; McCoy, 1981). However, excessively high levels of humidity may be unfavourable for the establishment of an entomophthoralean infection; the conidia for example should never be sprayed as a water suspension (Latgé & Papierok, 1988). Temperature requirements are highly dependent on the ecological niche of the fungus. Deuteromycetes, which are most frequently found in tropical or subtropical areas, germinate optimally in vitro at temperatures above 25 °C (Hall, 1981; McCoy, 1981; Ignoffo, 1981). In contrast, Entomophthorales with a predominantly temperate distribution, have a colder optimum (Latgé & Papierok, 1988). Temperature requirements may vary between species and also at the intraspecific level. Strains of *Erynia neoaphidis* originating in the subtropical or tropical zones of Mexico and Brazil have an optimum of 26-28°C, whereas strains of the same species isolated in Northern Europe germinate best at 20-23°C. Temperature optima for germination can also be related to the role of the spore in the life cycle of the fungus. For example, early spring temperatures (12-16°C) are optimal for inducing the germination of the azygospores of *C. obscurus* from temperate Northern Europe, but ballistospores of this species germinate best at 19-24°C (Perry & Latgé, 1982; Sampedro et al., 1984). The former spores are able to survive the winter and infect the first aphids arriving in the crop whilst the latter represent the active infective propagule present in aphid populations installed in the crop later in the season. The role of light in the germination of entomopathogenic fungi has been poorly studied (Uziel et al., 1981). All the meteorological data relating to germination are macroclimatic and have not accounted for microclimatic variations or other factors such as evapotranspiration from the plant or the insect or specific location of the spore on the insect cuticle (intersegmental folds compared with rough exocuticle). The sophistication and miniaturisation of thermal and hygrometric probes and the use of computers now allow a modelling of the true microclimatic conditions encountered by the fungus in its natural microenvironment (Perry & Whitfield, 1984). Such techniques would permit a better understanding of unexplained infection experiments: for example, Ferron (1977) found that conidia of *B. bassiana* would infect *Acanthoscelides obtectus* irrespective of the relative humidity; similarly, sustained temperatures around 35°C did not harm *V. lecanii* in the greenhouse but were lethal to the fungus in vitro (Hall, 1981).

Germination of entomopathogenic spores is also dependent on the nutritional environment. In the case of *N. rileyi* or *B. bassiana*, the conidia germinate very poorly in the absence of nutrients (Woods & Grula, 1984; Boucias & Pendland, 1984) whilst the quality and quantity of nutrients available to *C. obscurus* will influence its mode of germination (Sampedro et al., 1984). Dillon & Charnley (1986) have shown that germination of *M. anisopliae* is initiated by water but progress to the first overt stage of germination (swelling) is dependant on a suitable exogenous nutrient. In vivo, germination of entomopathogenic spores is highly dependent on both the level and the type of chemicals present on the epicuticle. The surface of the corn earworm larvae contains sufficient nutrients, primarily amino acids, to allow germination and growth of *B. bassiana* (Woods & Grula, 1984). It is interesting to note that surface nutrients readily removed with water from the cuticle reappear after a few hours, demonstrating the continuous secretion of these nutrients through the epicuticle. Chloroform extracts of *Anticarsia gemmatalis*, containing an heterogenous mixture of lipids, increase the rate and level of conidial germination of *N. rileyi*; the most active lipid classes being the sterol diacylglycerol and polar lipids (Boucias & Pendland, 1984). Lipid cuticular extracts can influence not only the total level of germination but also the mode of germination. Such examples have been found in the Entomophthorales. Cuticular lipids of the adults of the house fly, *Fannia canicularis*, induce vegetative (i.e. infective) germination of *Erynia variabilis*, but extracts of pupae (qualitatively identical to the adult ones but 5 times less concentrated) promote only the formation of secondary spores. The specificity of *E. variabilis* to adult dipteran hosts has been postulated to be due to the characteristics of the fatty acids in this order of insects. The adult stage contains sufficient amounts of oleic acid to induce vegetative germination and low levels of linoleic and linolenic acids which have a deleterious effect on *E. variabilis* germination (Kerwin, 1982, 1984). Aggressive strains of *C. obscurus*, in contrast to non-aggressive ones, were able to form germ tubes in the presence of aphid cuticular extracts. Two types of extracts were found to stimulate germination: a water soluble extract containing free hexoses and amino acids, originating from the honey dew excreted by the aphid; and a chloroform extract, in which the hydrocarbon fraction was the most active component (Latgé et al., 1987b). However, the influence of cuticular components is probably less specific than has been previously thought. Aphid cuticular lipids are able to stimulate the germination of non-aphid pathogenic strains of *N. rileyi*, whilst extracts from *A. gemmatalis* will induce the formation of germ tubes of primary spores of *C. obscurus* unable to infect the velvet bean caterpillar (Boucias & Latgé, 1988a). Factors more specific than these cuticular nutrients must be involved, therefore, in inducing germ-tube penetration.

An inhibitory effect of some cuticular compounds has also been reported. Removal of the cuticle lipids from various insects, either chemically or mechanically, can increase their susceptibility to different entomopathogenic fungi. The lipids extracted have been shown to inhibit the germination of fungal spores (Koidsumi, 1957; Evlakhova & Chekourina, 1962; Fargues, 1981). The inhibitory effect has been considered to be due to free short chain fatty acids (FA) (C_6 to C_{12}) often detected in larvae of Lepidoptera (Smith & Grula, 1982; Saito & Aoki, 1983). However, all the in vitro germination tests using short chain FA were carried out at concentrations 10^2 to 10^3 times higher than those present in the cuticle (Latgé et al., 1987b). It has been known for a long time that short chain FA have a non-specific antagonistic effect on a wide range of fungi, at the doses tested with entomopathogenic fungi (Teh, 1974). Moreover, short chain FA are not found in all insect orders: they are absent from Coleoptera and Homoptera (Latgé & Fargues, unpubl.; Brey et al.,

1985; Champlin et al., 1981) and consequently could not be the cause of the absence of germination on non-susceptible hosts pertaining to these orders. These results demonstrate the necessity of relating in vivo and in vitro conditions and show that the role of these short chain FA in pathogenesis is questionable and may have previously been exaggerated. The role of lipids in germination may be more subtle. For example, removal of hydrocarbons in the stimulatory cuticular extract will make this extract inhibitory (Latgé et al., 1987b). Changes in the structural conformation of the lipid layer may interfere with chemotactic factors involved in germination (Latgé et al., 1987b).

Compounds other than lipids can have an antagonistic effect on the germination of entomopathogenic spores. In particular, the role of phenols and other polar components of the cuticle should be considered more carefully. Significant amounts of unoxidized diphenols and by-products, left over from the sclerotization process are trapped in the small spaces between the macromolecules of the cuticle (Andersen, 1985). Similar phenolic compounds have been recognized as fungitoxic amongst various plant and human pathogens (Touzé & Esguerre-Tugayé, 1982). Cuticular dihydric phenols and their oxidized by-products are able to inhibit the germination of spores of *Cordyceps militaris* (Latgé, unpubl.). A 50% ethanolic extract of the cuticle of pea aphids resistant to *C. obscurus*, prevents the germination of spores of this pathogen (Latgé, unpubl.). Microbiological factors can also influence the behaviour of germinating spores. The saprophytic microflora associated with the insect cuticle can either stimulate or inhibit the germination of the spore in vivo (Schabel, 1978; Fargues, 1981).

Penetration of the host integument

Successful germination on the host cuticle is not always synonymous with infection. Milner (1982) reported that *Erynia neoaphidis* conidia were able to germinate on both susceptible and resistant *Acyrthosiphon pisum* biotypes but that penetration was inhibited in the 'resistant' aphids. The arrival of fungal elements inside the insect body, a requisite to infection, depends on the ability of the germ tubes to penetrate the external epicuticle and the internal procuticle. The cuticle, because of its hardness and thickness, represents in fact the main barrier to fungal infection and it has been known for a long time that wounded insects become susceptible to weak fungal pathogens or even saprophytes, such as *Mucor* and *Fusarium* (Vey, 1971a). The chemistry of the epicuticle has been described above. The procuticle, the main part of the integument, is composed of chitin,

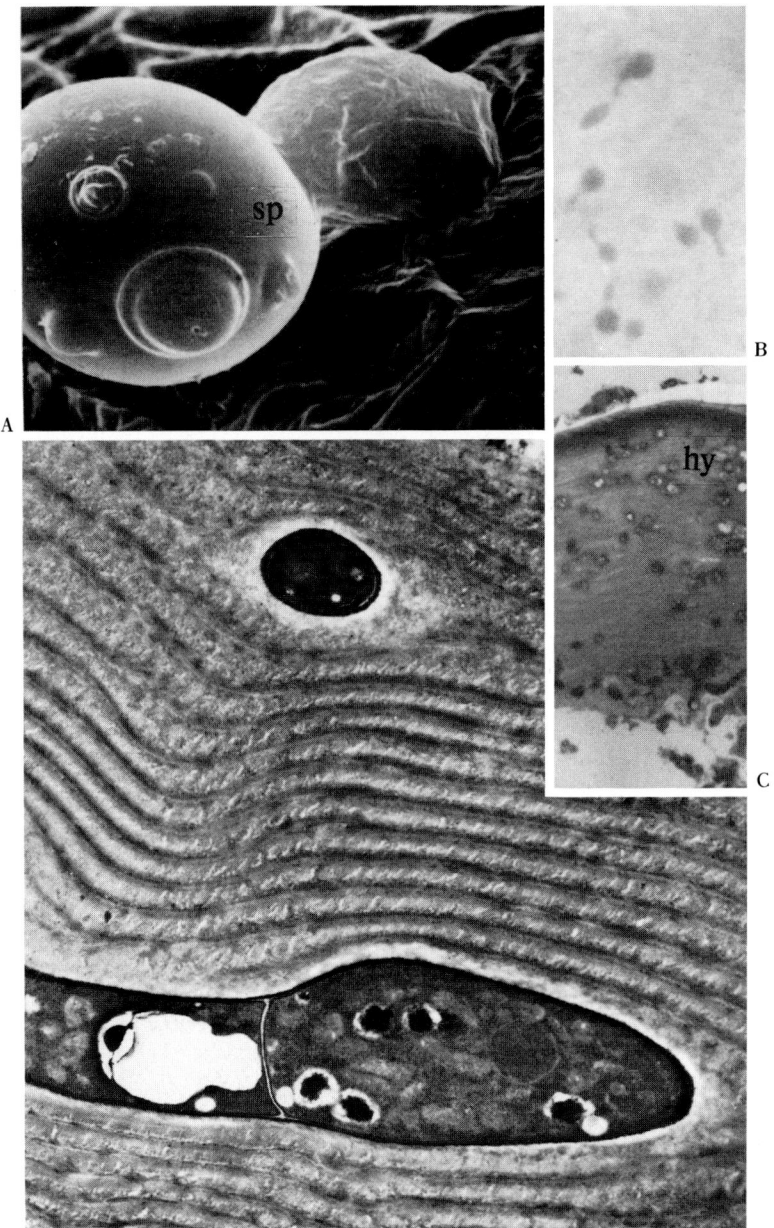

FIGURE 4-4. A Penetration of the aphid cuticle by a germ tube of *Conidiobolus obscurus* primary spore (sp) (× 750). B-D Penetrating structures of *Beauveria bassiana* in the cuticle of *Galleria mellonella*: B appressoria (× 1200), C penetrating hyphae (hy) (× 900), D inflexion of cuticular laminae and cuticular lysis during penetration (× 40.000).

embedded in a matrix of proteins fibres, running parallel to the surface with slight and progressive changes in the fibre orientation (Livolant et al., 1978), producing under the electron microscope an image of parabolically-arranged filaments. The procuticle may be differentiated into sclerotized exocuticle and soft endocuticle, depending on the nature of the proteins associated with chitin and the type of links between the proteins and chitin. During sclerotization diphenolic compounds (N-acetyldopamine and N-alanyldopamine) are enzymatically oxidized to reactive intermediates which combine with proteins and with each other (Andersen, 1979; 1985). Penetration of the tough integument should involve mechanical and enzymatic activities in the developing germ tube (fig. 4-4, A). Inflexion of the cuticular laminae has been demonstrated, mainly under the penetration pegs produced by appressoria, suggesting that mechanical forces are involved (Zaccharuck, 1970 b; Nyhlen & Unestam, 1975; Travland, 1979a) (B). However, the cuticular alterations observed during fungal penetration (Lambiase & Yendol, 1977; Pekrul & Grula, 1979; Michel 1981; Zaccharuck, 1981; Brey et al., 1986) have shown the important role played by the enzymes released by the germ tube during perforation of the insect cuticle. (C,D). Ultrastructural results are confirmed by chemical studies showing that digestion of the integument requires a sequential lipase-protease-chitinase treatment (Samsinakova et al., 1971; Smith et al., 1981; St. Leger et al., 1986b). In some instances, lipase, protease and chitinase levels have been correlated with the aggressiveness of entomopathogenic fungi. For example, chitinase-negative and lipase-negative strains of *B. brongniartii* are not able to infect *Melolontha melolontha* (Paris & Ferron, 1979). Virulence of *V. lecanii* has been associated with high extracellular chitinase activity (Jackson et al., 1985). Aggressive strains of *B. bassiana* on *Heliothis zea* excreted the highest amount of elastase (Grula et al., 1984). Studies on *M. anisopliae* by St. Leger et al. (1986a-d) have demonstrated that all virulent strains of this species produce high amounts of proteases, amongst which one chymoelastase and one trypsin-like enzyme have the highest activity. The involvement of these proteases in penetration of the cuticle is confirmed by the fact that application of a protease inhibitor to the cuticle surface caused a significant delay in mortality when compared with the control (St. Leger et al., 1986b). In some host-pathogen associations no correlations exist between the enzymatic activities and the chemical nature of the cuticular components. For example, aggressive strains of *M. anisopliae* on *Culex pipiens* have the highest α-glucanase activity (Al-Aidroos & Seifert, 1980) and in many cases no clear relationship has been established between the levels of in vitro-produced enzymes and the aggressiveness of a fungal strain (Samsinakova et al., 1977; Bajan et al., 1979; Grula et al., 1979; Champlin et al., 1981; Fargues, 1981; Latgé et al., 1984 c). Several explanations could be advanced:

1 The success of most entomopathogens depends on limited breakdown of the cuticle rather than on extensive destruction of the integument. Therefore, it is conceivable that most of the isolates produce limited quantities of enzymes just sufficient to perforate the integument; the most aggressive strains being able to excrete the enzyme quicker than the non-aggressive ones. This suggests that the kinetics of enzyme production may be more important than the total amount of enzymes produced. Similar results have been found amongst phytopathogenic fungi where no relationships have been demonstrated between the various enzymatic activities and the compatibility or incompatibility of the host-parasite interaction (Cooper et al., 1981).

2 Almost all enzymatic studies have been undertaken in vitro with mycelium instead of germinating conidia, when in fact the metabolic activities of these two fungal stages are often very different (Trinci, 1978). The few enzymatic studies performed on germ tubes have been qualitative histochemical studies which have demonstrated the presence of proteases, aminopeptidases, lipase, esterase and N-acetyl-glucosaminidase produced by the germ tube in the integument (Ratault & Vey, 1977; Michel, 1981; St. Leger et al., 1986d). In vitro studies should also be accompanied by in vivo experiments. Compatibility between enzymes of *M. anisopliae* produced in vitro and in the insect cuticle has been demonstrated using rabbit antisera (St. Leger et al., 1986d). However, as with phytopathogenic fungi, enzymes of entomopathogenic fungi produced in vitro could differ qualitatively (at the isozyme level) from those formed in vivo (Hancock, 1976).

3 Most of the enzymatic studies so far have employed non-specific substrates such as casein or vegetable oil, instead of true cuticular polymers. It is now known that cuticular ghosts from *Anticarsia gemmatalis* and *Astacus astacus* stimulate much greater protease production in *N. rileyi* and *A. astaci* respectively, compared with artificial proteolytic substrates (Boucias, unpubl.; Persson et al., 1984). Defined enzymatic substrates such as synthetic chromogenic peptides (p-nitrophenylesters or blocked peptide nitroanilides) can give a much better indication of the specificity of the enzymes involved and their mode of action at a molecular level (Grula et al., 1984; Persson et al., 1984; St. Leger et al., 1986d).

4 Finally, the role of the enzymes may not be limited to degrade the cuticle but also to liberate monomers that can be metabolized by the germ tube in order to continue to grow into the integument. In these conditions, a non-pathogenic strain may produce the same type and amount of enzyme as the pathogenic one but defects in permease systems may inhibit its use of the enzymatic by-products. Such a phenomenon has been suggested in *C. obscurus* strains, in respect of the chitinase-N-acetyl glucosamine complex (Latgé et al., 1984c). Failure to establish an infection results not only from the inability of the fungus to lyse the integument but also from the active role played by the integument to stop the invading fungus. Moulting, and the subsequent discarding of associated fungal cells with exuviae, is regarded as an important host resistance mechanism (Vey & Fargues, 1977). The presence of increasing phenoloxidase activity in the integument during infection will result in the oxidation of available phenols, such as protein-bound tyrosine residues, to reactive quinones and then melanines deposited around the invading germ tube (Nyhlen & Unestam, 1975; Michel, 1981). In certain instances, this reaction can be powerful enough to block the infection process. In other cases the penetrating germ tubes do not appear to be melanized (Michel, 1981), suggesting the secretion of a phenoloxidase or proteinase inhibitor by the penetrating germ tube. The active role of the cuticle in the defence against fungi is also illustrated by the pathogenesis of *Aphanomyces astaci* in the American crayfish, *Pacifastacus leniusculus* which is more resistant to this pathogen than the European crayfish species. By depleting the haemocyte number in *P. leniusculus*, the fungus can invade and grow without melanin deposition in the cuticle and eventually the crayfish dies from the fungal infection (Persson & Söderhäll, 1983). Therefore, it appears that a communication exists between the cuticle and the haemocytes which is probably essential in the defence reaction elicited by the cuticle. Protease inhibitors which have been demonstrated in the crayfish cuticle (Häll & Söderhäll, 1983) and the substrates for cuticular phenoloxidase are likely to be transported from the haemocytes to the cuticle.

Development of the fungus inside the host

Although arthropods have a primitive immunological system, particularly lacking immunoglobulins, they are able to discriminate between the self and the non-self and to react to the entrance of a fungal pathogen inside the body cavity (Boucias & Latgé, 1986b). In some instances, the internal reaction of the host towards the pathogen can be strong enough to eliminate the pathogen. For example, several cases of recovery have been reported in Lepidoptera larvae after intra-haemocoelic injection of entomopathogenic Hyphomycetes, indicating a powerful host defence reaction (Fargues, 1981; Ignoffo et al., 1982).

Cellular and/or humoral defence mechanisms have been reported (Götz & Boman, 1985; Lackie, 1986). Phagocytosis has rarely been described (Vey & Fargues, 1977), probably because the large size of the fungal spore (compared with bacteria) will make engulfment by haemocytes very difficult. However, the presence of toxins, such as destruxins, seems to favour phagocytosis (Vey, unpubl.) The main cellular response of the insect is a multihaemocytic encapsulation of the fungal element following initial recognition of the fungus by haemocytes (Vey & Vago, 1971). The putative receptors in the host and fungal cells have been inadequately investigated. The involvement of lectins in the recognition of fungal particles has been proposed recently. Lectins specific for carbohydrates commonly present in the cell wall of entomopathogenic fungi have been detected in arthropod blood. Most of the lectins found in the insect haemolymph are galactose-specific (Renwrantz, 1983; Pendland & Boucias, 1985, 1986 b; Lackie & Vasta, 1986; Natori, 1986). In *Sarcophaga*, the galactose lectin participates in the elimination of foreign substances introduced into the body cavity since antibodies active against *Sarcophaga* lectin or galactose injected with the foreign material block the defence reaction (Natori, 1986). In *A. gemmatalis*, a direct correlation has been observed between the production of a galactose-agglutinin and the presence of fungal cells in the larval haemolymph (Pendland & Boucias, 1985). It is possible that lectins adsorb to microbial walls as indicated by removal of haemagglutination activity after incubation of fungal cells in the lepidopteran haemolymph (Pendland & Boucias, 1987). Such adsorption could render the fungal cell more susceptible to agglutination and consequently encapsulation (Yeaton, 1981). However, the specificity of the lectins is not always correlated with the surface carbohydrates of the potentially pathogenic fungus (Pendland & Boucias, 1984b; 1986a), suggesting that lectins have another role to play than helping recognition such as carrying toxins or causing growth inhibition of the cell (Yeaton, 1981; Natori, 1986).

Another recently proposed mechanism involved in the recognition phenomenon in crustaceans and insects is the prophenoloxidase activating system (Ashida & Dohke, 1980; Söderhäll, 1982; Söderhäll & Smith, 1986; Andersson et al., 1987; Harmstorf & Götz, 1987). For the majority of arthropods, it is located in specific cells, viz: granulocytes in some insects and semi-granular cells in crayfish (Söderhäll & Smith, 1983; Smith & Söderhäll, 1983b; Ratcliffe et al., 1984). The proPO cascade plays an essential role in phagocytosis, adherence during nodule formation, microbial killing and cell communication (Smith & Söderhäll, 1983ab, 1984; Ratcliffe et al., 1984; Johansson & Söderhäll, 1985; Leonard et al., 1985). However, the various proteins of the proPO cascade, most of them having enzymatic activities, remain to be isolated and chemically characterized.

The mechanics of the fungal encapsulation process in insects has been studied (Vey, 1971b, 1977; Vey & Götz, 1975; Vey et al., 1975) and is similar to that observed in bacteria (Lackie, 1980; Ratcliffe & Rowley, 1979). In less than 30 minutes, a rapid recruitment and adhesion of haemocytes follows the initial recognition at the site of contact with the fungal propagule. The granular cells (containing the proPO system) are disrupted and release the components of the proPO system, provoking the appearance of electron dense material in the intercellular spaces. This initial triggering of the activation of the proPO system is due to water-soluble or insoluble glucans composed of $\beta(1\rightarrow 3)$ linked glucopyransoyl residues. Such polysaccharides are frequently found amongst entomopathogenic Entomophthorales and Deuteromycetes (Latgé et al., 1984a; Latgé & Boucias, unpubl.). Isolated semi-granular cells from Crustaceans have been found to respond to $\beta(1\rightarrow 3)$ glucans by degranulation and subsequent lysis resulting in the release of the proPO system from these cells (Johansson & Söderhäll, 1985); $\beta(1\rightarrow 3)$ glucan receptor of proPo-activating system has been recently isolated and purified (Yoshida et al., 1986). In certain arthropods, the proPO system can be activated spontaneously by low calcium concentrations. A calcium-dependent clotting process can also be observed (Söderhäll, 1981, 1983; Durliat, 1985; Ratcliffe, 1985). Encapsulation is always associated with the activation of the phenoloxidase resulting in melanization of the fungal propagule. Several proteases seem to be involved in the conversion of proPO to phenoloxidase. One of them, a calcium-dependent serine protease appears to be common to most arthropods (Ashida & Dohke, 1980; Ashida et al., 1983; Söderhäll, 1983; Söderhäll & Smith, 1986). After being activated, the phenoloxidase system becomes sticky and attaches to different types of foreign substances (Söderhäll et al., 1979; Ashida & Dohke, 1980; Ashida et al., 1982). The coating of the foreign material by coagulated haemocytes stimulates the recruitment and adhesion of plasmatocytes (containing little or no phenoloxidase) which form multicellular sheets around the coated fungi (Ratcliffe et al., 1984) (fig. 4-5 A). The nature of all the attaching components present in the coagulated haemocytes, which are eventually released into the haemocoel, is unknown (Söderhäll et al., 1984). In crayfish, a cell adhesion factor located in the cells granules has been recently characterized (Johansson, unpubl.). Haemocytes continue to accumulate in compact layers around the engulfed fungal propagules to form, after 1-3 days, a granuloma which can reach 300 to 400 μm in diameter (fig. 4-4 C). The haemocytes in contact with the fungus are always altered whilst the external ones maintain a normal morphology (fig. 4-5 B) (Vey, 1971b; Vey & Vago, 1971). As the capsule forms, the recruitment of haemocytes decreases. This indicates a de-

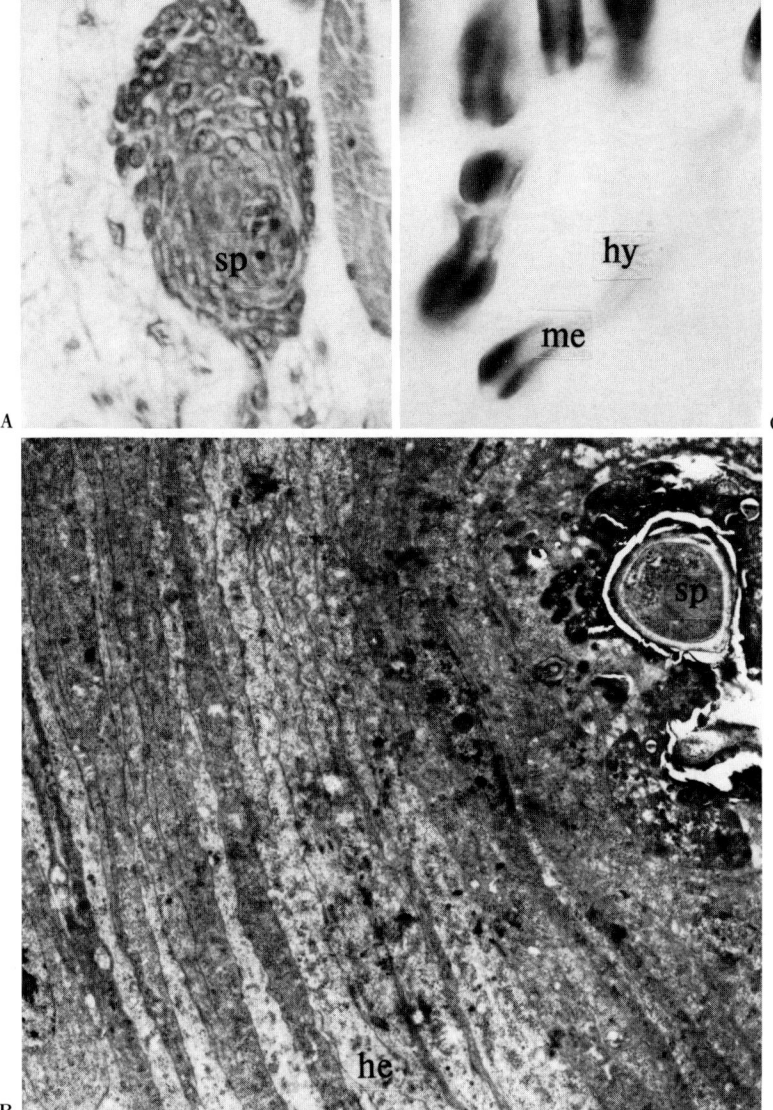

FIGURE 4-5. A Recruitement of *Galleria mellonella* haemocytes (he) around spores of *Cordyceps militaris* (sp), already melanized ($\times$ 500), B multicellular sheets of haemocytes around melanized spores of *Aspergillus niger*; note the completely destroyed haemocytes near the fungus ($\times$ 4000), C layer of melanin (me) covering the hyphae (hy) of *C.militaris* escaping from a granuloma ($\times$ 2000). (B: courtesy of A. Vey).

crease in the release or the diffusion of the opsonizing substances (Lackie, 1980), or the liberation of a promotor disconnecting the attachment process. Biochemical systems such as serine-protease or subtilisin inhibitors, which may block the phenoloxidase activation, have been described in various arthropods (Hall & Söderhäll, 1982; Söderhäll, 1983; Hergenhahn et al., 1986; Boucias & Pendland, 1986). In addition, Hergenhahn & Söderhäll (1985) have found a macroglobulin-like activity in the blood of crayfish, which might serve to sequester proteases from the circulation. These regulating systems could be under the control of a gradient of concentration of the chemical effectors and would protect the arthropod against a massive intravascular coagulation and melanization, suggesting that some communication exists between arthropod cells, the exact nature of which remains obscure.

Inside the granuloma, the fungus is always covered by a thick layer of melanin (fig. 4-5 C). Quinones and melanins produced from insect phenols are antagonistic to fungal growth (Söderhäll & Ajaxon, 1982) and may block or retard the development of certain fungi in non-susceptible insect hosts. Treatment of the insects with glutathion (a phenoloxidase inhibitor) increases the level of lethal infection by reducing the melanization of the fungus inside the host without modifying the pattern of cell aggregation (Vey, 1977). The formation of haemocytic granuloma is not the only type of reaction encountered in arthropods. In the aquatic larvae of mosquitoes and chironomids, melanin-like substances are deposited at the surface of the fungus without direct participation by the blood cells (Vey & Götz, 1975). In other insects, such as the pea aphid, no apparent defence reaction, either humoral or cellular, has been detected. The absence of a host reaction in the pea aphid has been correlated with the lack of circulating haemocytes and a low level of phenoloxidase activity (Latgé & Brey, unpubl.).

The defence system of arthropods against fungal invasion has been insufficiently studied. However, based on the few models, which have been investigated so far in invertebrate immunology, it appears that some of the defence reactions developed by invertebrates are similar to various cellular and humoral immunological reactions present in vertebrates. In vertebrates, including humans, IgM and the complement are involved in the multicellular defence reactions. In invertebrates, the role played by coagulogen may correspond to the IgM. The prophenoloxidase system comprises a complex enzyme cascade, possessing opsonin and antimicrobial properties, and could represent a primitive form of the complement which in contrast to the complement systems of mammals, is contained within the circulating blood cells (Söderhäll & Smith, 1986).

Fungi have developed several mechanisms to escape the insect defence reactions. The most intriguing has been shown by several species of Entomophthorales. These species can spontaneously form protoplasts in the haemocoel of the insect, and are able to grow in vivo and in vitro in this protoplast stage (Tyrrell & MacLeod, 1972; MacLeod et al., 1980; Butt et al, 1981; Carruthers et al., 1985; Nolan, 1985; Latgé et al., 1987a). Protoplast formation is not induced by an enzymatic lysis of the hyphal tip but results from the influence of specific media components inducing a reversible inactivation of the fungal polysaccharide synthetases (Beauvais & Latgé, 1988). The protoplasts of *Entomophthora muscae* and *Entomophaga aulicae* do not expose $\beta(1 \rightarrow 3)$ glucans on their membranes and thereby avoid recognition by the haemocytes. Thus, they are able to invade the haemocoel without inducing a defence reaction (Dunphy & Nolan, 1980, 1982; Latgé, Beauvais & Vey, 1986a). Membrane glycoproteins with β galactose residues also seem to play an important role in the inhibition of the haemocoelic encapsulation (Dunphy & Chadwick, 1985). Protoplasts have also been shown to be produced by fungi of the genus *Coelomomyces* (Powell, 1976).

Another way for a fungus to overcome the defence reaction of the insect is by the multiplication of infective propagules. It is well known that a resistant insect ceases to be resistant above a certain dose of injected propagules. Most entomopathogenic fungi, especially Hyphomycetes, are able to form yeast-like propagules by simple hyphal fission. These yeast-like cells are quickly disseminated throughout the blood stream making it difficult for the circulating haemocytes present in limited numbers to engulf all the fungal elements. Moreover, hyphal bodies are not as antigenic as mycelium (Pendland & Boucias, 1986b). Most entomopathogenic fungi are also able to excrete toxins; the best detailed examples being the cyclodepsipeptides produced by *Beauveria* and *Metarhizium* (Roberts, 1981). Toxins alter insect cells which are not invaded by the fungus, particularly at the mitochondrial or endoplasmic reticulum levels (Zaccharuck 1981; Vey et al., 1973) and consequently can weaken the insect multicellular reaction. The fact that pathotypes of *M. anisopliae* remain encapsulated in heterologous hosts while the specific strains readily escape from the granuloma has been correlated with toxin production. Recently, Huxham et al. (1986) have demonstrated in an in vitro assay, that metabolites of the fungus *M. anisopliae* including destruxins, can reversibly suppress activation of prophenoloxidase in *Schistocerca gregaria* and *Periplaneta americana*. However, the role of toxins has been established only from metabolites produced in vitro and these products may not be found in vivo or may not be as active. Injection of haemolymph of *N. rileyi*-infected silkworm into healthy larvae had the opposite effect to injections of in vitro produced culture filtrates (Roberts, 1981). Moreover, enzymes produced during infection, which degrade the insect tissues, have often been confused with toxins (Roberts, 1981).

The biochemical changes occurring in infected insects has been poorly studied (Domnas, 1981). Such studies should be undertaken to determine the cause of insect mortality. Is the main cause of death a physiological starvation caused by the deprivation of insect metabolites by the pathogen? This has been suggested by Domnas (1981) in the case of *Culex* larvae infected with *Lagenidium giganteum*, where protein and sugar levels drop considerably in the insect following the development of the fungus inside the host. True starvation has been observed as a consequence of gut wall paralysis of *Heliothis zea* by *B. bassiana*, the pathogen inducing cessation of feeding and loss of weight (Cheung & Grula, 1982). In these cases, death would occur without any toxin emission. Toxins have been claimed also to be responsible for the death of numerous insects infected by Deuteromycetes (Ferron, 1981). For most Entomophthorales, however, death seems to occur only when all the tissues of the insect have been substituted by the fungus. Living infected aphids have often been observed with the fungus already emerging from the body.

Host death marks the end of the parasitic phase of fungal development. The mycelium then grows saprophytically, producing antibiotics antagonistic to the intestinal bacterial flora. When the environmental conditions are favourable, the fungus grows outwards through the integument and develops conidiogenous structures. Under unfavourable climatic conditions, most fungi are able to produce resting propagules (chlamydospores, zygospores or oospores) which allow the fungus to overwinter or to withstand adverse conditions in absence of the host (Latgé et al., 1978a; McCoy, 1981; Pendland, 1982)

Conclusions

Amongst the three phases recognized in the establishment of a fungal infection in an arthropod host, the first two phases: the adherence and germination of the fungal spore on the integument and the penetration of the cuticle by the germ tube, seem the most critical. Usually, the multicellular defence reaction of the haemocytes and/or the humoral reaction, if present, seem effective only against weak pathogens or very low doses of aggressive pathogens. It has been shown, for example, that fast-growing saprophytes, like *Aspergillus*, *Fusarium* and *Mucor* can easily infect an insect after direct injection into the haemocoel or after a wound in the cuticle (Vey, 1971a). Immune reactions occurring in invertebrates, may help to clarify the phylogeny of the more sophisticated mammalian immunological system. For example, many infectious diseases, such as malaria, are transmitted by insects and the causal organisms develop and migrate in their host insect without developing any cell reaction (Ratcliffe et al., 1984).

The reasons for such unexpected phenomena, considering the strong cellular reaction of the haemocytes to fungi, may be elucidated by a thorough study of the nature of the multicellular reaction. How are lectins involved in insect immunity? What is the exact role of each haemocyte category in the defence reaction? What are the mediators regulating the cell to cell communication? Will it be possible to find in the insect chemotactic factors similar to the leukotrienes or lymphokines active in humans? What is the exact role of the proPO system in the defence reaction? How is the specificity of the immune response determined since, for example the antibacterial cecropins were only detected after bacterial infection and wounding, but not after a fungal infection? All these questions need to be answered before the process of granuloma formation in arthropods can be elucidated.

Our understanding of the germination process of the fungal spore on the cuticle and the penetration of the integument by the germ tube is limited in comparison with phytopathogenic fungi. Research should be centered on the biochemical and genetical factors that could explain the aggressiveness of a fungal strain or the resistance of a given insect to a fungal isolate. The availability of strains with variable susceptibility should facilitate such studies. Another promising approach may be a study of the mechanisms involved in non-host resistance, particularly to understand if general resistance and specific varietal resistance mechanism are comparable or not (Matta, 1982). This research would be more meaningful if the few entomopathologists involved in this area of research would work on the same insect host. One has to remember that most progress in immunology, especially at the genetic level, or in molecular biology has only been made possible by the selection of common laboratory organisms by the scientific community, viz: *Mus domesticus* and *Escherichia coli*. The question is: which insect and which pathogen should be selected? Obviously, a large host such as *Manduca sexta* and a fast growing fungus would provide very useful models.

Many fundamental questions remain to be answered in order to understand the fungal infection process in insect pathology. The first one is: what are the molecules involved in the reciprocal recognition of an insect by a fungus and could we consider, as in plants, that specific recognition for incompatibility and compatibility is the result of non-mutual recognition of the pathogen and host gene products (Ellingboe, 1982)? Based on analogies with plant and human immunology, the most likely candidates for these recognition events should be found amongst carbohydrate-containing proteins not only surface localized but also present intracellularly and subsequently excreted (Heitefuss, 1982). Another important question concerns the understanding of the biochemical mechanisms of the resistance or susceptibility of an insect to a fungus. What is the active role of the integument in the resistance to a fungal pathogen? Is the insect able to secrete substances similar to the phytoalexins produced by the plant in response to a fungal infection or can it activate killing factors playing a role similar to the complement in vertebrate immunology? If so, what fungal polymers are able to elicit the synthesis of these fungicidal metabolites? The identification of these products would facilitate genetic studies of the molecular interactions in the host-pathogen couples. Will it then be possible to induce cross protection of the insect to a fungus, similar to that obtained in phytopathology or to the immunization in mammalian systems, or is resistance to a fungus a non-specific unenhanceable phenomenon? Finally, it would be very interesting to know the genetic status of the virulence or aggressiveness of a fungus in entomopathology since it is now recognized for plants and animals that the interaction between a host and a parasite is governed by similar basic genetic patterns (interorganismal genetics); the genes in the host and the parasite determining the phenotype of the interaction (Ellingboe, 1982). The answer to this question will have a prerequisite: the discovery of resistance genes in the host. One would then be able to see if resistance in insects is the result of a gene for gene relationship similar to the situation observed most often in plant pathology (where it is called vertical resistance): in this case, a host line with an R-gene cannot be resistant to a pathogen unless the pathogen has the corresponding P-gene for virulence. But aggressiveness of a fungal isolate (horizontal resistance in plants) can also have a polygenic determination. In conclusion, relatively little is known about fungal pathogenesis in arthropods, particularly the determinants of the infection process, and more fundamental studies should be undertaken in this field. A consideration of the analogies with phytopathology and vertebrate mycology would be very constructive.

Chapter 5

Natural control: ecology and biology

Entomogenous fungi in nature cause a regular and tremendous mortality of many pests in many parts of the world and do, in fact, constitute an efficient and extremely important natural control factor. E. A. Steinhaus (1949)

Introduction

A greater knowledge of the role that entomopathogenic fungi play in arthropod population dynamics and the conditions which govern their activities in natural ecosystems is considered to be an essential prerequisite if man is to employ them to the best possible advantage in agriculture, forestry and vector control. In this chapter, observations relating to the occurrence and distribution of entomopathogenic fungi and the behavioural patterns of their hosts will be analysed in diverse habitats. These habitats can be divided roughly into two contrasting ecosystems: those exploited by man and those which have remained relatively untouched by his activities. Most studies have been undertaken in the former, typically relating to epizootics amongst populations of arthropods of agricultural importance. Relatively little attention has been paid to these fungi in other, non-agricultural habitats, which we here designate as primary ecosystems.

Primary ecosystems

AQUATIC HABITATS

The entomopathogenic fungi associated with these habitats belong predominantly to the lower fungi (Mastigomycotina, Zygomycotina) because of their aptitude to produce motile spores. Amongst the Oomycetes, *Lagenidium giganteum,* produces strains highly pathogenic to mosquito larvae (Umphlett, 1973). This ecologically obligate parasite can survive saprophytically for considerable periods and appears to have adapted to transient or semi-permanent aquatic habitats, producing thick-walled oospores probably to enable the fungus to survive in the absence of water. However, relatively little is known about the impact of this species on natural populations of mosquitoes. Similar information is also wanting in respect of the newly described Oomycete, *Leptolegnia chapmanii* (Seymour, 1984), which is pathogenic to *Aedes* and *Culex* larvae in various habitats (lakes, tree holes) in the USA.

Chytridiomycetes of the genus *Coelomomyces* are host specific endoparasites of aquatic larvae of biting flies, many of which have complicated life cycles, some dependent on alternate crustacean hosts to complete the cycle (Federici, 1981). Epizootics of *Coelomomyces* have been recorded in *Anopheles* populations in East Africa, with mortality levels of up to 95% in all larvae hatching out during the rainy season (Couch & Umphlett, 1963). The latter authors also quote examples from the USA, demonstrating that these fungi can in certain habitats exert a considerable natural control of mosquito populations. However, it has been concluded that, although *Coelomomyces* are well established in natural populations, they usually occur at a low or enzootic level and under normal conditions they are not considered to be significant control factors (Chapman, 1974). Both the requirement for an alternate host and the development of thick-walled (resting) sporangia would seem to be specific adaptations to the sporadic or seasonal occurrence of mosquito larvae. Adult flies can also be attacked by *Coelomomyces* and Couch (1972) is of the opinion that, although wind, birds and fish are implicated in dispersal, lightly infected adults carrying mature sporangia internally are the most important means of dissemination of the pathogen. The chytrid, *Coelomycidium simulii* is a highly specific pathogen of blackflies in Europe, Africa and North America (Strand et al., 1977). Infection rates are generally low, however, and epizootics appear to be uncommon. Fungal thalli are encountered typically in late instars of the fly host, the abdomen being filled with sporangia, but the overall life cycle is still poorly understood. The possibility of ovarial transmission of the pathogen has been suggested recently (Tarrant & Soper, 1986).

The aquatic Entomophthorales *(Erynia conica; E. rhizospora; E. plecopteri)*, which are particularly common on dipterans (Simulidae, Tipulidae) inhabiting fresh water ponds, streams and rivers in temperate regions (Thaxter, 1888), show remarkable adaptations to the habitat. It has been discovered recently that up to four spore types can be produced by a species to facilitate dispersal and infection, both aerially and aquatically, revealing a great plasticity in both conidial morphology and mode of germination (Descals et al., 1981; Descals & Webster, 1984). The basic life cycles of the above species are essentially similar, even though the hosts belong to three different insect orders (Diptera, Trichoptera, Plecoptera). Typically the infected insects die on the undersides of rocks and vegetation, usually firmly attached by rhizoids. The sequence of events is thus: the formation of 'aerial' primary conidia, which are unbranched and forcibly ejected, and which can replicate into shorter secondary conidia if a suitable host is not encountered; production of apically branched 'aquatic' primary conidia, which are passively released and which replicate into stellate secondary conidia. The latter are considered to be adapted for impaction on the underwater larval stages of the host. Secondary globose conidia were the only infective propagules contaminating adult simulids at their oviposition sites. No larval stages were infected by aquatic spores (Hywel-Jones & Webster, 1986). The mode of overwintering in these species is unknown since no resting structures have been reported. Another pathogen of mosquitoes, *Erynia aquatica*, infects adult insects by secondary spores discharged from primary ones floating on the surface of the water (Humber & Ramoska, 1986).

Few Deuteromycetes are associated with natural populations of aquatic insects, *Culicinomyces clavisporus*, for example, being recorded until recently only as a laboratory contaminant of mosquito larvae. Its distribution and impact on mosquitoes in nature is still not well documented (Goettel et al., 1984). *Tolypocladium cylindrosporum* has been identified recently as a significant natural control factor of *Aedes* larvae in brackish coastal regions of New Zealand and from oak tree holes in the USA, apparently in some instances causing high mortality (60-90%) (Weiser & Pillai, 1981; Pillai, 1982). Those larvae infected with the fungus that remain afloat produce abundant dry conidia from the aerial mycelium but, with both of these Deuteromycetes, the infection can also be passed on to the adults and this is probably the most important and effective method of long-distance dispersal.

Forest habitats

Gray (1858) prophetically stated that: 'The results obtained from the histories of the various entomophytes seem to show that if the countries having primaeval forests, or subjected to heavy rains, especially within the tropical portions of the world, were diligently searched by collectors, they would probably produce numerous examples of this curious phenomenon.' The primary rain forests of the tropics are probably the only really untouched ecosystems within this context. However, much of the managed coniferous forests, particularly in North America, probably reflect the stability and mimic the ecosystem which existed prior to man's intervention and hence are included within this section.

Humid, tropical forests have a rich and varied entomopathogenic mycoflora (Evans, 1982). Particularly well represented, are species of the genus *Cordyceps* and its many proven or suspected anamorphs (*Hirsutella, Hymenostilbe, Nomuraea, Paecilomyces, Verticillium*). This genus appears to have reached its evolutionary peak in such habitats, judging from the diversity of forms which reflects that of the higher plant and arthropod communities, all of which are closely interlinked. It has been postulated that entomopathogenic fungi are an integral factor within these habitats contributing to the stabilisation of arthropod populations and in turn being controlled by these very same population levels (Evans, 1974). The forest canopy buffers the understorey from extremes of temperature and humidity, creating a stable microclimate probably conducive to continual fungal activity even during severe dry seasons.

Observations in tropical forests in both Africa and South America indicate that the entomopathogenic mycoflora decreases in richness as the forests are exploited; whether this is due to the disappearance of the specific hosts or the loss of optimum conditions for infection, or a combination of both, is unknown. But certainly, in depleted forests and in agricultural land bordering these habitats, members of the genus *Cordyceps* are scarce. Conversely, fungal genera most familiar to agricultural entomologists, such as *Beauveria* and *Metarhizium*, are poorly represented in tropical forests (Evans, 1982). There are few data relating to the fungal diseases of arthropods in this ecosystem, although some studies have been made of entomopathogenic fungi attacking forest-dwelling ants (Andrade, 1980; Evans & Samson, 1982; 1984). Ants are the dominant arthropods in lowland tropical forests (Elton, 1973) and not surprisingly, therefore, are the most affected quantitatively, by entomopathogenic fungi, although as noted by Petch (1925) scale insects are also heavily attacked. Because the ant is infected at the adult stage, and the hard exoskeleton is not colonised by the fungus, sufficient salient taxonomic features are present to enable host identification at the generic or species level, allowing for an accurate assessment of host-pathogen associations which can only be speculated upon in most other arthropod groups.

Indeed, in the case of *Hypocrella* infections of scale insects, the entire host body is replaced by the fungal stroma, making host identification difficult, to such an extent that these fungi were initially assumed to be parasitic on the plant. However, evidence from *Cordyceps*-infected ant collections strongly suggests that there is a high degree of specificity within these associations, a *Cordyceps* species typically being confined to a single genus or tribe of ants. Nevertheless, there are odd exceptions to the rule, or unexplained jumps, since seemingly morphologically indistinguishable *Cordyceps,* which appear within the Formicidae to be restricted to a single tribe, can occur on a similarly narrow range of hosts within an unrelated insect order.

Investigations of a disease complex on a myrmicine ant, *Cephalotes atratus,* involving two species of *Cordyceps* were carried out over a two-year period in Amazonian rain forest (Evans & Samson, 1982). Monthly counts of infected ants showed some fluctuations but generally, mortality was relatively constant over the sampling period indicating that the disease was at an enzootic rather than an epizootic level, if the latter term is defined as a temporary increase in the incidence of an infectious disease, supporting the results of earlier studies on entomopathogenic fungi in similar habitats in Ghana (Evans, 1974).

Gregarious or social insects, which would be especially prone to an infectious disease, must have evolved defence mechanisms for reducing the impact of pathogens on the insect colony. In the case of *Cephalotes atratus,* it was observed that infected workers change their normal foraging routine, thereby avoiding contamination of the nest. Diseased ants, diagnosed by their erratic often uncoordinated movements, tended to aggregate on or around certain trees, away from any detectable ant trail or nest, characteristically hiding beneath bark tissues and epiphytic plants. This evasive behaviour had been noted previously in other ant genera infected with various *Cordyceps* species in both Old and New World forests, the ants seeking refuge beneath leaf litter or herb and shrub leaves. The essentially ground-dwelling ponerine ant of Africa, *Paltothyreus tarsatus,* invariably climbs shrubs when attacked by *Cordyceps australis* and dies on the branches totally exposed (Evans, 1974; Samson et al., 1982). Most *Camponotus* species infected with *Cordyceps unilateralis* also react in a similar manner. There are general indications, however, that arboreal ants tend to descend when infected and actively seek hiding places whilst non-arboreal species develop the urge to climb. The motives and physiological changes governing the alterations in behavioural patterns of ants infected with entomopathogenic fungi can only be speculated upon until more is known about the movements of healthy ants.

Observations of diseased ants suggest that the pathogen affects, physically or chemically, the central nervous system. Spasmodic twitching, excessive grooming, rasping or biting of the substrate and removal of ant cadavers in the vicinity, loss of orientation and balance are all external signs, diagnostic of fungal disease. The final activity of the ant involves gripping the substrate with mandibles and legs. The death grasp fixes the host in a particular selected or favoured niche. If the latter is a tree crevice or leaf litter for example, the evolution of hyphal outgrowths and specialised phototrophic structures becomes a necessity. The function of the *Cordyceps* stromata and anamorph synnemata is obviously to lift the spore-forming tissues into a position favourable for the dissemination of inoculum. Dolichoderine and formicine ants, attacked by *Stilbella* species in forests and old cocoa farms in Ghana, tunnel beneath tree bark, and the synnemata subsequently emerge from the abdomen or anus of the host and grow towards the light source. The length and complexity of the structure varies with the depth of the 'burrow' (Samson et al., 1981). *Hirsutella* synnemata on hidden ants in Amazonian forests creep through leaf litter or over tree bark, forming holdfast-like rhizoids on these substrates and produce aerial fertile heads. These tenacious structures remain long after the host has been removed by scavengers or by natural weathering and provide a durable inoculum source. Such outgrowths perform a similar function in *Cordyceps* associated with wood-boring larvae, or passively-hidden hosts, of which there are many examples in tropical forests (Evans, 1982).

The complete life cycles of many of the tropical forest *Cordyceps* still require elucidation. At what stage and how the different hosts (as diverse as tree borers, bagworms, soil-buried egg cases and trapdoor spiders) become infected by obligate pathogens remains a mystery. Massee (1895) was of the opinion that, because some *Cordyceps* anamorphs develop readily in culture, these fungi are also saprophytic in nature, independent of the arthropod host, thus considerably broadening their survival base. However, this is thought to be unlikely for host specific *Cordyceps* species but for non-specialised pathogens, such as *Beauveria* and *Metarhizium,* there are indications that they survive saprophytically on chitinous debris in the soil. Recently, hyphal migration of *B. bassiana* through soil has been demonstrated (Gottwald & Todders, 1984).

Entomopathogenic fungi are also well represented on plant-sucking homopterans in tropical forests and Petch (1925), with reference to the jungles of Sri Lanka, commented upon their impact: 'No one who has systematically collected these fungi in the tropics, especially those which attack scale insects, can fail to be impressed by the number of insects which are killed by them under certain conditions.'

Species of the genera *Hypocrella* (or their *Aschersonia* anamorphs), *Torrubiella*, *Hirsutella* and *Verticillium* are common on coccids and whiteflies colonising understorey shrubs and trees and can also be collected in the litter layer on leaves fallen from upperstorey trees (Mains, 1958; Evans, 1974; Evans, 1982). Invariably, the entire population is infected, the host being replaced by the typically bright-coloured fungal stromata. As a consequence, host identification in the majority of cases is rudimentary and thus the specific insect-fungal association has not been determined. There is a similar problem of assessing host specificity amongst members of the genus *Gibellula*, apparently obligate pathogens of free-living spiders, which are prevalent in forest habitats throughout the tropics (Samson & Evans, 1973; Evans, 1982). A number of distinctive species with elegantly-adapted synnemata have been collected recently in primary forests in Brazil and Indonesia, with or without *Torrubiella* teleomorphs (Evans, unpubl.), but the importance of these pathogens (especially the ubiquitous *G. pulchra*) in the natural regulation of spider populations is unknown.

There are also reports of disease outbreaks caused by entomopathogenic fungi in forest ecosystems in both subtropical and warm temperate regions, principally involving *Cordyceps* species, almost invariably *C. unilateralis* on formicine ants. Chen (1978) observed that in Taiwan. '... not uncommon to collect hundreds of specimens at one location', whilst Kobayasi (1941) described similar epizootics in oak forests in Japan. Although members of the Entomophthorales are rare in tropical forest, being found predominantly on small biting dipterans beneath herb leaves (Evans, 1982), they are commonly reported as the causal agents of disease of forest pests in temperate forest habitats (Burges, 1981). Natural epizootics of Entomophthorales have been noted in forest insect populations in tropical countries, but usually in high-altitude habitats with subtropical to warm temperate climates. Katerere (1983) found that *Entomophthora planchoniana* is a major controlling factor of the pine needle aphid (*Eulachnus rileyi*) in Zimbabwe. It was considered that the disease is dependent on a critical host population density; epizootics were recorded when host density was over four aphids per needle, concomitant with a drop in temperature (below $15\,°C$) and heavy rainfall (70–100 mm). From ecological studies of populations of the woolly pine needle aphid (*Schizolachnus pini-radiatae*) in Canadian forests, it was concluded that the major controlling factor was an entomophthoralean fungus (Grobler et al., 1962), later identified as *Erynia canadensis*. In a detailed epizootiological investigation subsequently undertaken by Soper & MacLeod (1981), it was discovered that, under Ontario conditions, disease incidence was not limited by climatic factors but was dependent upon inoculum levels and host population density. Almost a 100% kill of the aphids was noted in one season followed by a faster spread of the pathogen at lower population densities in subsequent years.

Two species of Entomophthorales, *Entomophaga aulicae* and *Erynia radicans*, are considered to play an equally important role in the decline of the eastern hemlock looper (*Lambdina fiscellaria*) also in North American forests (Otvos et al., 1973), causing severe mortality (up to 90% infected larvae). Analyses of epizootics have demonstrated that fungal infections build-up over a 2-year period and cause the collapse of looper infestations in the third year. The thick-walled resting spores, which develop during the latter part of the infection cycle, ensure survival of the fungus during the 9-month period that the host is absent. These spores germinate in spring coinciding with first instar emergence, showing perfect synchrony with the host life cycle. Following infection, the pathogen spreads rapidly by means of abundant 'summer conidia'. It is probable that in the natural forest ecosystem, rather than in managed forests, the looper was held in check by a number of factors, including entomopathogenic fungi, and never achieved pest status.

Perry & Whitfield (1984) and Perry & Régnière (1986) reported that the spruce budworm (*Choristoneura fumiferana*) in Canadian forests was infected by seven entomopathogenic fungi: *B. bassiana*, *H. gigantea*, *M. anisopliae*, *P. farinosus*, *V. lecanii*, *E. radicans* and *E. aulicae*. These authors adopted a system approach to model the life cycle of *E. radicans* and *E. aulicae*. The model accurately depicted the sequential stages of larval development and fungal infection. The most important variable for *E. radicans* was temperature since this critically affected not only the activity of the resting spore but also the type of conidial germination. Natural infection occurred only in the fifth and sixth instar larvae, even though earlier larval stages were proven to be susceptible in laboratory and field tests, and this was shown to be related to the overwintering requirements of the fungal resting propagules. Resting spore dormancy is broken at a constant $4 \pm 2\,°C$ over a 4-month period and maximum germination occurs after a 10-month cold treatment; the life cycle of the pathogen having evolved a temperature-induced synchrony with that of the host. High host density and accentuated movement in the fifth instar larvae coincided with *E. aulicae* epizootics. High levels of precipitation during secondary infections favoured epizootic development, which was insensitive to temperature. The type of sporulation is a function of the host-fungus interaction. In vivo, *E. aulicae* isolates produced conidia, whereas *E. radicans* developed resting spores suggesting that these species possess enzootic (resting spore-producing) and epizootic (conidial-forming) phases (Perry & Régnière, 1986).

Macleod (1956, 1963) described epizootics amongst populations of the forest tent caterpillar (*Malacosoma disstria*) caused by *Erynia megasperma* in maple woods in Canada and Balazy (1985) reported the regular occurrence of entomophthoralean diseases on aphids in deciduous or mixed forest stands in Poland. These insects in the same habitat were also occasionally attacked during the summer and autumn months by *Hirsutella aphidis*, the affected aphids being found on plants on the forest floor.

It is perhaps significant that several species of the Entomophthorales (*Erynia formicae*, *E. myrmecophaga*) also attack ants in temperate European forests (Balazy & Sokolowski, 1977; Humber, 1981). Like the *Cordyceps*-infected tropical ants, the behaviour of the temperate *Formica* species attacked by Entomophthorales is apparently altered, and histopathological studies have revealed that the brains of infected insects are invaded by fungal hyphae, possibly provoking abnormal host reactions (Loos-Frank & Zimmermann, 1976). Marikovsky (1962) elaborated on the behaviour of diseased ants in the USSR and postulated that infected workers never die in the nest but move away to avoid being eaten and climb herbs and grasses near ant trails, grasping the stems with their mandibles and legs before dying in this exposed position. He argued further that the actions of the diseased ants are determined by fungal toxins and serve to create ideal conditions for dispersal of the fungus. The identity of the pathogen involved remains in doubt so analogies cannot be drawn with other ant-fungal complexes.

Non-forest habitats

Most of these habitats are difficult to define as primary ecosystems but are included here in a non-agricultural context.

Some unusual host-pathogen relationships have been observed in natural habitats as for example that of large, burrow-constructing, root-feeding caterpillars (Hepialidae, Lepidoptera) and a number of *Cordyceps* species in Australasia. *C. robertsii* occurs on larvae beneath tree ferns in New Zealand and river banks and creeks in Australia (Willis, 1959). Some early reports state that the stromata develop from living larvae (Gray, 1858; Cooke, 1892), but this seems extremely doubtful. *C. gunnii*, *C. hawkesii* and *C. taylori* also attack similar hosts in similar habitats in the humid subtropical areas of Queensland and Tasmania, occasionally in epizootic proportions. Larvae infected by *C. gunnii* build extensive burrows, up to 90 cm deep, and apparently the *Cordyceps* stipe emerges after the infected larvae have moved upwards towards the mouth of the burrow.

Cicadas with their specialised life cycles are susceptible to a range of entomopathogenic fungi but certainly the most unusual are those belonging to the genus *Massospora*. These species attack gregarious cicadas, which occur in large numbers, and grow only in the abdomen of the host (Soper, 1974). The fungal life cycle is closely adapted to that of the host, having had to evolve special mechanisms to cope with the extended absence of adult cicadas, for periods of up to 17 years. Soper et al. (1976) analysed an epizootic of *M. levispora* on *Okanagama rimosa* in a blueberry-sweetfern habitat in Canada and found that the initial infection rate was constant, the adults being exposed to infection from resting spores only briefly prior to emergence. The infected cicadas remain alive and active, even though the abdominal integuments become detached, and mix with the healthy population. Indeed, the male cicadas are still able to call and attract females and it is thought that most infection results from this contact. If the conidia liberated from the ruptured abdomen infect a cicada then only resting spores are produced and there is no secondary conidial cycle. The function of the conidia, in this case, is to create a rapid build-up of the disease during the short time that the host is exposed to infection. Diseased cicadas, carrying resting spores, which return to the soil appear to provide the inoculum for long-term survival and subsequent infection of the next generation of emerging adults.

A similarly gruesome entomophthoralean fungus is *Entomophaga kansana* which grows on calyptrate flies of the families Calliphoridae, Muscidae, Sarcophagidae and Tachinidae in the USA (Hutchinson, 1962). Fungal epizootics were observed in fly populations along moist stream beds, the infected insects walking or flying with missing abdominal integuments. A gelatinous substance released from the abdomen eventually glued the unfortunate flies to the substrate and it was conjectured that healthy individuals were subsequently attracted by a distinctive odour emanating from the dead and dying insects. The interval from initial fungal infection to sporulation was as short as four days, during which time the insects became sluggish and displayed 'spastic, jerky movements'. Recently, it has been demonstrated that pathotypes of *E. grylli* infecting the grasshopper *Melanopus*, and forming only resting spores in this insect, are capable of producing ballistospores directly from hyphal bodies or immature resting spores only when the insect body has been disrupted and the inner fungus exposed to the air (Humber & Ramoska, 1986). Caves offer an unusual habitat for entomopathogenic fungi, but there appears to be a highly adapted mycoflora associated with troglobiotic dipteran and coleopteran hosts. *Hirsutella* and *Stilbella* species are reported to be common on such hosts in European caves (Samson et al., 1984).

It should be noted here that the presence of entomopathogenic fungi in non-agricultural habitats can have an effect on nearby agricultural ecosystems. In Switzerland, Keller & Suter (1980) showed that epizootics of species of Entomophthorales in aphid populations in various crops could be initiated from infected aphids on weeds in adjacent non-cultivated land. Indeed, monitoring the insect fauna and the associated fungal pathogens in non-agricultural habitats can provide useful data about the natural host preferences of insect pests and the seasonal fluctuations and specificity of the different fungi involved. For example, in the Normandy area of France, the aphid *Myzus ascalonicus* on *Achillea millefolia* was found to be attacked mainly by *E. phalloides* (77-98% infection), whilst *M. ornatus* occupying the same niche was infected almost exclusively by *E. neoaphidis* (71-90% mortality) (Fargues & Remaudière, 1977; Remaudière et al., 1981).

Agricultural ecosystems

Reports of epizootics caused by entomopathogenic fungi in diverse agricultural habitats are much more numerous, particularly in temperate regions, than those in primary ecosystems, probably because populations of both entomologists and arthropods are significantly higher. Studies in agricultural crops have also been more quantitatively biased than those in primary ecosystems because of the necessity to accurately determine the economic threshold of the pest and the precise impact of the pathogen on the host population. Recently, mathematical models have been constructed in order to predict the evolution of the pest population as a function of abiotic and biotic factors. Two types of models have been developed with predictive or explicative goals: deterministic models simulating the development of the pest population; regression models which assess the interrelationships between the various abiotic and biotic factors. Space does not allow for a comprehensive treatment of this subject but selected case examples from a range of habitats are analysed.

Tree crops

Natural control of pests by entomopathogenic fungi appears to be well substantiated. Burges (1981) highlighted the importance of fungi in controlling populations of the green apple sucker in Canadian orchards as shown by the resurgence of the pest following the application of fungicides, leading him to conclude that such ecosystems are stable '... similar in this respect to the forest'. Much earlier, Thaxter (1888) had commented upon the occurrence and importance of *Erynia radicans* on leafhoppers in apple orchards in the USA.

Cocoa farms in Africa and South America tend to have less pest problems when the conditions reflect those of the natural habitat, the multi-storeyed tropical forest. Epizootics have been reported on lepidopteran larvae (Samson & Evans, 1982); membracids (Evans & Samson, 1977) and ants (Samson et al., 1981), caused by a range of genera, including *Beauveria, Paecilomyces, Sporodiniella* and *Stilbella*. Typically, heavily sporulating species are involved in these disease outbreaks and only infrequently are the teleomorphs recorded: a reverse situation to that observed in the primary forest. The so-called 'friendly fungi', a term covering those entomopathogenic fungi naturally occurring in citrus orchards in Florida (e.g. *Aschersonia aleyrodis, A. goldiana, Myriangium duriaei, Podonectria coccicola, Nectria auranticola*), were much investigated during the early part of this century. Natural control of citrus pests, principally scale insects and whiteflies, was considered to be important in regulating pest populations, enforced by observations that the use of fungicides resulted in an increase in these pests (Fawcett, 1944). A similar situation has recently been observed in Cuba and Mexico, where whiteflies and mites are controlled naturally by *Aschersonia* and *Hirsutella* species respectively (Latgé, unpubl.). Most of the current research has been concentrated on one of the pathogens, *Hirsutella thompsonii*, which regularly causes epizootics in populations of the citrus rust mite and which also has been recorded as the main natural regulator of *Eriophyes guerreronis*, a serious mite pest of coconuts in Central America and the Caribbean (Cabrera Cabrera, 1977: Hall & Espinosa Becerril, 1981). McCoy (1981) concluded that: 'Where citrus rust mite is an economic pest *H. thompsonii* is a key factor in its natural control, but the fungus is sometimes suppressed by pesticides.' An analysis of the epizootics in Florida showed that the mites move from the old to the new foliage or fruit in April and increase during May. The disease can be detected at a low level by mid-June on mites colonising the new growth. As the mite populations increase, epizootics lasting 2-3 weeks develop and further pest build-up during autumn and winter is prevented. Thick-walled chlamydospores formed within the host probably provide the seasonal carryover.

In fruit orchards in Israel, it appears that entomophthoralean fungi are important in controlling pest populations as this group is well represented in such habitats (Ben-Ze'ev et al., 1984), whilst Byford & Ward (1968) observed epizootics of various members of the Entomophthorales on aphids *(Phorodon humuli, Brachycaudus helichrysi)* in plum orchards in the UK. These authors also noted behavioural changes in infected aphids which were highly correlated with the fungal life cycles. A proportion of newly-infected aphids became agitated and crawled into bark crevices, where the fungus produced mainly resting spores. Conversely, those insects which died on leaves in the tree canopy formed only conidia. There was no apparent relationship with environmental factors since both spore types could be found at the same time of year. They suggested that the conidia from resting spores germinating in the spring would be in an ideal position to infect the nymphs newly hatched from eggs laid on or near the tree buds the previous autumn. The climatic conditions promoting germination of the resting spores are probably similar, therefore, to those necessary for hatching of the aphid eggs. The host and pathogen life cycles would appear to be in perfect synchrony, both spatial and temporal.

Arable crops

The first detailed ecological study of entomopathogenic fungi in relation to the pests of arable crops must be accredited to E. Metchnikoff in Russia in the 1870's who was impressed by the fluctuating populations of the wheat cockchafer *(Anisoplia austriaca)*. He concluded that *Metarhizium anisopliae* was one of the principal factors in the natural control of the beetle. Later, he investigated its effect on the sugar-beet weevil *(Cleonus punctiventris)* and estimated that the fungus was responsible for 40% mortality in natural populations. Much work on the muscardines and related Hyphomycetes, particularly their role in pest population dynamics in arable habitats, has followed since, although the environmental factors favouring natural outbreaks of these pathogens have not been critically evaluated. Generally, muscardine epizootics develop during warm moist conditions (Ferron, 1978; 1981), with relative humidity near saturation point and temperatures between 23-28°C. However, some recent investigations have indicated that high macroclimatic humidity is not essential for the development of hyphomycete epizootics and that optimum infection can occur during dry periods, the fungi reacting to favourable microclimatic factors (Johnson et al., 1984). Hyphomycete conidia are less dependent, therefore, on the presence of free water compared with those of the Entomophthorales.

Ignoffo (1981) summarised the epizootiology of *Nomuraea rileyi* on the velvet bean caterpillar *(Anticarsia gemmatalis)* in soyabean fields in the southern USA according to the following sequence of events:

1 transmission of soilborne conidia in early spring to seedlings (e.g. by rainsplash);
2 infection of larvae feeding on contaminated plants;
3 dispersal and death of infected larvae within plants;
4 sporulation and creation of infection foci;
5 repetitive infection and inoculum increase throughout late spring and summer, wind dispersal of spores to initiate epizootics in late summer;
6 elimination of susceptible larvae in late summer and early autumn;
7 contamination of soil by conidia;
8 overwintering and survival of conidia in soil.

A mathematical model established by Kish & Allen (1978), confirmed that the factors which determine when the epizootic peak is reached depend upon: the amount of viable soil inoculum; the early availability of hosts; environmental conditions; and amount of early season dissemination of conidia. By constructing such models it should be possible to accurately predict the occurrence of epizootics, and to rely on the natural regulation of lepidopteran pest populations, without resort to premature control measures. Fuxa (1984) later studied the dispersion and spread of the pathogen in soyabean fields and concluded that heavy rainfall actually reduces disease levels, infection declining from 16.5 to 0.6% and that infected hosts were more aggregated than non-infected insects.

Although Hyphomycetes such as *Paecilomyces, Nomuraea, Beauveria* and *Metarhizium,* usually do not have specialised resting or overwintering structures, the relatively thin-walled conidia have obviously adapted to a soil survival phase, essential in the natural life cycle of these pathogens as well as to their potential use as biological control agents (Madelin, 1966). In fact, Hyphomycetes are almost the only fungi found on soil insects such as *Melolontha melolontha, Otiorhynchus sulcatus* and *Leptinotarsa decemlineata* (Zimmermann, 1981; Keller, 1986a). In this context, Gottwald & Tedders (1984) considered that the ability of *B. bassiana* to spread from infected pecan weevil larvae in the soil to colonise alternative energy sources made it a much more effective pathogen against soilborne pests than *M. anisopliae*. Johnson et al. (1984) also linked the soil phase of *B. bassiana* to its disease cycle on the Egyptian alfalfa weevil. The highest incidence of the pathogen was registered during dry periods and peak incidence was correlated with harvesting, which opened-up the crop canopy and exposed the larvae to soilborne inoculum. Rainfall, therefore, was not a critical component of natural control. Pioneering work on the modelling of insect diseases was carried out by Ullyett & Schonken (1940) in South Africa, who conducted an intensive study of the natural control of the cabbage pest *Plutella maculipennis,* with particular reference to the fungal pathogen *Erynia radicans*. They found that, although the fungus caused an immediate and significant reduction of the *Plutella* population when environmental conditions were favourable (high humidity, high temperature), it was adjudged to be responsible for an overall increase in the average host density, resulting subsequently in economic crop damage. To explain this anomaly, they proposed that the sporadic intervention of the fungal pathogen (regarded here as a density-independent and temporary mortality factor, because of its requirements for unusual local weather conditions) seriously disturbed the natural balance of the host population, effectively replacing the density-dependent and permanent mortality factors. As the conditions became drier and the fungus disappeared, the host population was able to recover more rapidly than its parasites and predators and consequently achieved a higher density level than before the fungal epizootic.

Similar abiotic and biotic factors were recognised as determining the natural spread of Entomophthorales, particularly from studies of aphid and weevil populations, in arable crops in Europe and the USA. The most critical environmental factor is relative humidity. Epizootics usually develop only during periods of high humidity or rainfall (Wilding, 1981). Los & Allen (1983) found that *Erynia phytonnomi* was effective in suppressing populations of the alfalfa weevil *(Hypera postica)* below economic threshold levels only during periods of high rainfall. Johnson et al. (1984) investigated the same pathogen on the Egyptian alfalfa weevil *(H. brunneipennis)* in California and showed that abnormally heavy rainfall early in the season initiated an earlier and longer epizootic and, in such instances, *E. phytonomi* was a major biotic mortality factor. The fungus survived dry periods in mummified host larvae, termed resting larvae. Conversely, as previously mentioned, *B. bassiana* achieved its highest incidence in the same ecosystem during dry weather. Rainfall can be directly correlated with sporulation of this entomophthoralean pathogen since it has been demonstrated that when intracanopy relative humidities exceed 91% for three hours, the infected weevil cadavers discharge abundant spores called 'showering' (Millstein et al., 1982). Although these workers postulated a relationship between accumulated relative humidity (humidity hours) and conidial dispersal, and hence increased infection, they did not prove that microclimatic humidity is the driving variable in the system.

Studies of the population dynamics of the aphid *Sitobion avenae* on winter wheat in Western France, revealed that the increase in infection due to Entomophthorales is directly related to the number of rainfall days in May (Pierre & Dedryver, 1984). In England epizootics occurred only in those years with six consecutive days of rain and with aphid densities above two per tiller (Wilding et al., 1986a). Temperature as well as moisture requirements are important and, although these are generally broad, they may vary significantly from one fungal species to another. Of the three species of Entomophthorales which regularly cause epizootics amongst aphid populations in France, *E. neoaphidis* occurs predominantly during cool, humid periods whereas *E. planchoniana* and *Neozygites fresenii* are most common in warm, moist weather (Remaudière et al., 1981). The latter pathogen is the only one encountered in tropical climates. In addition to environmental factors, other variables such as insect phenology and behaviour, the amount and distribution of fungal inoculum, influence the spread of disease in aphid populations. Most long-term studies in arable ecosystems have revealed that entomophthoralean epizootics are density dependent, occurring only when the host population is high (Remaudière et al., 1981; Wilding, 1981). Entomophthorales attacking bean aphids *(Aphis fabae)* appear earlier in the season in colonies with a high density when compared with smaller populations, *N. fresenii* being more host density dependent than *E. neoaphidis* (Rabasse & Dedryver, 1982). Similarly, Nordin et al. (1983) found that the natural termination of entomophthoralean epizootics in populations of the alfalfa weevil occurred when the host density fell below a critical threshold of 1.7 larvae per stem.

In temperate arable crops, the Entomophthorales have a biological cycle similar to that described earlier in North American forest habitats (Otvos et al., 1973; Perry & Whitfield, 1984), which is typically thermoregulated, coinciding with host development (Latgé & Papierok, 1988). In the case of the aphid pathogen, *Conidiobolus obscurus*, infected insects filled with resting spores fall to the soil where they remain viable during the winter months, undergoing a period of obligatory dormancy. Early spring temperatures are optimal for inducing asynchronous germination of the resting spores which produce infective conidia over a period of several weeks. These infect the first aphids appearing on the crop. Optimal temperatures for conidial germination correspond to midspring-early summer, sufficient to initiate epizootics amongst the already high aphid populations (Latgé, 1983).

Detailed epizootiological studies have also been undertaken of other insect pest-Entomophthorales associations in arable crop situations. Le Ru (1986) investigated the role of *Neozygites fumosa* in the regulation of the cassava mealy bug *(Phenacoccus manihoti)* in West Africa and found that insect populations fell at the beginning of the rainy season and remained at a low level until the following dry season. The fungus was considered to be the main controlling factor; being favoured by a high relative humidity ($>90\%$) over several days and temperatures above $20\,^{\circ}$C. Harper et al. (1984) reported that populations of soyabean loopers *(Pseudoplusia includens)* in the USA can be reduced to levels below those causing economic damage following epizootics of *Erynia gammae*. *Entomophthora muscae* apparently plays a similar important role in the control of the onion fly *(Delia antiqua)* in North America. Natural infection levels of almost 100% were recorded in early (first or spring generation) and late (third or autumn generation) instar populations (Carruthers et al., 1985). A number of abiotic and biotic variables were compared but host and pathogen density were the only factors which elicited a significant response. Apparently, microclimatic variables are not the limiting or controlling components in this disease system. The disease cycle in the onion agroecosystem comprises: primary infection of the flies emerging from the soil, by overwintering resting spores; host death and production of either conidia or resting spores. Flies which produce conidia (ca. 90% of total infected flies) typically die in elevated positions in the late afternoon and the conidia are formed rapidly and discharged throughout the night when environmental conditions are optimal for germination. If resting spores are produced (ca. 10%), the host dies on the soil surface, the abdomen becoming brittle and subsequently fracturing to release the spores into the soil. Both methods of sporulation occur simultaneously through the season and thus the pathogen must exist in two distinct physiological states: one ensuring that the host moves to an exposed situation (epigeal), suitable for aerial dispersal of conidia; the other reversing the behaviour of the fly so that it moves downwards to the soil (hypogeal), the most suitable niche for overwintering and ultimately for infection of flies as they emerge from soil-based pupae. The behaviour of entomophthoralean-infected insects has also been well documented in other arable habitats. Harper (1958) observed that sugar-beet aphids, *(Pemphigus betae)*, which normally live and feed below ground, when infected by *Erynia nouryi* crawl up the plants and die on the soil surface or on the crown and leaves of the crop. The clinging habit of the aphids on the substrate was compared to that of grasshoppers attacked by *Entomophaga grylli*. The latter fungus is considered to be a significant natural control agent of locusts in Africa and North America and this has recently been confirmed by monitoring population changes of locusts *(Oxyahyla)* in ricefields in Asia (Weiser et al., 1985). Characteristically the infected insects climb up grass stems and die freely exposed (Brady, 1979). Similarly, carrot flies *(Psila rosae)* infected with *E. muscae* will never die within the crop but always at the top of the trees in neighbouring hedgerows (Eilenberg, 1986). In particular, the females alter their egg-laying habit, reducing the chances of survival of the eggs which are are normally laid on or near the host plants. The death of infected insects in elevated postions will undoubtedly favour the dispersal of the fungus (King & Humber, 1981). Personal observations in Indonesia on ricehoppers *(Cofana spectra)* indicate that insects infected with Entomophthorales die with their wings outstretched at the apices of the rice leaves, whilst those attacked by *Metarhizium* tend to occur much lower down the rice culm with wings appressed, suggesting in this case that it is the type of fungal infection which determines host behaviour.

Entomophthoralean epizootics have also been noted amongst populations of *Hylemya* flies (Anthomyiidae), serious pests of vegetables, in the USA and Europe. Berisford & Tsao (1974) reported on outbreaks of *Entomophthora muscae* and studied disease symptomatology. The infected flies gathered on shrubs and fences, exhibiting sluggish movements and grasping the substrate, finally becoming permanently attached by glue-like substances exuded from the labellar lobes and by fungal rhizoids emerging from the proboscis. This 'summit' disease syndrome fixes the sporulating host in a position ensuring efficient dispersal of the fungus. Humber (1976) also described a disease caused by a member of the Entomophthorales on the same hosts. Anthomyiid flies infected by *Erynia (Strongwellsea) castrans* show no changes in their behavioural patterns even though the fungal hyphae penetrate the host nervous system. This is an interesting difference from the dramatic effects of the majority of the entomopathogenic fungi on host behaviour. *E. castrans* is a highly adapted parasite which produces its spores internally and disseminates them through a hole in the abdomen of the living insect. When the host dies, the fungus also dies shortly afterwards.

PASTURES

Pasture grasses are attacked by a number of damaging pests but relatively little information is available on natural control of these by entomopathogenic fungi. Mathieson (1949) investigated an epizootic on the pasture cockchafer *(Aphodius howitti)* in Australia caused by an undescribed species of *Cordyceps (C. aphodii)* and observed that infection can occur in active larvae at any time from hatching in February until August. A conidial state developed from April to August followed by the teleomorph stromata, which matured by October. However, the author was uncertain of the function of the ascospores liberated during this period since susceptible larvae were absent, but he did find that old stromata which survived the winter produced a new crop of spores following the February rains, thereby acting as a primary inoculum source for subsequent infection of the emerging larvae. This cycle was essentially confirmed by Coles (1978), during a survey of the populations of *Aphodius* species in pastures in Southern Australia; the fungus occurred in a wide range of soils in the wetter areas with an average annual rainfall of over 750 mm. Keller (1986a) recorded populations of the European cockchafer *(M. melolontha)* in Switzerland over a number of years and found that *B. brongniartii* is an important natural control factor, being positively associated with population decline.
Recent studies of the eriophyid mite, *Abacarus hystrix*, a vector of the ryegrass mosaic virus in pastures in the UK, have shown that a

variety of *Hirsutella* species is involved in natural mortality, accounting for about 16% of the mite population (Minter et al., 1983). The occurrence of entomopathogenic fungi on these small hosts was only verified by examining grass leaves under the dissecting microscope and undoubtedly many more fungi remain to be found using microscopical rather than field observations.

Epizootics on spittlebugs, caused by an Entomophthorales species, have been reported recently in tropical pastures in Ecuador (Evans, 1982) and in Mexico, where the froghoppers *Aeneolamia albo-fasciata* and *Prosapia simulans* were infected by *E. neoaphidis* and *C. apiculatus* respectively (Latgé, unpubl.). *E. radicans* appears also as an important bioregulator of *Empoasca* leafhopper in Brazil (Wraight, unpubl.)

Annual epizootics of *E. grylli* in populations of the grasshopper (*Camnula pellucida*) have been monitored in high altitude rangelands in Arizona (Carruthers et al., 1986). Whilst these are dependent on high host density and humidity, they do appear to play a primary role in regulating grasshopper populations.

General thoughts and speculations

Natural control of arthropods by entomopathogenic fungi is widespread in both primary and agricultural ecosystems in a range of habitats throughout the world. Members of the Clavicipitales are particularly well represented in tropical and subtropical regions and dominant in arthropod communities in lowland rain forests. Conversely, the Entomophthorales increase in importance away from the tropics and are most commonly found in temperate zones in both forest and agricultural habitats. It would appear that the ability to invade and exploit living arthropods arose independently within these two major groups and that the fungal pathogens have co-evolved with their hosts over a considerable period of time.

Most of the entomopathogenic representatives of the Entomophthorales and Clavicipitales have restricted host ranges and this specificity is reflected by sophisticated morphological adaptations to the host life cycle. In contrast, many of the Deuteromycetes are opportunistic species which attack hosts in a wide range of insect orders. Some specificity can be circumstantial as in the case of the green aphid pathogen, *Neozygites fresenii*, which will also switch to the black aphids in the summer when its normal hosts are absent and the fungus is highly active biologically. A further example is that of *Beauveria brongniartii* which has many potential hosts but is found predominantly on insects in soil habitats, where the fungus displays its highest pathogenicity (Fargues & Remaudière, 1977).

It is generally accepted that absolute parasitism is more highly evolved than the essentially pathogenic habit shared by all the fungi included in this book. In evolutionary terms, it is significant that no basidiomycete fungus is obligately entomopathogenic – it should be noted here that until relatively recently the genus *Hirsutella* was still considered to have basidiomycetous affinities (Steinhaus, 1946) – and the entomogenous Septobasidiales apparently exist in a mutually beneficial symbiotic association with scale insects and can not be said to be true parasites. Humber (1984) concluded that in the Entomophthorales there is a shift away from rapid killing of the host, either through toxin release or histolysis, towards absolute parasitism, in which the host is not directly killed by the activities of the fungus. However, no member of this group has perfected this relationship. Some species of the genera *Erynia* and *Entomophthora* grow and sporulate only in the living host, whilst those of the genus *Massospora* species complete most of their development in the still active host, and thus are regarded as the most highly adapted or evolved of the entomophthoralean fungi.

It would be expected, therefore, that the most primitive forms rapidly kill their hosts and evidence for high toxin activity in the least specialised of the Entomophthorales has been put forward (Humber, 1984). However, it could be argued that such species (e.g. *Conidiobolus coronatus*) are bad pathogens since, although the fungi rapidly kill their hosts, they do not appear to have evolved any mechanism to prevent or limit secondary colonisation of the cadaver which often develop bacterial septicemia, effectively preventing fungal sporulation. Toxin development in the higher fungi appears to have been linked with additional safeguards (viz: production of antibiotics; Ferron 1978, 1981) to ensure successful colonisation of, and subsequent sporulation on, the host. Nevertheless, the ability to produce insect toxins is not commonly reported in the Entomophthorales and the majority of species appear to kill the host only after extensive tissue colonisation, which strongly suggests that toxins are not intimately involved in pathogenesis. Indeed, Roberts (1981) is of the opinion that the lower fungi overcome the host primarily by utilisation of the available nutrients in the haemocoel rather than by low molecular weight toxins. None of the supposedly more highly evolved entomopathogenic Clavicipitales exhibits a prolonged parasitic phase and host mortality is rapid, almost certainly due to toxin release, and there is no evidence of this group having advanced towards true parasitism. Humber (1984) further supposes that the first Entomophthorales grew saprophytically in the leaf litter and that all the genera radiated from this soil saprophytic habit. Certainly, the possibility exists that some of the facultative entomopathogenic hyphomycete genera evolved from a soil niche to colonise weakened or damaged arthropods and then adapted enzyme and toxin systems which enabled them to invade and kill healthy individuals. Observations in tropical forests of *Beauveria bassiana* growing on insect remains in the soil and studies of the same pathogen on pecan weevil larvae would seem to support this hypothesis (Evans, 1982; Gottwald & Tedders, 1984).

The Entomophthorales have a biological cycle perfectly adapted to a pathogenic existence on insects. Specialized resistant structures are present to ensure survival of the fungus in the absence of a suitable host either during a short (capillispore) or a long (resting spore or mycelial sclerotium) period of time. These fungi ensure successful dispersal of their spores either by the activities of the living host or by a violent discharge mechanism which carries the spores away from the still air layer around the host and is effective in the case of very dispersed insect colonies. In case of aggregative colonies (such as black aphids or some mealybugs), capillispores with a sticky apex are produced and are much more adapted for local transmission from one insect to a 'close neighbour', functioning like the sticky conidia of homopteran-colonising Hyphomycetes (*Verticillium lecanii*, *Hirsutella* spp.). From the evidence gathered during studies of Entomophthorales epizootics in insect populations, it would seem that host behavioural patterns vary with pathogen morphogenesis, within any one fungal species. Different substances may be produced by the fungus at different growth stages which specifically affect host movement to favour the production and dispersal of a particular spore type. In no species are specialised sporogenous outgrowths formed, although Madelin (1966) considers that the presence of rhizoids, which serve to anchor the host to the substrate, represents an advanced feature for a Zygomycete: '... because few of these have the capacity for production of pseudotissues by coordinated hyphal growth'. As we have emphasized previously, this is a crucial attribute of those pathogens adapted to hosts which seek refuge when infected. It seems logical, therefore, that this type of host reaction would be extremely disadvantageous for an entomophthoralean pathogen lacking highly-organised, phototropic, sporogenous structures. It has been postulated that arthropods attacked by Entomophthorales become positively phototropic and climb, eventually succumbing to infection and dying in an exposed position which favours subsequent spore dissemination. Is the 'summit disease' syndrome entirely governed by the fungus, in whatever the host, or were the Entomophthorales only able to successfully complete their life cycles on hosts which reacted by moving to prominent death positions? The subject is ripe for speculation and, hopefully, future investigation.

It is worthwhile trying to analyse the development of the tropical forest entomopathogenic Clavicipitales because they probably represent the end products of an ancient co-evolution with arthropods, this association having been buffered from environmental disturbances unlike those in most other habitats. In the case of the genus *Torrubiella*, the perithecia occur directly on the host and/or around it on the substrate. The arthropods are invariably freely exposed, albeit mainly on the undersides of leaves, so there would have been no pressure to develop aerial sporogenous structures. As previously mentioned, a high proportion of the forest *Cordyceps* are found on hosts which seek shelter when infected. Obviously, therefore, it became essential for fungi attacking this type of host to develop stromatic outgrowths for repositioning the perithecia. Forest *Cordyceps* species with *Hirsutella* and *Hymenostilbe*-type anamorphs have highly differentiated, well-organised stromata with a tough, sterile stipe region and a lateral, subterminal or terminal fertile area, with the perithecia typically buried and arranged in a multilayered packing tissue. The anamorph hymenium may also occur on this structure, either laterally or terminally. Such species are invariably difficult to establish in culture and it is thought that they are highly developed, host specific, obligate pathogens. They appear to have adapted both to the avoidance mechanism of many of the forest arthropod hosts and to their low population density by producing a durable, highly-organised fruitbody which liberates a succession of spores at a relatively low intensity but over an extended period of time. Effectively, these species have opted for a slow-growing structure of long-term survival value. Observations of forest ants infected with *Cordyceps* have revealed that old or damaged stromata can successfully regenerate new fertile heads and thus considerably prolong the sporulation period. In addition, some of these species also produce autonomous, similarly highly differentiated synnemata which may bear several distinct conidial types: dry spores for long-distance air dispersal; slime spores for short-distance rainsplash. Darkening of the mucilage around the latter is thought to increase their longevity by reducing ultraviolet radiation, another adaptation to the long-term strategy for spore dispersal. This complexity of conidial form is perhaps analogous to that reported for the Entomophthorales associated with various insects, especially simuliid flies, in natural aquatic habitats in temperate regions (Descals & Webster, 1984). The brightly-coloured, fleshy *Cordyceps* species, usually found on forest soil- or log-buried larvae, have the perithecia either organised in discrete heads or scattered laterally, partly or completely buried in the fleshy tissues with no definable stipe region. The anamorphs are not usually formed on these structures and, if present, are to be found covering the buried host. Typically, such *Cordyceps* are easy to establish in culture from ascospore inoculum and produce *Paeci-*

lomyces or *Verticillium*-like anamorphs. This may indicate that these species do, in fact, have the ability to survive saprophytically, independently of the arthropod host, as proposed by Massee (1895). This attribute would be extremely important to these fungi, in view of the fleshy, non-durable nature of the stromata, for ensuring their long-term survival. Conceivably, the ascospores which fail to hit a susceptible host could germinate in the soil and colonise chitinous debris, living saprophytically before producing air- or splash-dispersed conidia. Indeed, Mathieson (1949) was puzzled by the autumnal production of ascospores in *C. aphodii*, since host larvae were not available for infection until the following spring. The entomopathogenic Hyphomycetes common in agricultural ecosystems may represent the remnants of this apparently primitive *Cordyceps* group.

In common with many *Cordyceps* and *Torrubiella* species, the other Clavicipitales pathogenic on scale insects in tropical forests are also brightly coloured. The evolutionary advantages of this striking colouration, however, are unknown. In the genus *Hypocrella*, both the ascospores and the conidia of the *Aschersonia* anamorph are well protected within stromatic tissue and the spores exude in mucilage. Indeed, the entomopathogenic hyphomycete fungi associated with plant-sucking homopterans in general, such as *Fusarium coccophilum* and *Verticillium lecanii*, also have their spores enveloped in mucous, there being very few examples of dry-spored anamorphs on these hosts. It would appear that the mass infection of homopteran colonies, which is a characteristic of these fungi, is due to the efficient dispersal of the slime spores on the leaves and within the plant foliage or canopy, probably in run-off and rain-splash during tropical downpours.

One reason for the disappearance of *Cordyceps* species from disturbed habitats is that their fruitbodies, particularly on hosts which die in the soil or leaf litter, would be readily destroyed by the farming practices employed in intensively-cropped agricultural ecosystems. Thus, the occurrence of *Cordyceps* can be said to be inversely related to the exploitation of the habitat explaining why primary tropical forests are rich in *Cordyceps*. This may also indicate why *Cordyceps* species are relatively well represented in pasture habitats, compared with arable situations (Mathieson, 1949; Evans, unpubl.) where cropping is non-seasonal. Such a relatively stable ecosystem, in which there is no annual removal or burial of pathogen reservoirs, would contribute to the maintenance of high inoculum levels over extended periods. Those species with the capability of rapidly producing a heavily-sporulating anamorph would be at an advantage in disturbed habitats where host populations would be seasonally high but erratic. This investment of food reserves in a massive and explosive production of conidia would be at the expense of the teleomorph which would probably be irrevocably lost from the life cycle. Survival between crops and host populations would then be dependent upon conidia adapted for soil persistence and/or a saprophytic mycelial existence.

Some of the findings relating to the factors which promote the development and maintenance of epizootics in members of the Entomophthorales appear to be contradictory. Studies of pest-pathogen complexes in forest ecosystems have shown that in certain cases, the critical factors are fungal inoculum levels and host population density, with climatic variables being of secondary importance (Soper & Macleod, 1981). In contrast, the macroclimate (particularly humidity) is often considered to be the most important factor in arable habitats (Wilding, 1981; 1982). It is possible that this is related to differences in buffered (forest) and non-buffered (agricultural) ecosystems.

In summarising natural control, we conclude that entomopathogenic fungi are significant factors in long-lasting, stable ecosystems, where they exert a steady (enzootic) control of arthropod populations. Logically, there is a high degree of synchrony with the host as its elimination would mean self-destruction of the pathogen. Natural control by entomopathogenic fungi in agricultural ecosystems is unstable and unpredictable but may frequently assume epizootic proportions, usually after the host has reached a critical population density. In other words after it has achieved pest status and caused significant economic damage. The rationale behind biological control is to utilise these fungi in the most efficient manner. Premature inducement or advancement of epizootics would seem to be the most meaningful approach by increasing natural inoculum levels and by modifying the macro- or microclimatic environment.

Chapter 6

Biotechnology

Introduction

As mentioned previously, studies of natural epizootics of entomopathogenic fungi during the latter part of the nineteenth century stimulated man's interest in employing them as mycoinsecticides to control agricultural pests. Mass production of the selected fungus is a necessary prerequisite for any large scale field application and the methodology involved was developed at an early stage to suit a number of different pest-pathogen situations (Krassilstschik, 1888; Snow, 1896; Rorer, 1913). However, this technology stagnated as disillusionment in the practical value of fungi as biological control agents of arthropods became widespread following inconclusive or negative reports of their efficacy against a range of pests in various parts of the world (see Chapter 7) and the initial overwhelming success of chemical pesticides.

The revival of interest in mycoinsecticides over the past 15 years has led to the large scale production of several promising candidate fungi and to the marketing of the first commercial mycoinsecticides, Mycotal and Vertalec, based on formulations of *Verticillium lecanii*. Despite these recent encouraging developments, little information is available on the biotechnology of entomopathogenic fungi and their industrial production is still relatively unsophisticated. This chapter will review some of the recent progress in the field of fungal biotechnology and at the same time will attempt to answer those questions likely to be asked by anyone interested in, or encharged with developing mycoinsecticides: what fungi and which propagules can be most suitably exploited in order to produce a cheap, safe and efficient alternative to conventional insecticides; what are the factors governing the growth and sporulation of these fungi; what is the most appropriate fermenter technology to mass-produce these fungi?

What to produce?

Which organism?

The selection of a fungal strain, or of a species in the case of a target pest susceptible to several pathogens, is critical since the aggressiveness of a fungus is highly strain dependent. The assessment of pathogenicity is usually based on the results of pest, or progeny (in the case of fast reproducing insects), mortality obtained from laboratory tests (Papierok, 1982; Hall, 1984). It should be stressed, however, that bioassays conducted under laboratory conditions invariably optimise the potentialities of the fungus and thus the data should be interpreted carefully. For example, bioassays are run in the absence of any microbial competitors, in ambient conditions ideally chosen for the pathogen and with in vitro reared insects, the physiology of which may be different from that of the wild types. These bioassays, which are useful to compare strains or species, should represent, therefore, only the preliminary step before field experimentation. The latter is essential in order to determine if the microclimatic requirements of the pathogen and the host coincide and thereby to assess the true potential of an entomopathogenic fungus as a biocontrol agent. The pathogenic stability of a strain during repetitive transfers should also be checked. The media used for these transfers and for subsequent production should be chosen with care since it is known that nutrients can markedly influence conidial viability (Goral, 1979; Fargues, 1981). Preferably, the strain selected should not have too narrow a specificity, being commercially advantageous

---

This chapter has been written by J.-P. Latgé and dr. R. Moletta. The last author's address is: INRA, Bioconversion, Narbonne, France.

if a product has a relatively wide host range within an insect group containing several pest genera. However, the host range cannot be too wide and obviously must exclude beneficial insects as well as other invertebrates and vertebrates. Experiments have been carried out with several entomopathogenic fungi to test their effects on vertebrates, specifically to evaluate any allergic, irritation or toxic properties. Only minor allergic responses have been detected amongst a few of the entomopathogenic fungi screened so far (Lisansky & Hall, 1983).

Strains with the highest sporulation capacities should be selected (Papierok, 1982; Jackson et al., 1985) since variations both in the amount of spores produced and in their mode of production have been reported (Kononova, 1979; Hall, 1982; Roberts & Sweeney, 1982; Latgé & Sanglier, 1985). For example, in the case of *Hirsutella thompsonii* under the same culture conditions, four different strain types have been identified: strains unable to sporulate; strains sporulating exclusively on solid media; strains producing conidia in shake culture either on conidiophores arising from mycelial filaments or on conidiophores originating from conidia through a microcycle (Latgé et al. 1988a). Industry will also screen for strains with the simplest nutritional requirements.

Most strain selections have been made from wild isolates of entomopathogenic fungi from naturally-infected hosts. However, recent developments in fungal genetics suggest that the natural properties of a strain can be improved through genetic manipulation. In the past, mutagenesis has been used to enhance the virulence or sporulation of *Metarhizium anisopliae* (Al-Aidroos & Roberts, 1978; Al-Aidroos & Seifert, 1980). Conceivably, this process could also be exploited to produce mutants resistant to the pesticides normally encountered in the crop habitat of the target pest. Recombination of selected strains, obtained by mutagenesis and characterized by convenient genetic markers, has been attempted amongst the Deuteromycetes. Parasexual recombinants from heterozygous diploids, produced by hyphal anastomosis, have also been identified in *M. anisopliae* (Al-Aidroos, 1980; Magoon & Messing-Al-Aidroos, 1984; Riba et al., 1980; Bergeron & Messing-Al-Aidroos, 1982). Recombination by protoplast fusion has been investigated in *Beauveria brongniartii*, *M. anisopliae*, *Nomuraea rileyi* and *V. lecanii* (Paris, 1980; Heale, 1982; Jackson & Heale, 1983; Samuels et al., 1985; Boucias, unpubl.). One of the problems typically encountered in recombination is the poor stability of the recombinants and attempts either to increase the virulence of a highly sporulating strain or to widen its host spectrum have failed. The genetic mapping of entomopathogenic fungi is totally unknown. However, in entomopathology, as in phytopathology, the aggressiveness or the sporulation potential of a fungal strain should be controlled by numerous genes. Thus, genetic improvement, either by the classical techniques or the new methods of genetic engineering, is fraught with difficulties. Nevertheless, it is feasible that the genes responsible for toxin excretion could be cloned and that their reinsertion into the genome of another strain or species would be a method of increasing the efficiency of a mycoinsecticide.

Which propagule?

All entomopathogenic fungi are characterized by a biphasic biological cycle: a mycelial vegetative phase and a reproductive phase. Two spore types are usually found: asexual spores, for promoting rapid dissemination of the fungus and resting spores (sexual spores or vegetative chlamydospores), responsible for survival of the pathogen during adverse conditions or in the absence of suitable hosts. Theoretically, any of these fungal propagules could be considered for the production of mycoinsecticides.

Because of their primary role in the infection process, spores have been considered since the beginning of biological control history as the most adapted fungal propagule to produce. With the exception of the Oomycetes and the Zygomycetes, where large scale production of infective spores – motile zoospores (Oomycetes) or discharged ballistospores (Entomophthorales) – looks unattainable, spores still remain the most viable proposition at the present time. Conidia of the Deuteromycetes are readily mass-produced on solid media under aerated conditions. However, the sporulation cycle in vitro is relatively long (1-3 weeks) (Couch, 1982). Conidia can also be obtained in liquid media, being produced on typical conidiophores arising from hyphal filaments (Goral, 1979; Kononova, 1979; Roberts & Sweeney, 1982) or directly from the spore through a sporulation microcycle (Latgé et al., 1988a). This microcycle, typically induced by nutrient and/or temperature manipulation, has been developed as a model to study the biochemical events occurring during sporulation of the conidial fungi (Smith, J.E. et al., 1981). This potential for microcyclic sporogenesis has been of particular interest in the case of entomopathogenic fungi since it shortens the culture time and increases spore yields (Latgé et al., 1988a).

The other type of mass producible spore is the resting spore of the Mastigomycotina and Zygomycotina. Because of their role in disease carryover, these spores offer the advantage of being highly resistant and can survive for several months both in vitro and in nature (Jaronski, 1982; Perry & Latgé, 1982; Kerwin et al., 1986). They can also be produced in liquid culture as well as on solid media (Latgé et al., 1977a; Latgé, 1980; Brey, 1985; Kerwin & Washino, 1986a). However, these spores, like the mycelium, are not directly infectious; their pathogenicity being dependent upon their potential to produce infective spores by germination.

The production of ascospores, in the case of Ascomycetes such as *Cordyceps,* or of chlamydospores, often reported amongst Deuteromycetes (McCauley et al., 1968; McCoy, 1981; Pendland, 1982), has never been attempted.

The production of mycelium has been also contemplated especially in the case of Oomycetes and Zygomycetes for reasons explained above. Mycelial propagules of entomopathogenic fungi are non-infective and thus the successful use of a mycelial formulation in biological control is dependent upon the ability of the mycelium to sporulate under natural conditions. On solid media, a continuous segmented mycelium is usually produced, whilst in shake-liquid cultures, as in the insect body, fungal development is most often characterized by the formation of yeast-like cells able to reproduce by fission. The terminology applied to these propagules depends on the fungal group under consideration, being termed: blastospores (Deuteromycetes); hyphal bodies (Entomophthorales, Coelomomyces); hyphal segments or subthalli (Lagenidiales). The yeast-like cells are most often produced by hyphal constriction and thus their wall structure is mycelial. The fungal dimorphism exhibited by entomopathogenic fungi needs to be investigated more thoroughly in vitro since the multiplication of yeast-like forms would facilitate not only mass production but would also be an indication of the virulence of a strain as these cells are responsible for the rapid colonisation of the insect haemolymph. In medical mycology, several factors control fungal dimorphism: temperature, carbon dioxide concentration, availability of nutrients such as hexoses, sulphydryl compounds, salts and vitamins (Cole & Nozawa, 1981). Similar studies have not been undertaken with entomopathogenic fungi, although it seems that the factors promoting the formation of yeast-like cells are species dependent. In *V.lecanii,* for example, blastospore production is carbon dioxide dependent (Hall & Latgé, 1980), whilst in *N.rileyi,* the key factors appear to be the presence both of a high concentration of yeast extract and of tension activators such as Tween 80 (Riba & Glandard, 1980). Some of these Deuteromycetes, do produce pathogenic yeast-like cells in vitro and consequently the use of these propagules has been contemplated (Catroux et al., 1970; Fargues et al., 1979) even though many species of this order readily produce asexual spores in vitro.

A further disadvantage of using mycelial propagules from in vitro cultivation, or even those formed in vivo, is their short viability in comparison with spores (Ferron, 1978). Hyphal bodies of Entomophthorales enclosed in an aphid cadaver or sclerotia of *N.rileyi* can easily overwinter in nature but the mycelium produced in liquid culture is short-lived by comparison (Latgé, 1982; Ignoffo, 1982; Keller, unpubl.). However, formulation can significantly improve the longevity of the in vitro mycelium (Fargues et al., 1979; Soper & Ward, 1981; Mc Cabe & Soper, 1985). Additional research should be undertaken to determine if the greater longevity of in vivo mycelium is related to morphogenetic changes, such as thickening of the wall and the accumulation of lipid droplets. The formation of resting mycelia may be governed by critical environmental conditions or the presence of certain chemical stimulants, particularly sulphur-containing compounds or polyphenols and polyphenoloxidases, as in the case of mycelial sclerotia of *Claviceps purpurea* (Willets, 1978).

Growth factors

Nutritional requirements

The growth requirements of most entomopathogenic fungi have been poorly studied despite the fact that this information is essential for mass production. In particular, such knowledge may permit simplification of the medium, cutting costs without affecting yield. The choice of industrial nutrients will obviously be directly related to the nutritional requirements of the selected fungus. For example, Entomophthorales do not metabolise sucrose (Latgé, 1975); as a result all industrial media for the production of these fungi should exclude carbon sources extracted from sugar beet or sugar cane, which have high sucrose contents, and should be based on corn residues rich in dextrose.

Entomopathogenic fungi require oxygen, water, an organic source of carbon and energy, a source of inorganic or organic nitrogen and additional elements amongst which are minerals and growth factors. The carbon source is usually dextrose but can be replaced by polysaccharides (such as starch) or lipids. Nitrogen can be supplied in the form of nitrate, ammonia or organic compounds such as amino acids or proteins. Other essential macronutrients are phosphorous (as phosphates), potassium, magnesium and sulphur, the latter supplied either in an inorganic form (as sulphate) or organic (cysteine or methionine). Essential microelements usually include calcium, copper, iron, manganese, molybdenum, zinc and water soluble B-complex vitamins, especially biotine and thiamine. All these micronutrients usually occur in the raw materials included in industrial media but can be supplied as protein hydrolysates or yeast extract (Berry, 1975; Garraway & Evans, 1984).

The nutritional requirements of entomopathogenic fungi vary with the fungal species or even the fungal strain under consideration

(Fargues, 1981; Latgé, 1981). Deuteromycetes typically have low requirements and substantial growth of *B. bassiana* and *M. anisopliae* can be obtained in media containing only dextrose, a nitrate and a macromineral solution (Kononova, 1979; Fargues, 1981). However, semi-defined media, including protein hydrolysates, or natural undefined substrates rich in starch (rice, oatmeal, potato) have proved to consistently give the highest yields (Samson, 1982). Nevertheless, certain Deuteromycetes, particularly those belonging to the genera *Hirsutella* and *Gibelulla*, may have more specific requirements than the majority of this group since some species of these genera have failed to establish in culture (MacLeod, 1960; Samson & Evans, 1973).

The entomopathogenic Mastigomycotina and Zygomycotina have the most complex growth demands. These fungi require an organic form of nitrogen to attain substantial growth and cannot metabolise nitrates (McInnis, 1971; Latgé, 1981). Often vitamins and oligominerals have to be supplied to the medium (McInnis, 1971; Latgé & Sanglier, 1985). In these subdivisions of the fungi, usually considered phylogenetically less advanced than the Deuteromycotina, the interconnections between the anabolic pathways are limited and the precursors of all the essential metabolic pathways, or even the intermediary or final products exploited by the fungus in the host, have to be added to the medium. Their absence in the culture medium is probably responsible for the lack of success in establishing in vitro cultures of any of the *Coelomomyces* species and many of the Entomophthorales and Trichomycetes. Most of the Mastigomycotina which have been cultured have been grown on complex media such as coagulated egg-yolk or mixtures of protein hydrolysates with a lipid or sugar source (Unestam, 1965; Latgé et al., 1978b; Latgé, 1981; Brey, 1985). Nevertheless, some species can be grown in relatively simple defined media (Latgé, 1981).

Until recently, the nutritional requirements of entomopathogenic fungi were determined by studying the influence of the variations of one factor on the growth response, all the other factors being constant. This procedure is unsatisfactory because of the possibility of erroneous results and the excessive number of replicated experiments necessary to avoid such errors. The use of factorial designs is more appropriate since such models permit rapid evaluation of the controlling factors under a range of different conditions but with a minimum number of experiments (Latgé & Sanglier, 1985). This methodology is commonly employed in industry and the selection or improvement of a medium is usually accomplished in shake culture. It is composed of two stages: in the first stage, the essential nutrients and the size of the experimental domain are determined; in the second stage, the proportions of the selected components are optimalized.

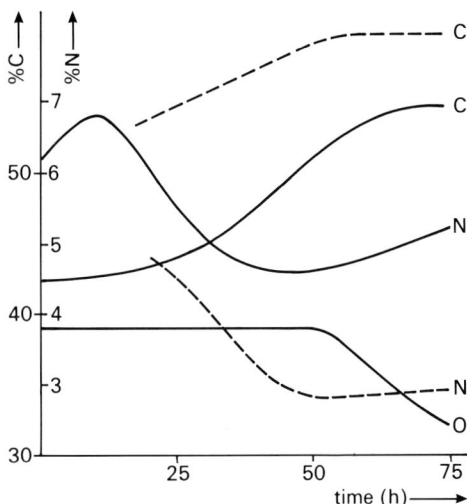

FIGURE 6-1. Variations over time of carbon (C), nitrogen (N) and oxygen (O) content of the mycelium of *Conidiobolus obscures* (—) and *Verticillium lecanii* (– – –).

Growth kinetics

Fungal growth is an autocatalytic process, the rate of which depends on the individual organism and the physicochemical condition of the culture medium, which will influence both the mass transfer kinetics, especially the gas-liquid interface, and also the type of growth (yeast-like, mycelial floc or pellet).

Fungal fermentations can be characterized by growth parameters most often ignored by entomopathologists. This chapter will present some of the most important basic criteria that should be determined to completely understand and thus to control the fermentation of entomopathogenic fungi. In any fermentation experiment, the quantities of substrate to be added to the fermenter in order to obtain the desired concentration of product should be initially assessed, and for this, stoechiometric equations of growth can be used. The elementary composition of a microorganism can be described by a pseudomolecular formula on the basis of elemental analysis of C, H, O and N (which represent over 90% of the fungal dry weight). For example, *C. obscurus* grown on a dextrose-yeast extract medium can be represented by $C_{3.55} H_{6.56} N_{0.42} O_{2.34}$ (Latgé & Moletta, 1983). Considering that only the substrate (in this case the dextrose) will react with oxygen and ammoniac to produce the cells, water and carbon dioxide, a stoechiometric equation of growth can be estimated by:

$$a\, C_6 H_{12} O_6 + b\, O_2\, c\, NH_3 \rightarrow d\, C_{3.55} H_{6.56} N_{0.42} O_{2.34} + e\, H_2O + f\, CO_2 \qquad (1)$$

The equations relating to the balance of each component (e.g. for carbon 6 a = 3.55 d + f) and the values of the growth yield for the particular substrate are employed to calculate the different stoechiometric coefficients of the reaction and to estimate the minimal substrate requirements to achieve the desired production. The equation (1) will vary every time since the molecular formula of the fungus can change with the reaction time (fig. 6-1).

After the requirements of the fungus have been defined, microbial growth kinetics can be studied either in solid or in liquid media (Righelato, 1975). The data will be more meaningful if they are analysed in term of growth parameters the most important of which are presented in the following paragraphs.

When a microorganism grows over a small time interval (dt), the biomass increases of a value (dX) proportional to the amount of microorganism present (X) and to the length of dt:

$$dX = \mu X dt \tag{2}$$

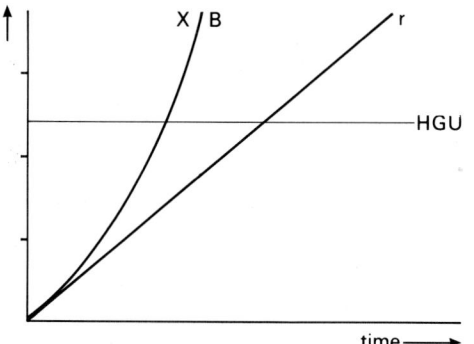

FIGURE 6-2. Schematic characteristics of mycelial growth in solid media. r: length of the colony radius (mm), X: biomass (g/l), B: number of branches, HGU: hyphal growth unit (mm) = total mycelium length / number of branches.

The specific growth rate μ (h^{-1}) is the basic measure for all fermentations. When μ is constant, integration of equation (2) gives:

$$\ln X = X_o + \mu t \; (X_o = \text{initial biomass}) \tag{3}$$

The biomass doubling time ($t_g = \dfrac{0.69}{\mu}$) is calculated from equation (3) by putting $X = 2 X_o$. Other important parameters of the growth depending on substrate (S) uptake are the growth yield $Y = (\dfrac{dX}{dS})$ the metabolic quotients q $(= \dfrac{1}{X} \cdot \dfrac{dS}{dt} = \dfrac{\mu}{Y})$.

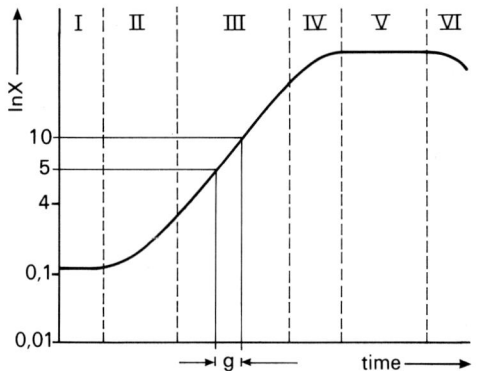

FIGURE 6-3. Batch growth curve with six phases; g = generation time.

GROWTH KINETICS IN SURFACE CULTURE. Fungi are the easiest microorganisms to grow on solid substrate in the quasi-absence of free water. The minimal percentage of free water necessary to obtain any biological activity is 12%, the availability of free water depending on the nature of the substrate employed (Golveke, 1977, Moo-Young et al., 1983).

The kinetics of growth of a fungal colony were determined by Trinci (1971). The fungal thallus can be regarded as an aerial disc and hyphal growth outwards from the original inoculum is from the tip only. The length of the growing tip varies with the species from 0.1 to 10 μm (Caldwell & Trinci, 1973). Initially the growth of the mass of the colony on a surface is exponential, i.e.:

$$X = X_o e^{\mu t} = (H \pi d r_o^2) e^{\mu t} = H \pi d r^2 \tag{4}$$

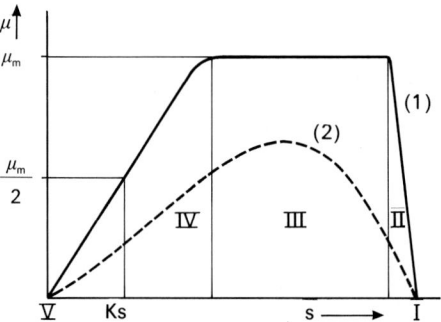

FIGURE 6-4. Specific growth rate (μ) plotted as a function of substrate concentration I, II, III, IV, V: growth phases (1) exponential growth and linear growth, μ_{max}: maximum specific growth rate, Ks: saturation constant.

H = disc height; r = radius of the colony (r_o = initial radius); d = microorganism density.
Taking logarithmus, equation (4) becomes

$$2 \ln r = 2 \ln r + \mu t \qquad (5)$$

$$\text{or } r = r_o e^{\frac{\mu t}{2}}$$

However, after a certain time of growth, the nutrients become limiting in the centre of the colony and only a peripheral annulus A, situated at an average of r_m from the initial inoculum will grow exponentially. The biomass of this annulus X_A can be characterized by the following equations:

$$\frac{dX}{dt} = \mu X_A \qquad (6)$$

$$X_A = A \pi H d 2 r_m \ (A \ll r_m) \qquad (7)$$

From equations (4), (6) and (7),

$$\frac{dr}{dt} = \mu A \text{ or } r_m = \mu A t + r_o \qquad (8)$$

A and μ being constant, the radius of the colony will increase linearly. However, in a fungal colony, the branches of the mycelium increase in number exponentially, at a specific rate similar to the increase in mycelial length (Trinci, 1978). The hyphal growth unit, i.e. the ratio $\frac{\text{total mycelium length}}{\text{number of branches}}$ stays constant. Consequently, the biomass of the fungal mat increases in an exponential way (Latgé & Moletta, 1983) (fig. 6-2).

GROWTH KINETICS IN SUBMERGED CULTURE. The use of stirred fermenters with automatic control of the culture environment is the most suitable technique to evaluate fungal kinetics. Cultures can be discontinuous (batch culture) or continuous.
The batch growth curve of a microorganism can be divided into six phases (fig. 6-3): lag phase; accelerating growth; exponential growth; decelerating growth; stationary phase; lytic decline phase. Growth has been often represented by mathematical models (Pirt, 1975). In the case of substrate limitation, the equation of Monod is most often used (fig. 6-4)

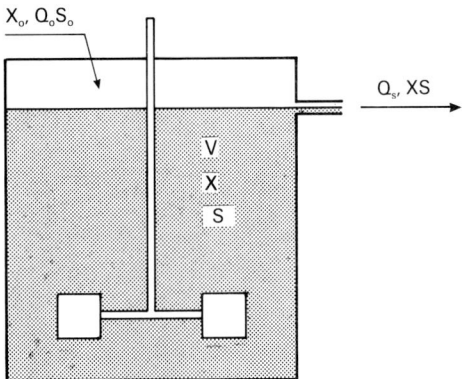

FIGURE 6-5. Diagram of a continuous culture. V: vessel volume, Q flow rate at the entrance (Q_0) and exit (Q_s) of the fermenter, X: biomass at the entrance (X_0) and exit (X) of the fermenter, S: substrate at the entrance (S_o) and exit (S) of the fermenter.

$$\mu = \mu_m \frac{S}{K_s + S} \qquad (9)$$

μ_m = μ maximum specific growth rate; K_s = saturation constant.
If $K_s << S$, $\frac{S}{K_s + S} \sim 1$ and the growth is exponential; if the value of K_s is relatively high in comparaison to S, when S decrease, a decelarating growth phase is reached. Values of μ are dependent on the combination substrate-fungus but are most often comprised between 0.1 and 0.4 h^{-1}.

Values of K_s a can be determined by plotting $\frac{1}{\mu}$ against $\frac{1}{S}$; the (equation (9) becoming $\frac{1}{\mu} = \frac{K_s}{\mu_m} \cdot \frac{1}{S} + \frac{1}{\mu_m}$.
In the case of non exponential growth K_s value can be approximated from the curve $\mu = f(s)$ (fig. 6-4).
With fungi in batch culture, it is difficult to observe exponential increases in biomass for more than five doubling times (Righelato, 1978). After a certain growth time, the fungus will eventually modify the physicochemical condition of its environment and corresponding growth slows down due to limitation in nutrient concentration or oxygen transfer or accumulation of staling products. Morphologically, linear growth can be correlated with a blockage of the branching whereas branching of the mycelium or the fermentation of yeast-like cells induce an exponential increase of the biomass (Latgé & Moletta, 1983).
Continuous flow cultures permit continuous production of the fungus by the controlled addition of fresh medium to the culture vessels so that the volume remains constant (fig 6-5). The biomass balance,

i.e. the increase in biomass = growth in the culture vessel - output from the culture vessel can be written for a small time interval dt:

$$V dX = Q_o X_o\, dt + \mu X\, dt V - K_D X V dt - Q_s X\, dt \qquad (10)$$

V = vessel volume; Q = flow rate of the broth at the entrance (Q_o) and exit (Q_s) of the fermenter; X_o = biomass at the entrance; K_D = death rate.

In the absence of feed back of biomass X_o = O, and in a stationary regime $Q_o = Q_s$, if the death rate is nul, the equation (10) can be written:

$$\frac{dX}{dt} = X(\mu - \frac{Q}{V}) \qquad (11)$$

At the steady state, i.e. when $\frac{dX}{dt}$ = O, the rate of wash out of the biomass will exactly balance the growth rate of the microorganism and the dilution rate $\frac{Q}{V}$ is equal to μ.

The wide use of continuous culture techniques, either in the laboratory or in industrial plants, has led to the design of different systems adapted to the particular problem to be resolved: plug flow culture; chemostat; turbidostat single or in series with or without biomass feedback (Pirt, 1975).

RHEOLOGY AND DIFFUSION LIMITATION. In liquid media, fungi can grow as unicellular forms (hyphal bodies or blastospores in entomopathogenic fungi) or filamentous hyphae infrequently branched in flocs or tightly branched in pellets. Exopolysaccharides of a more or less viscous nature, can also be excreted into the culture medium. The growth form of the fungus influences the rheological characteristics of the medium and consequently the mass transfers (especially the gas transfer at the interphase gas-liquid-fungus) (Pace, 1980; Pace & Righelato, 1980). The rheological properties of the fungal suspension is of primary interest since they will determine the metabolic rates and even routes and influence the design and operation of the fermentation process (Reuss et al. 1982). Rheology is normally expressed in terms of shear stress (τ). In cultures behaving like water, τ increases with the shear rate γ, i.e. the speed gradient at which the fluid is moved.

$$\tau = \mu^* \gamma \qquad (12)$$

Fluids characterized by this law are Newtonian fluids; the viscosity of the fluid, μ^* staying constant whatever speed is used. Cultures containing spherical microbial bodies, yeast-like cells or pelleted mycelia, exhibit a more or less Newtonian flow behaviour; the viscosity of the culture usually being low and function of the culture medium viscosity. Conversely, fungi growing as long mycelial hyphae or producing exopolysaccharides exhibit a pseudoplastic behaviour; the apparent viscosity of the culture being a function of the mycelial concentration (Righelato, 1978).

The transfer of oxygen into the cultures is markedly influenced by the broth rheology; the oxygen transfer falling with increasing viscosity. In contrast to other substrates which can be added in excess to the liquid medium, oxygen is usually a crucial factor because of its low solubility in liquid media ($\sim 7 \times 10^{-3}$ mgl^{-1}) and the high demand of the fungus (100-250 mg O_2 g^{-1} cell h^{-1}) (Sinclair & Mavituna, 1983). The rate of transfer of oxygen dS_{O_2} is characterized by the gas-transfer coefficient $K_L a$ defined by equation (13).

$$\frac{d S_{O_2}}{dt} = K_L a\, (S_{O_2}^* - S_{O_2}) \qquad (13)$$

K_L = specific resistance of O_2 to diffusion; a = specific interfacial area; $S_{O_2}^*$ = maximal O_2 concentration; S_{O_2} = real O_2 concentration. Different methods have been described to measure $K_L a$ in culture media free of microorganism or in actively respiring fungal systems (Pirt, 1975; Sinclair & Mavituna, 1983). When a culture is growing in a filamentous form or produce exocellular polysaccharides, the apparent viscosity of the medium increases and $K_L a$ decreases as a result of transfer resistance in the fermenter. To avoid this drawback, the agitation must be increased. However, increase in speed remains limited because of the risk of the mycelium being damaged by the impellers. In contrast to filamentously growing cultures, pellet fermentation is not linked to an increase in the medium viscosity and to low $K_L a$ values. Nevertheless, oxygen transfer to cells within a pellet is always diffusion-limited. In the case of *Penicillium chrysogenum*, for example, pellets more than 0.8 mm in diameter are completely lysed in the centre due to oxygen depletion. The presence of oxygen could not be detected more than 0.2 mm from the pellet surface (Wittler et al., 1984).

These few data show that the optimization of fermentation should be a compromise, taking into account: the characteristics of the fungal propagule to be produced, the nutritional requirements of the fungal strain; the composition of the culture medium, the configuration of the fermenter vessel and the type of fermentation process selected.

Factors governing sporulation

It is well-documented that conditions favouring spore formation are usually different and more restricted than those controlling mycelial growth (Hawker, 1966; Smith, 1978). Although spores are the main target propagule in the production of mycoinsecticides, the environmental factors governing the sporulation of entomopathogenic fungi have been poorly studied. In particular, the optimum conditions for sporulation and for enhancing the pathogenicity or viability of the spores produced need to be determined, since it has been established that the environmental factors during fermentation influence both the aggressiveness of the spores and their survival (Aoki, 1967; Goral, 1979; Fargues, 1981; Kerwin & Washino, 1986a). Despite the scarcity of basic studies on sporulation, entomopathogenic fungi do not seem to differ significantly from related fungi and thus general data on fungal sporulation can probably be applied to entomopathogens.

A period of vegetative growth invariably precedes sporulation (Couch, 1982; Latgé, 1982) and the production of a large number of spores normally requires a well-nourished mycelium. The nutrient concentration and quality that favour sporogenesis are often highly specific. Exogenous sterols are necessary for the production of zoospores and oospores of the mosquito pathogen *Lagenidium giganteum* (Domnas et al., 1977; Kerwin & Washino, 1983, 1986c; Kerwin et al., 1986). In *Conidiobolus obscurus*, biotin, leucine, proline and thiamine specifically stimulate the formation of azygospores without interacting with vegetative growth (Latgé & Sanglier, 1985). A study of the effect of nitrogen and carbon sources on the sporulation of the muscardine fungi in submerged culture has shown that the yield of conidia is dependent upon the composition of the medium. For *B. bassiana*, maximum sporulation is attained with glucose and vegetable oil plus either glutamine, lysine or serine (Goral, 1979). Nitrate favours conidial production whereas peptone enhances blastospore formation (Kononova, 1979). Sporulation of *N. rileyi* is optimal in a Sabouraud maltose agar supplemented with 1% yeast extract (Couch, 1982). Starvation or reduction in food supply usually stimulates sporulation, nitrogen being the first nutrient to be exhausted, which appears to be a defence against autolysis or the formation of non-viable spores (Smith, 1978; Belova, 1979; Latgé, 1980). However, sporulation can occur without any starvation of the mycelium, the production of conidia of *H. thompsonii* and blastospores of *V. lecanii* in batch culture being parallel to mycelial growth (Latgé et al., 1986c) (fig. 6-6, 6-7).

Studies relating to the production and utilisation of the sexual spores have been restricted almost exclusively to the Mastigomycotina. Al-

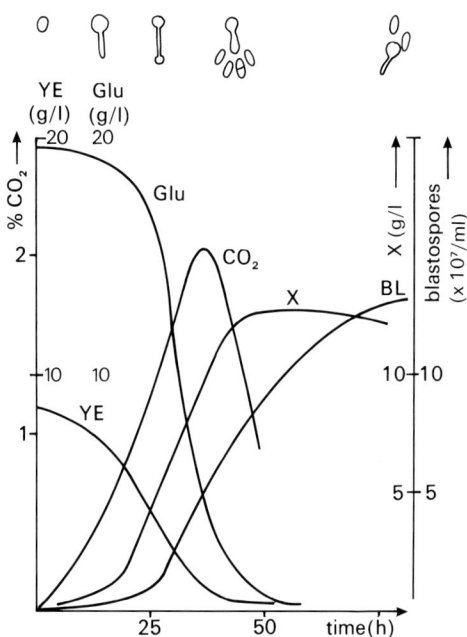

FIGURE 6-6. Batch culture of *Verticillium lecanii* in a glucose (Glu)-yeast extract (YE) medium. X: biomass, BL: number of blastospores.

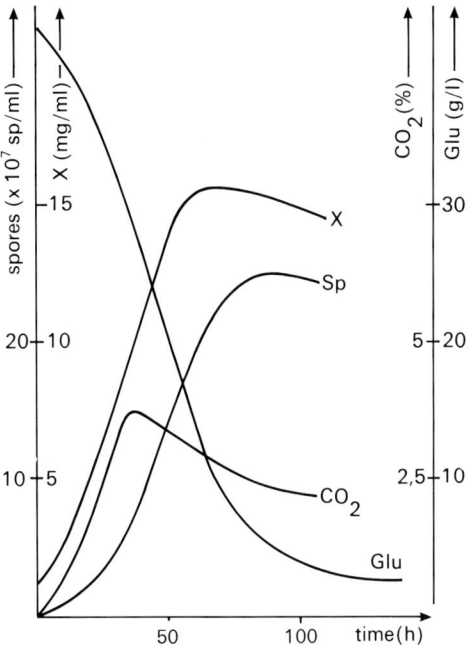

FIGURE 6-7. Spore (Sp) production of *Hirsutella thompsonii* in batch culture in a glucose-yeast extract medium.

though zygospores of *Conidiobolus thromboides* and oospores of *Lagenidium giganteum* have been obtained in shaken liquid culture (Latgé et al. 1977b; Kerwin & Washino, 1986a), conjugation of the gametes, and consequently sporulation, may be favoured by the adsorbtion of mycelium on solid substrates. The involvement of sexual hormones has been proposed for these fungi (Van den Ende, 1978; Warner & Domnas, 1981). The production of ascospores of entomopathogenic fungi has received little study, although Basith & Madelin (1968) reported on the formation of sterile perithecial stromata of *Cordyceps militaris* in a highly concentrated medium based on haemoglobin.

Asexual spores are formed predominantly in aerated conditions and this has led to the classic two-step production procedure for most of the Deuteromycetes: submerged culture of the mycelium and solid substrate fermentation for sporulation. Nevertheless, there are no rules concerning the degree of aeration in order to reach a satisfactory sporulation level amongst the entomopathogenic fungi. Conidia of *B. bassiana*, *C. militaris* and *H. thompsonii* can all be obtained in submerged culture (Carilli & Pacioni, 1977; Goral, 1979; Kononova, 1979; Van Winkelhoff & McCoy, 1984; Latgé et al., 1988a). However, it is felt that entomopathologists should take advantage of the possibility of using fermenter techniques in order to improve their knowledge of the factors controlling sporulation, especially in the Deuteromycetes. Of particular interest, would be a study of the transfer of oxygen and the relationships between growth rate, nutrient utilisation and sporulation (Smith, 1978). Thus, as with other filamentous fungi, stepwise control of sporulation in entomopathogenic fungi could be achieved in continuous culture by manipulation of the physical and chemical environment. In contrast to filamentous fungi, however, the fermentation process for the production of azygospores of *C. obscurus* is composed of three phases: a growth phase; a sporulation phase and; a spore maturation phase (Latgé, 1980) (fig. 6-8). Sporulation occurs only after growth has ceased. The maturation step can also be attained outside the reactor (Latgé & Perry, 1980). Two-step continuous culture could also be designed for this type of mass production.

In order to produce a spore inoculum, the concept of propagation in number and not only in biomass becomes essential. Spore or yeast-like cell production can be compared on a kinetic basis to the formation of microbial metabolites (Pirt, 1975). Two situations can be found. The production of spores is independent of the growth rate but proportional to the amount of biomass. This is the case for batch cultures, where the spores are produced after the end of the growth phase, concomitant with nutrient exhaustion, as for example in the formation of resting spores of *C. obscurus* (Latgé, 1980) (fig. 6-8). In this case the spore (sp) fermentation rate

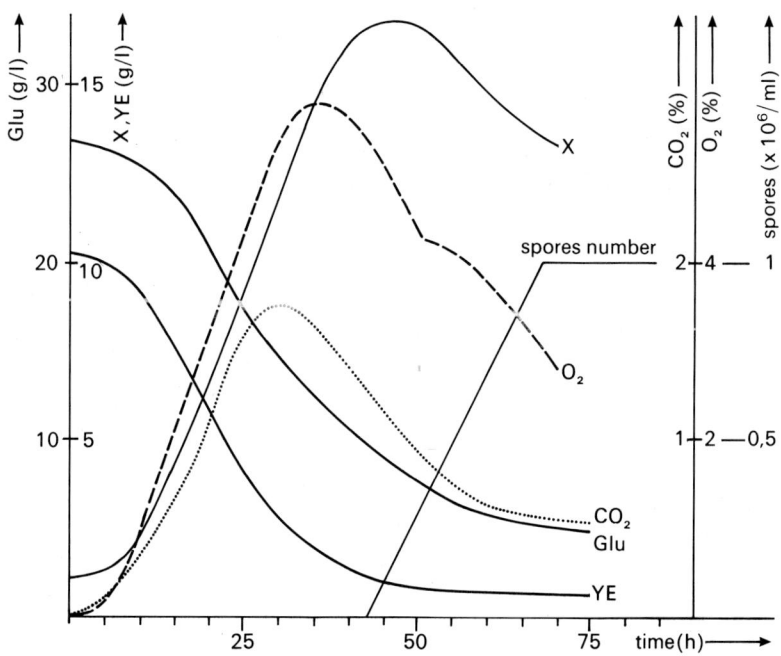

FIGURE 6-8. Azygospore formation of *Conidiobolus obscurus* in batch culture in a glucose-yeast extract medium.

$$\frac{dsp}{dt} \text{ is: } \frac{dsp}{dt} = \beta X \qquad (14)$$

The production of spores can also be dependent (at least partly) upon growth rate; the amount formed being directly proportional to that of the biomass. Examples of this type are given by the formation of blastospores of *V. lecanii* or the production of *H. thompsonii* conidia (Latgé et al., 1986c) (fig. 6-6, 6-7). In these cases, the specific rate of spore formation is:

$$\frac{dsp}{dt} = Y_{sp/x} \mu X \qquad (15)$$

$Y_{sp/x}$ number of spores recovered per unit of biomass. Equation (15) becomes:

$$\frac{dsp}{dt} = Y_{sp/x} \mu X + \beta X \qquad (16)$$

for spore formation partly growth linked and partly independent of the growth rate. In all these cases, the method of increasing sporulation is to increase biomass concentration. Such a goal could be achieved by increasing the concentration of the growth-limiting sub-

Table 6-1
Examples of industrial nutrients used in industrial fermentation

carbon source	nitrogen source	miscellaneous
sugarcane	$NaNO_3$	$NaOH$
bagasse corn syrup	NH_4NO_3	H_2SO_4
sugar beet molasses	NH_4OH	Na_2CO_3
whey	$(NH_4)_2SO_4$	$CaCO_3$
vegetable oil (soy)	soyflour	K_2HPO_4
	cotton seed flour	$MgSO_4$
starch	cotton steep liquor	$FeSO_4$
rice flour	brewer's yeast	$ZnSO_4$
	meat hydrolysate	yeast extract

Table 6-2
Present status of the mass-production of the most common entomopathogenic fungi

fungus	propagule produced*	mode of culture solid	mode of culture liquid
Lagenidium giganteum	Sp (oospores)		+
Conidiobolus obscurus	Sp (azygospores)		+
Entomophthorales (non-producers of resting spores in vitro)	M (hyphal bodies)		+
Beauveria brongniartii	M (blastospores)		+
Beauveria bassiana	M (blastospores)		+
	Sp (conidia)	+	+
Metarhizium anisopliae	Sp (conidia)	+	
Hirsutella thompsonii	Sp (conidia)	+	+
Verticillium lecanii	M (blastospores)		+
	Sp (conidia)	+	
Nomuraea rileyi	M (blastospores)		+
	Sp (conidia)	+	
Aschersonia aleyrodis	Sp (conidia)	+	
Culicinomyces clavisporus	Sp (conidia)		+

*Sp = spore, M = mycelium

strate whilst at the same time removing the staling substances. Multi-stage continuous culture, with some form of biomass feedback, could be efficiently employed.

Mass production

After the growth and sporulation processes have been thoroughly investigated and tested at the laboratory level, mass production of the fungus can be undertaken on an industrial scale with various raw materials (table 6-1). Until now, entomopathogenic fungi have been produced in liquid or solid media (table 6-2). Several fermenter designs, though rarely employed, could be used to attain a more efficient mass-production of these fungi.

Production in liquid media

Three types of reactors or fermenters are commonly used: the stirred tank; the tower and the loop fermenters (Kristiansen & Chamberlain, 1983). The stirred tank fermenter has the form of a vertical cylinder with the agitator mechanism centrally placed (fig. 6-9 A). These reactors produce a violent agitation of the culture medium with good homogenization of the broth and a high gas transfer coefficient whilst at the same avoiding mycelial aggregation and subsequent pellet formation. One drawback is damage to the mycelium on contact with the stirring mechanism. Stirred tank fermenters have been employed for the production of both mycelium and yeast-like cells of all the common entomopathogenic fungi. Blastospores of Deuteromycetes, conidia of *B. bassiana* and *H. thompsonii* and resting spores of *C.obscurus*, *C.thromboides* and *L. giganteum* have also been produced in such tanks (Latgé et al., 1977b; Belova, 1979; Latgé, 1980; Kerwin et al. 1986; Latgé et al., 1986c). The tower fermenter is a vertical cylinder with a height/diameter ratio greater than six and lacks any mechanical agitation (B). Nutrient mixing is promoted by the injection of gas at the base of the reactor. Most fungi produce

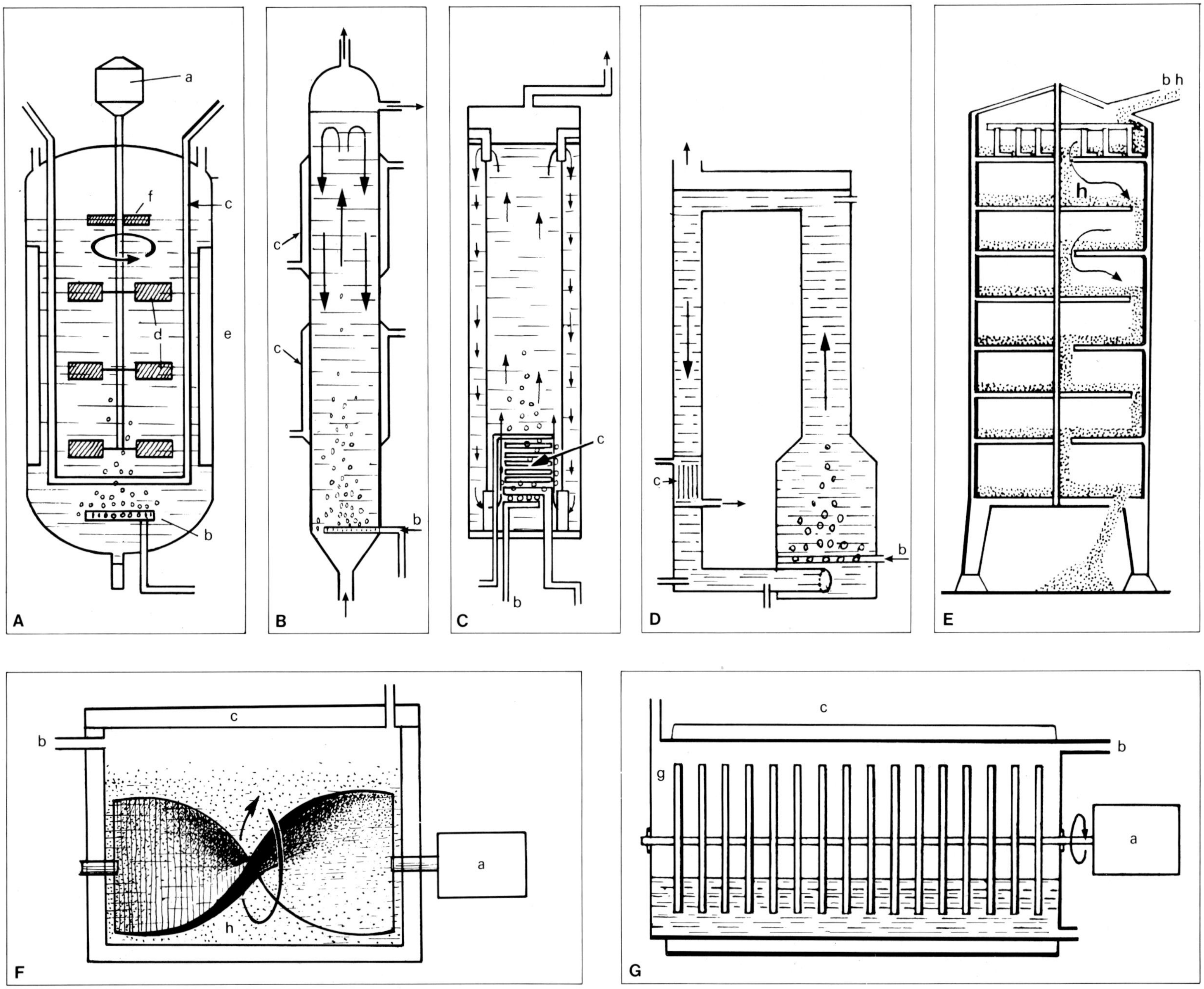

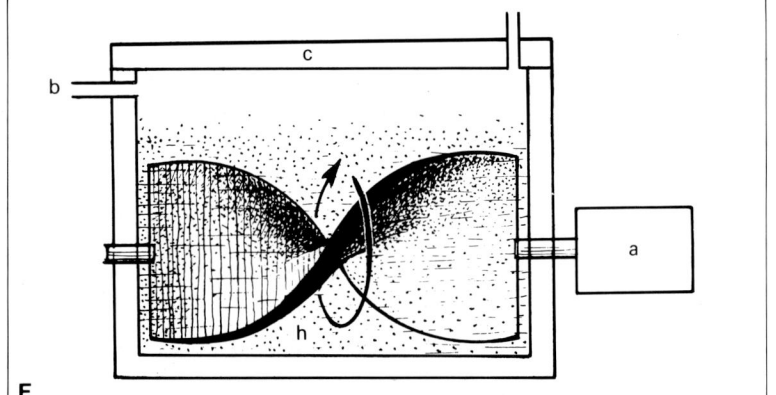
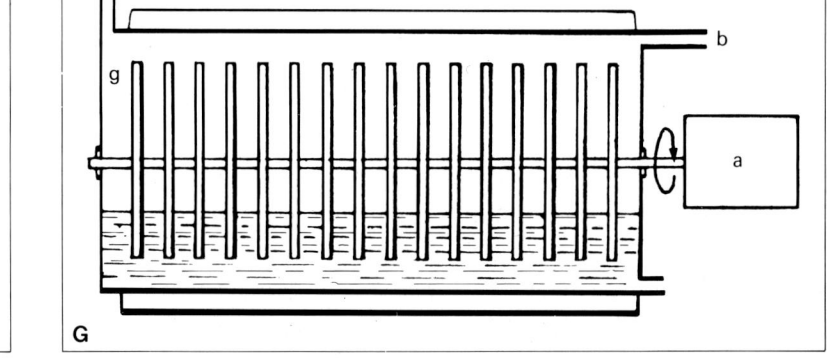

FIGURE 6-9. Diagrams of fermenters suitable for the production of entomopathogenic fungi. A Stirred tank fermenter, B tower fermenter, C and D loop fermenter with internal (C) or external (D) recirculation of the medium, E tray reactor, F homogenous solid reactor, G rotating disc fermenter: a motor, b air sparger, c heating and cooling system, d impeller, e baffle, f foam breaker, g rotating discs, h granular substrate.

mycelial aggregates in this type of fermenter, which can also be used for those species producing spores by conjugation. The loop fermenter is a modification of the latter in which the culture medium is forced back down to the bottom of the reactor. The recycling of the medium is achieved by the incorporation of a draught tube (internal recirculation, C) or by a pipe (external recirculation, D) in the design. The mass transfer at the gas-liquid interface can be as efficient as the stirred tank fermenter but with an important saving of energy. Resting spores of *C. obscurus* can be produced in this type of fermenter (Latgé, unpubl.).

PRODUCTION ON SOLID MEDIA OR SOLID SUPPORT

The mass production of entomopathogenic fungi is usually undertaken using primitive reactors, in which the solid substrate is stirred only weakly, or not at all. The substrates used industrially are mainly cereal grains, broken or not, supplemented with specific nutrients or inert clays, such as vermiculite, impregnated with the culture medium.

Tray reactors consist of several trays containing the substrate, with or without stagewise transport from one tray to another (E). Constant temperature and humidity are maintained by the addition of water or nutrient and the system can be illuminated if this is essential for sporulation. Conidia of *B. bassiana*, *H. thompsonii* and *N. rileyi* are produced in such tray fermenters, the inoculum liquid coming from stirred tank reactors (Couch, 1982). This technology has been adopted in Brazil for the large-scale production of *M. anisopliae*, where rice is fermented in autoclavable plastic bags (Marques et al. 1981). There are two major disadvantages of the tray reactors: because of the long fermentation cycle (3-4 weeks), the risk of contamination is high; substrate utilization is inefficient.

The homogeneous solid reactors are temperature-controlled vessels in which the medium is mixed by means of a simple rotating arm (F). The homogenization of the substrate is superior to that of the tray but the illumination is lower. Rotary fermenters have been used for the mass production of *V. lecanii*.

The rotating disc fermenter (G) allows the development of a fungal biofilm on a series of partially submerged discs which rotate slowly in a trough of medium, exposing the film to the nutrient solution and air. Film thickness is self-regulating since excess growth sloughs from the carrier surface (Anderson, 1983). These fermenters would be suitable for the large-scale production of sexual spores but costs would be high.

PRODUCTION STRATEGY

The type of production process selected is critical and the advantages and disadvantages need to be carefully assessed. In continuous culture, the investment in equipment is low, which is in contrast to the more expensive batch culture production system. However, in the latter, the risk of contamination is lower as also is the appearance of non-productive mutants.

Fermentation in liquid media has been well-documented and many fermentation parameters and fermenter designs can be used and controlled through automatic computing. Production is rapid, thereby reducing contamination problems, but the type of propagule formed may not be suitable for subsequent application in the field. In fermentation on solid media, the energy costs are low and all types of fungal propagules can be produced. Moreover, the survival of these propagules is relatively high and the fungus can utilise the solid substrate as an infection starter in nature. Control of fermentation, however, is difficult, although this can be achieved by monitoring the effluent gas, the biomass being evaluated by the dosage of a specific fungal metabolite such as chitin, sterol and ATP (Matcham et al., 1984).

RECOVERY

The recovery processes are relatively simple and depend upon the production system employed. Mycelium or spores produced in liquid media can be separated from the fermentation broth by centrifugation or direct filtration, which is facilitated by mixing with an inorganic carrier or ultrafiltration. The recovered biomass is usually very sensitive to dryness and should be kept wet until formulation. Spores produced on solid media can be recovered pure, and then air-dried under vacuum, or in a mixture of spores and substrate after grinding of the medium. Spores can be washed with a formulating agent but this processing will entail additional centrifugation and drying.

FORMULATION

Very little information is available on formulation because such technology tends to remain an industrial secret. In general, formulating agents should have spreader-sticker qualities and protect the organism from desiccation and solar degradation, before and during germination of the spore on the target insect (Most & Quinlan, 1986).

A typical formulation is a mixture of several products which have to be compatible. These include: the active ingredient, typically spores; a diluent and/or a dispersant; a wetting agent and a sticker. Dry or liquid formulations are available and most of them are directed against foliage-living insects. Three types of dry formulation exist: dusts, composed of 10% spores and 90% fine clay (particle size 40 µm) as a diluent; granules, made up of the active ingredient (5-20%), a carrier (80-95%) and a binder (1-5%); wettable powders, consisting of a high concentration of active ingredient (50-80%), a filler such as clay (15-45%), a dispersant (1-5%), a surfactant (3-5%) and a vehicle such as water or oil (35-65%). The choice of formulation type will depend upon the habitat of the pest (for example, granules are best suited for soil application) and on the chemical and physical characteristics of the spores, such as hydrophilic properties, size, shape, surface structure and specific gravity (Soper & Ward, 1981).

Storage

Storage is a key problem for industrial preparations of entomopathogenic fungi which may need to be stored for at least one year without any loss of viability. The survival capacity of a spore is directly dependent upon its moisture content: a slow dehydration of the spore being favourable for the conservation of Deuteromycete conidia. (Couch, pers. comm.). Cold temperatures (4-5°C), although uneconomic, are necessary for most long-term storage. Under such conditions, spores of *V. lecanii*, *H. thompsonii* and *C. obscurus* remain viable for at least one year (Latgé & Perry, 1980; Lisansky & Hall, 1983). At room temperature, most fungi rapidly lose viability, often in less than one month. Formulation can enhance survival and is essential for fragile fungal propagules such as mycelium and yeast-like cells produced in liquid culture. Mycelial preservation of the Entomophthorales has been made possible through washing of the mycelium with a 20% maltose solution and formulating the washed mycelium (McCabe & Soper, 1985).

Conclusions

Any scientist interested in developing a commercial product based on an entomopathogenic fungus that has proven to be efficacious under laboratory and field conditions, must work in cooperation with industry. Agrochemical companies will be able to select the most suitable fermentation and formulation technologies. The industry will also be competent to evaluate the economics of the project; the development of a product being dependent upon the production costs and the market size. At present, production systems are reasonably well-developed and relatively inexpensive for entomopathogenic fungi. The most important research which needs developing concerns the stability of the fungal preparation and reliable assessment of its effectiveness under field conditions. Fungal stability can be improved through a greater knowledge of formulating agents and by the selection of more resistant strains. Such goals can be achieved by genetic manipulation or by utilising pathogens which produce resting spores (chlamydospores, oospores and zygospores) ideally suited for long-term storage even under unfavourable conditions.

Chapter 7

Biological control: past, present and future

Introduction

'No fungus disease has ever exterminated an insect or prevented an epidemic. That such diseases do kill off large numbers of insects periodically and so exercise a considerable natural control is undoubted but it has not yet been possible to improve on nature in this respect. It is quite possible that a study of the insects in relation to the fungi might disclose facts which would throw some light on the conditions which govern the incidence of these diseases and that in consequence of such discoveries it might be possible to utilize them in controlling certain pests, but at the present time the available evidence is decidedly opposed to the idea that any practical use can be made of them'. (Petch, 1925).

Petch's scepticism was seemingly based on contact with non-scientists whom he considered enthusiastic but misguided and unfamiliar with the past failures in the field of biological control:' ... proposals of this nature are still periodically put forward with no better foundation than the discovery of another fungus on another insect.' This chapter chronicles the past attempts at utilising these fungi, pre- and post-Petch, and reviews present biological control programmes which incorporate or are attempting to manipulate fungal pathogens of arthropods. A discussion of the future role that entomopathogenic fungi could play in helping to combat arthropod pests of agricultural and medical importance concludes the chapter.

From conception to realisation

The concept of harnessing entomopathogenic fungi for man's benefit is not new and Steinhaus (1956) credits A. Bassi (ca. 1836) with the original idea and subsequently workers in France (L. Pasteur, ca. 1874) and the USA (J. L. le Conte, ca. 1873; H. A. Hagen, ca. 1879) who made similar but perhaps more clear-cut proposals. Traditionally, however, E. Metchnikoff (ca. 1879) is regarded as the practical founder of the doctrine of biological control of pests, not only elucidating the possible uses of entomopathogenic fungi but also carrying out the first scientific experiments to test his hypotheses. Following the example of Petch (1925), we approach the subject from the experiences concerning an individual fungus or a group of fungi, either related or intimately involved in a pest-crop complex.

METARHIZIUM ANISOPLIAE

Metchnikoff advocated the mass production of this fungus for the control of the wheat cockchafer in Russia and by 1884 a small production plant was in operation to supply inoculum for use principally against the sugar-beet weevil (Krassilstschik, 1888). Employing simple field application methods, it was estimated that the pathogen was responsible for 55–80% mortality of larvae but further reports indicate that the work was not continued. It has been suggested that, because of ignorance of basic epizootiology, the results were not consistent and the trials were suspended (Steinhaus, 1956). An elegant method for mass-culturing of *M. anisopliae* was developed by Rorer (1913) for the control of the sugarcane froghopper (*Aeneolamia saccharina*) in Trinidad. Production units were established in a number of sugar estates and large-scale field experiments were conducted with variable success. Petch (1925), however, was of the opinion that insufficient experimental data were analysed and that it was impossible to determine if froghopper mortality was due to natural or artificially-induced epizootics: ' ... the artificial distribution of spores does not appear to have made any appreciable difference to the froghopper blight.' Rorer was anticipating further experiments to test the fungus against froghopper nymphs early in the season, in effect inducing a premature epizootic, but interest in this method of control declined. Similar work was also underway during this period in the Far East, relating to the rhinoceros beetle of coconut palms. Promising results were obtained in Samoa by introducing inoculum of *M. anisopliae* into baited breeding sites but subsequent trials, particularly in Sri Lanka, were disappointing (Swan, 1974). The role of *M. anisopliae* is being re-evaluated in integrated control programmes with baculovirus in new hybrid coconut plantations (Bedford, 1980). Recent unpublished work in the Philippines and Tanzania gives grounds for optimism.

Petch in 1925 summarized the situation: 'Thus the green muscardine has been the subject of experiments at intervals extending over 40 years, but from the foregoing it is evident that in no case has any result been demonstrated, sufficiently successful to warrant its adoption as a means of destroying insects.' What significant advances have been made then in the intervening 60 years since that statement was issued? Ferron (1978, 1981), in assessing this fungus, and entomopathogenic fungi in general, considered that we now have a much better understanding of arthropod dynamics, especially the factors that affect disease incidence. In essence, we have appreciated at last, that entomopathogenic fungi are not a panacea – the solution to all pest control problems – but, sensibly used, we can still take advantage of them in integrated control programmes.

One really has to look to the entrepreneurial efforts of Brazilian scientists to see any significant developments in the use of *M. anisopliae* in biological control. In that country, a number of semi-commercial preparations are in vogue in the various regions against diverse pests. Perhaps the most publicised and successful usage has been against the sugarcane spittlebug, *Mahanarva posticata* in the north-east region. Although an economic evaluation of the fungus has not been feasible (Marques et al., 1981), it is considered to be a success in the sense that there is a more or less constant level of insect mortality (30-40%) in plots sprayed with the fungus, which has reduced significantly the reliance on insecticides: an ecologically, and probably economically, sound justification. Trials with *M. anisopliae* against pasture spittlebugs *(Zulia* and *Aeneolamia* spp.) have been recommended in the central region of Brazil, although apparently not taken up on a large scale. Recent reports from Brazil indicate, however, that research into formulation and application is still being pursued for the control of pasture pests.

In Australia, significant reduction of the pasture pests, *Aphodius tasmaniae* and *Wigeana* spp., have been obtained using contaminated baits. In recent trials in the Philippines, conidia and dry mycelium applications were successfully employed to control populations of rice insect pests (Roberts & Wraight, 1986). The black vine weevil, *Otiorhynchus sulcatus,* can also be controlled in greenhouses, and to a lesser extent in the field, after spraying of conidial suspensions (Zimmermann & Simons, 1986).

BEAUVERIA

Petch (1925) and later Steinhaus (1949) reported at length on the considerable efforts in the American Midwest to utilise *B. bassiana* to control the corn chinch bug, *Blissus leucopterus*. Natural epizootics had been noted as long ago as 1865, and from 1891 to 1897 fungal inoculum was distributed to farmers throughout Kansas. Similar distribution was carried out in other states for a longer period of time but enthusiasm waned and by 1909 a commission of enquiry was set up to evaluate the campaign and to conduct carefully planned laboratory and field experiments. From these results, it was concluded that: the previously reported successes with *B. bassiana* could probably be accounted for by natural epizootics, since with the original trials there were no proper controls to compare untreated and treated plots; disease incidence and spread of the fungus were entirely dependent upon environmental conditions and these could not be enhanced by artificial introduction of inoculum. Recently, in Northern USA, the Colorado potato beetle populations have been significantly reduced after treatment of soil and leaves with conidial suspensions (10^{13} spores/ha). Control of the cornborer, *Ostrinia furnacalis,* has been achieved by placing the fungus in the apices of young corn plants (Roberts & Wraight, 1986).

Near the turn of the century, Giard (1893) in France was experimenting with *Beauveria brongniartii* against *Melolontha* larvae, devising methods for large scale production of inoculum and subsequent application. The field results were encouraging but again infection was dependent on abiotic conditions, especially soil moisture. Recently, Ferron (1978) has resuscitated interest in this pest-disease complex. Following injections of spore suspensions of *B. brongniartii* into pasture soil in Eastern France, infection of larvae was rapid and remained chronic for more than 12 months until the end of the larval cycle when it became acute. The surviving adults proved to be vectors of the fungus, the disease reappearing in the next generation of the pest. He concluded that the epizootics developed more rapidly in treated soil because infection of the larvae was both early and abundant and that continual infection was ensured because of the overlapping host generations. More recently in Switzerland, Keller (1986b) was able to introduce the disease into healthy populations of *Melolontha* and obtain permanent control of the insect. A liquid blastospore formulation was applied to the adults gathering in the mating sites. The infected adults subsequently died on the soil surface and the conidia produced on the cadavers infected the larvae.

Within the last 25 years, the relatively unsophisticated and economic appeal of the entomopathogenic fungi for the control of arthropod pests has been appreciated by agriculturists in both the USSR and China; countries with low-input, cumbersome agricultural systems. *B. bassiana,* because of its amenability to mass production, has been adopted successfully to control a number of important pests. In the USSR, the product Boverin is used in integrated control programmes, typically in combination with low doses of insecticides, against the Colorado beetle *(Leptinotarsa decemlineata)* and the codling moth *(Cydia pomonella).* Host mortality of the former averaged 92% over a 4-year period (Ferron, 1981). In China, *B. bassiana* is widely employed for the control of the European corn borer *(Ostrinia nubialis).* Since commercial preparations were developed in 1971, over 1000 production units have been established and a 5-year application campaign in the Yangtse Valley reduced corn borer damage from 60 to 2% (Hussey & Tinsley, 1981). Soper (1982) also cites reports of its effectiveness against the pine caterpillar, *Dendrolimus punctatus,* in China where a single application has maintained control for 3 years. In order to get the spores into the forest canopy, novel and ambitious techniques have been perfected, including the impregnation of fireworks, mortar shells and landmines with fungal inoculum!

VERTICILLIUM LECANII

Petch (1925) had first-hand experience of the initial attempts to utilise this fungus for biological control. He summarily dismissed control schemes devised in Indonesia and Sri Lanka for use of *V. lecanii* against the green scale of coffee *(Coccus viridis)*: 'There is no doubt that *Cephalosporium lecanii* kills enormous numbers of green bugs in Ceylon. At the beginning of each rainy period the green bug on coffee will generally be found to be covered with the fungus, and it is surprising that any manage to survive. The fungus is so generally distributed that artificial distribution could not make any appreciable difference' – a familiar criticism. He was referring to the earlier practice of pinning leaves with fungal-infected scales in coffee bushes in the hope of inciting epizootics, which probably stemmed from the pioneering work of Parkin (1906). Similar methods were also being employed in attempts to control a variety of scale pests in the West Indies: introducing infected insects into mango plantations to control scale-related sooty mould; pinning and spraying *V. lecanii* against *Coccus viridis.* The results were said to be encouraging (South, 1910).

V. lecanii has since been tested experimentally in the field in a number of countries against a range of pests with varying results. It was employed in Brazil for control of the green scale of coffee (Viegas, 1939). However, it is only in recent years in glasshouse systems in Europe that the potential of the pathogen has been realised. Two commercial products, Vertalec and Mycotal, based on strains specifically selected for use against aphids and whiteflies, have been marketed after extensive research in the UK (Hall, 1981; 1985). The continued success of this biological control agent would seem to depend heavily on the enthusiasm and the ability of the grower to adhere to the strict application regimes required to obtain efficient and lasting control, within his existing pest management programmes.

THE CITRUS EXPERIENCE

The controversy surrounding the use of entomopathogenic fungi for the control of citrus pests goes back almost to the beginnings of biological control and still continues to the present day. A wide range of pathogens and an equally wide range of arthropods are involved in the citrus experience (Rolfs & Fawcett, 1908; Berger, 1921; Fawcett, 1944; Steinhaus, 1949, 1975).

Awareness of the value of natural control of pests of citrus, and fruit trees in general, began in Florida about the turn of the century when it was observed that orchards previously heavily infected with the exotic San Jose scale, *(Aspidiotus perniciosus)* staged a remarkable recovery (Rolfs, 1897). The cause of this dramatic pest decline was eventually attributed entirely to epizootics of *Fusarium coccophilum.* Methods were then devised to disseminate the pathogen artificially using branches infested with diseased scales or spray applications of culture-grown spores. Several years later, Rolfs & Fawcett (1908) published a detailed bulletin on the fungal diseases of scale insects and whitefly in citrus, with information for the grower on where to purchase cultures of several pathogens and how to apply them. By this time several private enterprises were involved in the commercialization of the various fungi. In addition to *F. coccophilum,* two *Aschersonia* spp. *(A. aleyrodis* and *A. goldiana)* were recommended for control of citrus whiteflies by the leaf-pinning technique and by spraying. The latter methodology involved washing leaves containing abundant, red or yellow *Aschersonia* pustules in water and applying the filtrate. The brown fungus of whitefly *(Aegerita webberi),* the morphology of which had not been elucidated at this time, was also shown to be efficacious when pot-grown trees bearing fungal-affected insects were placed in orchards in contact with pathogen-free trees. The former trees were apparently selected and sold specifically for this purpose by citrus nurseries in Florida.

In the meantime, however, an independent investigation was underway to determine the practical benefits of the 'friendly fungi'. After four years of field trials, it was concluded that many of the purported successes were based entirely on casual observations, and not on hard data, resulting in an over-exaggeration of the value of entomopathogenic fungi: 'All experiments have shown that it is useless to force nature and that fungi cannot be successfully introduced unless the weather conditions are such that the fungi are spreading naturally in infected groves' (Morrill & Back, 1912). Despite these damaging findings, the Florida Plant Board continued supplying *Aschersonia*, grown on sweet potato strips in wide mouth pint bottles at 75 cents per culture, to growers (Berger, 1921). The latter author stated that: 'the importance of fungi in keeping down scale insects, for instance, can easily be demonstrated by spraying a tree with fungicide, such as Bordeaux mixture. This destroys the fungi and the scales increase, a fact discovered by many citrus growers'. He also illustrated alternative methods of applying the fungi, involving tying twigs with the pathogen onto scale-infested plants. Later workers, however, cast doubt on the pathogenic status of the 'friendly fungi', suggesting that they invaded and colonised only insects weakened by environmental factors or by endoparasites and that the adverse fungicide responses were the result of residual effects.

Fawcett (1944) attempted to defuse the situation and reconcile these contradictory findings by analysing the complex of factors involved in pest-pathogen relationships: 'If the conditions for natural distribution of a fungus are such that there is the maximum number of spores capable of infecting the maximum possible number of insects under prevailing conditions, then no added results could be expected from artificial distribution without at the same time changing the conditions. A relationship of this kind may be known as the "saturation point" for insect and fungus. When conditions are such that natural distribution is inefficient [...] then artificial distribution would be expected to increase the degree of infection.' Such an interpretation could help to explain many of the past failures mentioned previously.

Interest is once again being shown in the use of *Aschersonia* pathogens to control not only citrus pests but those of glasshouse crops in temperate countries (Samson & Rombach, 1985). Successful introduction of various *Aschersonia* spp. for the control of citrus whiteflies (*Dialeurodes* spp.) has been reported from the Black Sea region of Turkey and the USSR: application of spore suspensions accounting for as much as 90% larval mortality (Ferron, 1978).

Since 1912, it has been noted that populations of citrus rust mites (*Phyllocoptruta oleivora*) in Florida fluctuate dramatically during the rainy season. Fungal mycelium was observed during initial investigations to determine the cause of host mortality but no postive diagnosis resulted (Speare & Yothers, 1924). The fungus involved was finally identified as an undescribed species of *Hirsutella*, *H. thompsonii* (Fisher, 1950), and was subsequently demonstrated to be an important control agent of a number of eriophyid and tetranychid mite pests (McCoy, 1981). *H. thompsonii* has since been developed as a commercial mycoinsecticide, Mycar, for use early in the season to induce premature epizootics in rust mite populations: 5000 acres of citrus were treated in 1981 in Florida (McCoy, 1982; 1986). Latest reports indicate that the product has been withdrawn from the market: the controversy surrounding the citrus experience lives on.

NOMURAEA RILEYI

Apparently the first suggestion of using this fungus as a biological control agent was made by Johnston (1915) in Porto Rico. However, the idea was not put into practice until 1955, according to Ignoffo (1981), when it was tested experimentally against *Heliothis virescens* on tobacco in Southern USA, with inconclusive results. Investigations since have concentrated mainly on its effects against noctuid defoliators of soybeans and optimistic reports suggest that it may be of value as a prophylactic agent (Ignoffo, 1981; McCoy, 1982). The contributions of the former author in relation to the natural infection cycle of the pathogen were outlined in a preceeding chapter, and, based on these data, he showed that disease patterns could be altered by the early application of conidia. Epizootics were advanced by at least two weeks, compared with untreated plots, and, significantly, the peaks occurred prior to and during those stages of crop growth most sensitive to defoliation. The use of *N. rileyi* as a microbial insecticide in several integrated pest management programmes is still being evaluated in the USA. The pros and cons are discussed fully by Ignoffo (1981).

ENTOMOPHTHORALES

The first documented case in which a member of the Entomophthorales was exploited as a biological control agent dates back to 1895 in South Africa. In common with many of these early attempts, it is spiced with erroneous conclusions. A fungus, isolated from diseased locusts (*Acridium purpuriferum*) and believed to be a species of the Entomophthorales was mass produced for sale to farmers at sixpence per tube. The apparently beneficial employment of the fungus to control locust infestations attracted the attention of various mycologists involved with colonial agriculture. Requested cultures were sent to Australia and the fungus identified as *Mucor racemosus* (McAlpine, 1900). In order to resolve the problem, cultures were also sent to Kew and reported to represent a new species of *Mucor* (Massee, 1901). In the interim period, the fungus had been tested against locusts and grasshoppers in India with negative results. Inoculation experiments were repeated in 1906, again without success. Petch's subsequent explanation of the events is plausible: he deduced that the original field pathogen was *Entomophaga grylli* but the initial isolations were contaminated and pure cultures of the contaminant were distributed. The reported field successes were in all probability natural epizootics of *E. grylli*: ' ... the death of the insects would, as usual, be attributed to the artificial distribution of the fungus ... ' (Petch, 1925).

A similar experience was repeated recently in Mexico. The national plant protection agency was distributing a formulation of *Conidiobolus coronatus* to the farmers for use against pasture spittlebugs *(Aeneolamia, Prosapia)*. Despite the apparent success, no data was forthcoming on host mortality. An ecological survey showed that the pests were being controlled naturally by two species of Entomophthorales *(Erynia neoaphidis* and *Conidiobolus major)*. The two pathogens sporulated on the hosts but were quickly over-run by *C. coronatus* (30 to 60 minutes after sporulation started), giving the impression that this was the primary causal agent of disease. Cottage industry production plants were designed but fortunately have not been put into operation because of the natural decline of the spittlebug problem during the last few years (Latgé, unpubl.).

Earlier, Cooke (1892) had remarked on the susceptibility of aphids to *Entomophthora* infections: ' ... which are so destructive and so readily communicated, that the proposition has been made to introduce some one or more of the most favourable kinds of moulds of this class amongst the *Aphides* in glasshouses, with a view to their destruction'. It is probably because many Entomophthorales have a temperate distribution that so much research, relating to their potential in biological control, has been conducted on them. However, a major stumbling block to utilising these fungi as biological control agents has been the difficulties encountered in growing them in vitro. Consequently, early attempts at control utilising the Entomophthorales involved the introduction of diseased insects into healthy populations. In one classic experiment, larvae of the brown-tail moth *(Nygmia phaeorrhoea)*, inoculated with *Entomophaga aulicae*, were distributed in pasture and woodlands in the USA and induced 60-100% mortality in caterpillar populations, especially in areas where apparently the disease was not naturally present. Introduction of the fungus into the autumn population was found to be particularly effective since sufficient inoculum survived the winter to infect the spring emergents (Speare & Colley, 1912). Similar early or exotic introductions of inoculum of Entomophthorales to control pests have been tried since in North America. Dustan (1924) conducted field trials with *Erynia radicans* against aphids in apple orchards in Canada. The leaf-pinning method was recommended for introducing inoculum into orchards early in the season. Comparable trials followed amongst which can be mentioned those involving: *Myzus persicae* on potatoes (Harris, 1948); the spotted alfalfa aphid (Hall & Dunn, 1959); the eastern hemlock looper in forests (Otvos et al., 1973). More recently, Wilding (1981) working in the UK, released laboratory-infected aphids *(Aphis fabae)* into *Vicia* bean crops over a period of several years and found that a mean of 31% insects became infected in treated plots compared with less than 1% in control plots, 12 days after the introduction of *Erynia neoaphidis* inoculum. As in the early field trials in North America, it was noted that mortality could be increased by introducing the pathogen earlier and more systematically than occurs naturally. However, aphid populations were not reduced sufficiently to prevent crop damage suggesting that the fungus was not acting quickly enough. More detailed results have been presented since (Wilding, 1986b) which demonstrate that irrigation during dry periods greatly increased aphid mortality. It was concluded that the type of inoculum needed to be improved in order to permit a rapid insecticidal treatment. However, in areas where the pathogens are absent, the situation may be different. Milner (1982) has reported on the recent introduction of *E. radicans* into Australia in a classical biological control programme against the spotted alfalfa aphid; spread has been spectacular with 70-90% mortality near the original release site. Similar releases of this fungus against the leafhopper, *Empoasca fabae*, in potato and alfalfa fields have been carried out in the USA (McGuire et al., 1986; Wraight et al., 1986) Epizootics, with up to 80% infection were subsequently recorded.

Since 1970, improved methods of in vitro cultivation and application of the Entomophthorales have led to intensified research to evaluate their potential as biological control agents, particularly against aphid pests of temperate crops. For the first time, glasshouse and field trials have been conducted with industrial-type formulations based on resting spores (Latgé, 1982). However, the results have been disappointing and control of aphid populations has not been attained. Several reasons have been advanced to explain these failures: insufficient inoculum; erratic and slow germination of the resting spores; unfavourable abiotic and biotic conditions for the simultaneous development of resting spores and epizootics and in particularly, the limited capacity of the strains employed to spread within the insect population. The use of the Entomophthorales is still being evaluated against aphids as well as other pests, such as the brown planthopper of rice *(Nilaparvata lugens)*, using a formulation based on dry mycelium (marcescent technique) which is expected to be a faster-acting propagule than the resting spore (Roberts & Wraight,

1986). However, until now all attempts using such formulations have failed (Wilding et al., 1986a).

VECTOR FUNGI

According to Chapman (1974): 'fungi that infect mosquitoes have been known since the eighteenth century'. If this is true then their pathogenic status would certainly not have been appreciated or recognized. More than 20 species of Entomophthorales are implicated in natural control of Diptera of medical and veterinary importance, often being responsible for widespread epizootics (Latgé & Papierok, 1982). Most of the species are specific to these vector insects but are primarily effective against adult flies only. This, together with problems of propagation in vitro and spore fragility, has restricted investigations on their use in vector control. Experimental data are scarce and their potential appears limited at the present time.

The oomycete *Lagenidium giganteum* has been known for almost 50 years but it is only since the 1970's that attempts have been made to test it as a biological control agent of vector insects. Its major advantage is its ability to survive saprophytically in nature even in the absence of water. Field trials in California have given variable but generally promising control of mosquito populations in rice and flooded pasture. Agar cultures of the fungus were introduced into lagoons in North Carolina and resulted in 28-100% infection of *Culex* larvae (Jaronski, 1982). Recently, techniques for the production of oospores in liquid medium have been developed and field trials using such spores have shown some potential (Kerwin & Washino, 1983; 1986a). The use of *Coelomomyces* seems impractical since no species of this genus has been cultured in vitro and the situation is further complicated by their heteroecious life cycles. Nevertheless, the few field experiments that have been conducted have given encouraging results (Pillai, 1982). *Coelomomyces*-infected mosquito larvae have been released into new areas and habitats in Southern USA, using techniques developed by Couch (1972), and found to establish successfully to the extent of accounting for up to 60% mean larval mortality. Concurrent laboratory tests showed that if the inoculum is added at the right time, so that the spores are discharged during the first to third larval ecdyses, infection can be as high as 100%. However, according to Federici (1981) no field trials have been undertaken since the discovery of the role of the alternate crustacean host in mosquito infection.

Both *Culicinomyces clavisporus* and *Tolypocladium cylindrosporum*, in contrast to most other entomopathogenic fungi, are of recent discovery and consequently studies on their use as biological control agents, particularly in field situations, are scarce and applied research is still at a rudimentary stage. *C. clavisporus* has been evaluated in laboratory and field trials in Australia (Sweeney, 1982), against *Aedes* mosquitoes breeding in artificial rock pools and natural populations of *Anopheles* and *Culex* in ponds. Test results were variable, in some instances larval numbers declined rapidly 3-4 days after treatment, in others the effect was not so apparent and required careful assessment for which it was necessary to devise special sampling and bioassay techniques. It was estimated, however, that the original larval population declined by almost 80%, 5 days after treatment. It was also discovered that in certain habitats no control was obtained due to unfavourable physical and chemical conditions. Following the application, *C. clavisporus* apparently recycles but not at a sufficiently high level to provide for an effective on-going control of mosquito larvae. One advantage that this fungus has over the other fungi discussed here, is its ability to produce highly infective conidia in submerged liquid culture. Although formulation can improve the survival and floatability of the inoculum (Sweeney, 1986), one of the biggest problems of this fungus is its inefficacy in water when temperature exceed 20°C (Pant, 1986). Laboratory experiments with *T. cylindrosporum* have shown over 95% mortality in larvae of *Aedes*, *Anopheles* and *Culex* mosquitoes (Pillai, 1982). Unfortunately, there are few data on field testing of this pathogen although the recent WHO publication (1984) indicates that deep fermentation methods for producing inoculum (blastospores and conidia) have been developed in addition to methods suitable for production at the cottage industry level. However, saline or brackish water are toxic to this fungus (Pant, 1986).

Other non-specific fungi have been screened as potential biological control agents of vector insects but, although conidial preparations of *Metarhizium anisopliae* have proved to be effective as larvicides against several mosquito genera, the application rates 'are considered to be too high to merit further development of this fungus at present' (WHO, 1984).

Progression or rejection?

Fawcett (1944) reflecting on the general lack of applied research with entomopathogenic fungi summarized the situation at the time: 'One apparent cause for neglect of the field has been the hasty generalization that because of failures to get outstanding results with certain fungi in initial trials, the whole field has little promise of practical results.' This 'hasty generalization' or pessimism is still shared today by many involved in crop protection, or agriculture in general.

The main reason for considering mycoinsecticides with scepticism is their poor field performance, particularly their instability and costs when compared, for example, with *Bacillus thuringiensis* products. Moreover, most of the succesful trials reported are not based on scientific data and do not take into account the impact of natural inoculum. The improvement of this situation, up until now too speculative, will consolidate the commercial exploitability of entomopathogenic fungi: without investment the science will not progress, with more failures the concept will be rejected. The future is delicately balanced. There are many problems concerning the production, application and marketability of mycoinsecticides which remain to be resolved and none of the few products which have reached the crop protection stage are household names and have been unable to compete with chemical insecticides. Nevertheless, the list of credits supporting the use of entomopathogenic fungi as biological control agents is impressive:

1 safe, in comparison with chemical control methods;
2 environmentally sound, since they do not upset the ecological balance;
3 generally persistent, possibly giving on-going control;
4 compatible with other control agents, being readily incorporated into integrated control programmes;
5 less possibility of pest resistance, compared with chemical insecticides;
6 genetically manipulative.

Host specificity is often quoted as a highly desirable character of most entomopathogens, compared with chemical insecticides. Perversely, however, this very specificity can be counterproductive in that commercial concerns generally will not be interested in developing a product with such a limited pest market.

A more profitable direction for these highly specialised pathogens may lie in the study of their biochemical and physiological properties. The only common temperate species of the genus *Cordyceps*, *C. militaris*, has yielded an antibiotic and insect toxin, cordycepin, which is used in molecular biology to block RNA synthesis. Other insect toxins have been identified in *Beauveria* and *Metarhizium* (Roberts, 1981). For the first time, toxins isolated from *M. anisopliae* have been reported to be active per os (Poprawski et al., 1986), and there is every reason to believe that many more novel compounds remain to be isolated from entomopathogenic fungi, some of which could prove to be important not only in pest control but also in medicine.

On the debit side, the production and commercialization costs are extremely high, while the safety testing of mycoinsecticides can be expensive. The difficulties involved with the patenting of organisms or their biotechnologies and the overall concept of biological control, with all its marketing difficulties and uncertain profits, are not compatible with the policies of most large companies, with short-term investors or shareholders to satisfy. The potential of such organisms in high-input, westernised agriculture would appear to be limited by market forces. However, as demonstrated in China and USSR, their use as biological control agents of crop pests seems both feasible and apparently successful, when there is a cooperative, non-commercially orientated production system operating at the factory or cottage-industry level. The latter could also be exploited in developing or Third World countries, as long as there is an organised extension service, and thus help to reduce their dependence on insecticides, often dumped because of environmental restrictions in the developed world. A surprising occurrence has been the acceptance of fungal biological control agents in the highly commercialized, competitive Brazilian market which probably owes more to the entrepreneurial skills of a few individuals rather than to a state or company-controlled policy. However, claims of the effectiveness of many of these preparations should be treated with caution.

As many workers in the field have appreciated, entomopathogenic fungi are not the solution to pest control, they form a part of the answer and should be used as such in integrated pest management programmes. Jaques & Morris (1981) consider that natural control of pests by entomopathogenic fungi is significant in many agricultural ecosystems and they provide examples of pest problems, which have been exacerbated following the excessive use of fungicides and the subsequent suppression of the beneficial mycoflora. Therefore, it is obvious that future crop protection programmes must make the best possible use of these organisms, through:

1 the judicious use of fungicides, to maintain the natural equilibrium, and here it would be necessary initially to identify all the factors that are involved in the natural control of a potential pest;
2 the manipulation of these natural populations to obtain maximum benefit from them, for example: the employment of irrigation or cover crops to increase humidity levels; the use of low doses of chemical insecticides; the use of mathematical models to forecast the appearance of natural epizootics;
3 the induction of epizootics, particularly for those fungi which are difficult to mass produce;
4 the application of mycoinsecticides, only feasible for those fungi which can be readily and cheaply mass produced.

Epizootiological studies will provide the information essential for the successful implementation of these measures, especially the latter two which depend so heavily on favourable climatic conditions that cannot be readily controlled except in covered ecosystems such as glasshouses.

The inclusion of low doses of chemical insecticides in mycoinsecticide spray mixes has shown promise in the USSR and research on this aspect should be pursued, particularly the synergistic effects with other products or control organisms. A similar ploy along these lines, would be to apply chemical insecticides during the driest part of the crop season or during periods of low humidity and then apply the mycoinsecticides during the wetter months. Modelling of pest populations would be essential, of course, in this type of crop protection system for forecasting both the timing of application with a view to eliminating the epizootic lag phase, and the quantity of product to use. It is imperative, of course, that the interactions of climate, pathogen and host are completely understood, 'otherwise the application of fungi in biological control of insect pests must continue to be largely empirical' (Madelin, 1966), and, therefore, doomed to failure and rejection by those involved in crop protection. Hopefully, our current awareness of pest-pathogen-abiotic interactions is now sufficiently advanced to avoid such an empirical approach in the future.

As a final thought, it is worth considering that we have tested only a fraction of the entomopathogenic fungi that occur in nature. Certainly our stocks need to be replenished and the genetic base broadened. More collecting surveys are needed particularly in tropical areas which have been relatively understudied in this respect, compared with temperate regions. The restriction of many *Cordyceps* species to undisturbed or primary ecosystems means that habitat destruction also results in the disappearance of fungal germplasm of potential value not only for biological control but also as a source of novel metabolites.

Bibliography

Ainsworth, G.C. (1956). Agostini Bassi, 1773-1856. Nature 177: 255-257.

Ainsworth, G.C. (1968). Fungal parasites of vertebrates. In: The Fungi. Vol. III. (Eds. G.C. Ainsworth & A.S. Sussman), Academic Press, London & New York, pp. 211-226.

Ainsworth, G.C. (1981). Introduction to the History of Plant Pathology. Cambridge University Press, 315 pp.

Al-Aidroos, K. (1980). Demonstration of a parasexual cycle in the entomopathogenic fungus *Metarhizium anisopliae*. Can. J. Genet Cytol. 22: 309-314.

Al-Aidroos, K. & Roberts, D.W. (1978). Mutants of *Metarhizium anisopliae* with increased virulence towards mosquito larvae. Can. J. Genet. Cytol. 20: 211-219.

Al-Aidroos, K. & Seifert A.M. (1980). Polysaccharide and protein degradation, germination and virulence against mosquitoes in the entomopathogenic fungus *Metarhizium anisopliae*. J. Invert. Path. 36: 29-34.

Andersen, S.O. (1979). Biochemistry of insect cuticle. A. Rev. Ent. 24: 29-61.

Andersen, S.O. (1985). Sclerotization and tanning of the cuticle. In: Comprehensive insect physiology, biochemistry and pharmacology. Vol 3 (Eds. G.A. Kerkut & L.I. Gilbert), Pergamon Press, Oxford, pp. 59-74.

Anderson, J.G. (1983). Immobilized cell and film reactor systems for filamentous fungi. In: The filamentous fungi. Vol 4 Fungal Technology (Eds. J.E. Smith, D.R. Berry & B. Kristiansen), Edward Arnold Publ., London. pp. 145-170.

Andersson, K., Sun, S.C., Steiner, H. & Boman, H.G. (1986). Purification and properties of a prophenoloxidase activating enzyme and prophenol oxidax from *Hyalophora* Dev. Comp. Immunol. 10:623.

Andrade, C.F.S. de (1980). Epizootia natural causada por *Cordyceps unilateralis* (Hyprocreales, Euascomycetes) en adultos de *Campanotus* sp. (Hymenoptera, Formicidae) na regiao de Manaus, Amazonas, Brazil. Acta Amazonica 10: 671-677.

Aoki, J. (1967). Some considerations on the infection mechanisms of insect pathogenic fungi: nitrogen utilization of *Beauveria bassiana*, *Isaria farinosa* and *Isaria fumosa-rosea*. Proc. US-Jap. Sem. Microbial Contr. Pests, pp. 107-113.

Ashida, M. & Dohke, K. (1980). Activation of prophenoloxidase by the activating enzyme of the silkworm *Bombyx mori*. Insect Biochem. 10: 37-47.

Ashida, M., Ishizaki, Y. & Iwahana, H. (1983). Activation of prophenoloxidase by bacterial cell walls or ß-1,3-glucans in plasma of the silkworm *Bombyx mori*. Biochem. Biophys. Res. Commun. 113: 562-568.

Ashida, M., Iwama, R., Iwahana, H. & Yoshida, H. (1982). Control and function of the prophenoloxidase activating system. Proc. Third Int. Colloquium Invert. Path., Brighton, pp. 81-86.

Bajan, C., Kalakova, S., Kmitowa, K., Samsinakova, A. & Wojciechowska, M. (1979). The relationship between infectious activities of entomophagous fungi and their production of enzymes. Bull. Acad. pol. Sci. 27: 963-968.

Balazy, S. (1985). Notes on *Hirsutella aphidis*. Trans. Br. mycol. Soc. 85: 752-756.

Balazy, S. & Sokolowski, A. (1977). Morphology and biology of *Entomophthora myrmecophaga*. Trans. Br. mycol. Soc. 68: 134-137.

Basith, M. & Madelin, M.F. (1968). Studies on the production of perithecial stromata by *Cordyceps militaris* in artificial culture. Can. J. Bot. 46: 473-480.

Beauvais, A. & Latgé, J.P. (1988). Glucan synthases in the Entomophthorales. Exp. Mycol. (submitted).

Bedford, G.O. (1980). Biology, ecology and control of palm rhinoceros beetles. A. Rev. Ent. 25: 309-339.

Belova, R.N. (1979). Development of the technology of Boverin production by the submersion method. In: Proc. First joint US/USSR conference on the production, selection and standardization of entomopathogenic fungi of the US/USSR joint working group on the production of substance by microbiological means (Ed. C.M. Ignoffo), pp. 102-119.

Ben-Ze'ev, I., Kenneth, R.G., Bitton, S. & Soper, R.S. (1984). The Entomophthorales of Israel and their arthropod hosts: seasonal occurrence. Phytoparasitica 12: 167-176.

Berger, E.W. (1921). Natural enemies of scale insects and whiteflies in Florida. Q. Bull. Fla St. Pl. Bd 5: 141-154.

Bergeron, D. & Al-Aidroos, K. (1982) Haploidization analysis of heterozygous diploids of the entomogenous fungus *Metarhizium anisopliae*. Can. J. Genet. Cytol. 24: 643-651.

Bergstrom, A.C. & Nicholson, R.L. (1981). Invertase in the spore matrix of *Colletotrichum graminicola*. Phytopathol. Z. 102: 139-147.

Berisford, Y.C. & Tsao, C.H. (1974). Field and laboratory observations of an entomogenous infection of the adult seedcorn maggot, *Hylemya platura* (Diptera: Anthomyiidae). J. Ga. entomol. Soc. 9: 104-110.

Berry, D.R. (1975). The environmental control of the physiology of filamentous fungi. In: The filamentous fungi I. Industrial mycology (Eds. J.E. Smith & D.R. Berry), Edward Arnold Publ., London, pp. 16-32.

Blomquist, G. J. (1984). Cuticular lipids of insects. In: Infection processes of fungi (Eds. D. W. Roberts & J. R. Aist). Publs Rockefeller Fdn, pp. 54-60.

Blomquist, G. J. & Jackson, L. L. (1979). Chemistry and biochemistry of insect waxes. Prog. Lipid Res. 17: 319-345.

Boman, H. G. (1980). A molecular approach to immunity and pathogenicity in an insect-bacteria system. Molec. Biol. Biochem. Biophys. 32: 217-218.

Boucias, D. G. & Latgé, J. P. (1986). Adhesion of entomopathogenic fungi on their host cuticle. In: Fundamental and applied aspects of invertebrate pathology (Eds. R. A. Samson, J. M. Vlak & D. Peters), Foundat. Fourth Int. Colloq. Invert. Path., Wageningen, Netherlands, pp. 432-434.

Boucias, D. G. & Latgé, J. P. (1988 a). Non-specific induction of germination of *Conidiobolus obscurus* and *Nomuraea rileyi* with host and non host cuticle extracts. J. Invert. Pathol. (in press).

Boucias, D. G. & Latgé, J. P. (1988 b). Fungal elicitors of invertebrate cell defense systems. In: Fungal antigens (Eds. E. Drouhet et al.). Plenum, New York & London (in press)

Boucias, D. G. & Pendland, J. C. (1982). Ultrastructural studies on the fungus *Nomuraea rileyi* infecting the velvet bean caterpillar, *Anticarsia gemmatalis*. J. Invert. Path. 39: 338-345.

Boucias, D. G. & Pendland, J. C. (1984). Nutritional requirements for conidial germination of several host range pathotypes of the entomopathogenic fungus *Nomuraea rileyi*. J. Invert. Path. 43: 288-292.

Boucias, D. G. & Pendland, J. C. (1986). Detection of protease inhibitors in the haemolymph of resistant *Anticarsia gemmatalis* inhibitory to the entomopathogenic fungus, *Nomuraea rileyi*. In: Fundamental and applied aspects of invertebrate pathology (Eds. R. A. Samson, J. M. Vlak & D. Peters). Foundat. Fourth Int. Colloq. Invert. Path., Wageningen, Netherlands. p. 455.

Boucias, D. G., Pendland, J. C & Latgé, J. P. (1988.) Non-specific factors involved in the attachment of entomopathogenic deuteromycetes to host insect cuticle. Appl. Environ. Microbiol. (submitted)

Brady, B. L. K. (1979). *Entomophthora grylli*. C. M. I. Descriptions of Pathogenic Fungi and Bacteria 606, Kew.

Brey, P. (1985). Observations of in vitro gametangial copulation and oosporogenesis in *Lagenidium giganteum*. J. Invert. Path. 45: 276-281.

Brey, P., Ohayon, H., Lesourd, M., Castex, H., Roucache, J. & Latgé, J. P. (1985). Ultrastructure and chemical composition of the outer layers of the cuticle of the pea aphid *Acyrthosiphon pisum* Harris. Comp. Biochem. Physiol. 82A: 401-411.

Brey, P., Latgé, J. P. & Prévost, M. C. (1986). Integumental penetration of the pea aphid, *Acyrthosiphon pisum* by *Conidiobolus obscurus*. J. Invert. Path. 48: 34-41.

Brobyn, P. J. & Wilding, N. (1977). Invasive and developmental processes of *Entomophthora* species infecting aphids. Trans. Br. mycol. Soc. 69: 349-366.

Burges, H. D. (1981). Strategy for the microbial control of pests in 1980 and beyond. In: Microbial control of pests and plant diseases 1970-1980 (Ed. H. D. Burges), Academic Press, London & New York, pp. 797-836

Butt, T. M., Beckett, A. & Wilding, N. (1981). Protoplasts in the in vivo life cycle of *Erynia neoaphidis*. J. gen. Microbiol. 127: 417-421.

Byford, W. J. & Ward, L. K. (1968). Effect of the situation of the aphid host at death on the type of the spore produced by *Entomophthora* sp. Trans. Br. mycol. Soc. 51: 598-600.

Cabrera Cabrera, R. I. (1977). Estudio en Cuba del *Hirsutella thompsonii* Fischer. Control biologico del acaro del moho (*Phyllocoptruta oleivora* Ashm.). Agrotechnica Cuba 9: 3-11.

Caldwell, I. Y. & Trinci, A. P. J. (1973). The growth unit of the mould *Geotrichum candidum*. Archs Mikrobiol. 88: 1-10.

Carilli, A. & Pacioni, G. (1977). Growth and sporulation of *Cordyceps militaris* (Linn. ex Fr.) Link in submerged culture. Trans. Br. mycol. Soc. 68: 237-243.

Carruthers, R. I., Haynes, D. L. & MacLeod, D. M. (1985). *Entomophthora muscae* (Entomophthorales: Entomophthoraceae) mycosis in the onion fly, *Delia antiqua* (Diptera, Anthomyidae). J. Invert. Path. 45: 81-93.

Carruthers, R. I., Soper, R. S. & Feng, Z. (1986). Epizootiology of *Entomophaga grylli* in populations of the clear-winged grasshopper, *Camnula pellucida*. In: Fundamental and applied aspects of invertebrate pathology (Eds. R. A. Samson, J. M. Vlak & D. Peters). Foundat. Fourth Int. Colloq. Invert. Path., Wageningen, Netherlands, p. 237.

Catroux, G., Calvez, J., Ferron, P. & Blanchère, H. (1970). Mise au point d'une préparation entomopathogène à base de blastospores de *Beauveria tenella* (Delacr.) Siemaszko pour la lutte microbiologique contre le ver blanc (*Melolontha melolontha* L.). Ann. Zool. Ecol. Anim. 2: 281-294.

Cerenius, L. & Söderhäll, K. (1984). Chemotaxis in *Aphanomyces astaci*, an arthropod-parasitic fungus. J. Invert. Path. 43: 278-281.

Champlin, F. R., Cheung, P. Y. K., Pekrul, S., Smith, R. J., Burton, R. L. & Grula, E. A. (1981). Virulence of *Beauveria bassiana* mutants for the pecan weevil. J. econ. Ent. 74: 617-621.

Chapman, H. C. (1974). Biological control of mosquito larvae. A. Rev. Ent. 19: 33-59.

Chen, Zuei-Ching (1978). Notes on new Formosan forest fungi VI. Genus *Cordyceps* and their distribution in Taiwan. Taiwania 23: 153-162.

Cherbit, G. & Delmas, J. C. (1979). Potentiels cutanés et points de moindre impédance chez *Oryctes rhinoceros* (Coleoptera, Scarabaeidae). C. r. hebd. Seanc. Acad. Sci., Paris, 289D: 1077-1080.

Cheung, P. Y. K. & Grula, E. A. (1982). In vivo events associated with entomopathology of *Beauveria bassiana* for the corn earworm (*Heliothis zea*). J. Invert. Path. 39: 303-313.

Cole, G. T. & Nozawa, Y. (1981). Dimorphism. In: Biology of conidial fungi, vol I (Eds. G. T. Cole & B. Kendrick), Academic Press, New York & London, pp. 97-133.

Cole, G. T. & Pope, L. M. (1981). Surface ultrastructure of *Aspergillus niger* conidia. In: The fungal spore: morphogenetic controls (Eds. G. Turian & H. R. Hohl), Academic Press, London, pp. 195-215.

Cole, G. T. & Samson, R. A. (1979). Patterns of development in conidial fungi. Pitman, London & Melbourne, 190 pp.

Coles, R. B. (1978). The biology of *Cordyceps aphodii* (Sphaeriales: Clavicipitales). In: Proc. Second Australasian Conf. Grassland Invert. Ecol. (Eds. T. K. Crosby & R. D. Pottinger), Palmerston, New Zealand., pp. 207-212.

Cook, R. J. & Baker, K. F. (1983). The nature and practice of biological control of plant pathogens. American Phytopathological Society, St. Paul, Minnesota, 539 pp.

Cooke, M. C. (1892). Vegetable wasps and plant worms. Society for Promoting Christian Knowledge, London, 364 pp.

Cooper, R. M., Wardman, P. A. & Skelton, J. E. M. (1981). The influence of cell walls from host and non-host plants on the production and activity of polygalacturonide-degrading enzymes from fungal pathogens. Physiol. Pl. Path. 18: 239-255.

Couch, J. N. (1938). The Genus *Septobasidium*. University of North Carolina Press, Chapel Hill, 480 pp.

Couch, J. N. (1972). Mass production of *Coelomomyces*, a fungus that kills mosquitoes. Proc. nat. Acad. Sci. 69: 2043-2047.

Couch, J. N. & Umphlett, C. J. (1963). *Coelomomyces* infections. In: Insect pathology: An Advanced Treatise. Vol. 2 (Ed. E. A. Steinhaus), Academic Press, New York & London, pp. 149-188.

Couch, T. L. (1982). Production of Hyphomycetes. Proc. Third Int. Colloquium Invert. Path., Brighton, pp. 188-191.

Descals, E., Webster, J., Ladle, M. & Bass, J. A. B. (1981). Variations in asexual reproduction in species of *Entomophthora* on aquatic insects. Trans. Br. mycol. Soc. 77: 85-102.

Descals, E. & Webster, J. (1984). Branched aquatic conidia in *Erynia* and *Entomophthora* sensu lato. Trans. Br. mycol. Soc. 83: 669-682.

Dillon, R. J. & Charnley, A. K. (1986). Germination physiology of conidia of *Metarhizium anisopliae*. In: Fundamental and applied aspects of invertebrate pathology (Eds. R. A. Samson, J. M. Vlak & D. Peters). Foundat. Fourth Int. Colloq. Invert. Path., Wageningen, Netherlands, p. 255.

Domnas, A. J. (1981). Biochemistry of *Lagenidium giganteum* infection in mosquito larvae. In: Pathogenesis of vertebrate microbial diseases (Ed. E. W. Davidson), Allanheld, Osmun Publ. Ottawa, New Jersey, pp. 425-449.

Domnas, A. J., Srebro, J. P. & Hicks, B. F. (1977). Sterol requirement for zoospore formation in the mosquito-parasitizing fungus *Lagenidium giganteum*. Mycologia 69: 875-886.

Dunphy, G. B. & Chadwick, J. M. (1985). Strains of protoplasts of *Entomophthora egressa* in spruce budworm larvae. J. Invert. Path. 45: 255-259.

Dunphy, G. B. & Nolan, R. A. (1980). Response of eastern hemlock looper hemocyte to selected stages of *Entomophthora egressa* and other foreign particles. J. Invert. Path. 36: 71-84.

Dunphy, G. B. & Nolan, R. A. (1982). Cellular immune response of spruce budworm larvae to *Entomophthora egressa* protoplasts and other test particles. J. Invert. Path. 39: 81-92.

Durliat, M. (1985). Clotting processes in *Crustaceas decapoda*. Biol. Rev. 60: 473-498.

Dustan, A. G. (1924). The control of the European apple sucker, *Psyllia mali* Schmidb. in Nova Scotia. Pamph. Can. Dep. Agric. 45: 13pp.

Eilenberg, J. (1986). Effect of *Entomophthora muscae* (C.) Fres. on egg-laying behaviour of female carrot-flies *(Psila rosae F.)*. In: Fundamental and applied aspects of invertebrate pathology (Eds. R. A. Samson, J. M. Vlak & D. Peters), Foundat. Fourth Int. Colloq. Invert. Path., Wageningen, Netherlands, p. 235.

Eilenberg, J. Bresciani, J. & Latgé, J. P. (1986). Ultrastructural studies of primary spore formation and discharge in the genus *Entomophthora*. J. Invert. Path. 48: 318-324.

Ellingboe, A. H. (1982). Genetical aspects of active defence In: Active defence mechanisms in plants (Ed. R. K. S. Wood), Plenum Press, New York & London, pp. 179-192.

Elton, C. S. (1973). The structure of invertebrate populations inside neotropical rain forest. J. Anim. Ecol. 42: 55-104.

Escoubas, P., Clément, J. L., Lange, C. & Ronzani, N. (1986). Epicuticular hydrocarbons: a chemical barrier against toxins and pathogens. In: Fundamental and applied aspects of invertebrate pathology (Eds. R. A. Samson, J. M. Vlak & D. Peters), Foundat. Fourth Int. Colloq. Invert. Path. Wageningen, Netherlands, pp. 421-422.

Evans, H. C. (1974). Natural control of arthropods, with special reference to ants (Formicidae), by fungi in the tropical high forest of Ghana. J. appl. Ecol. 11: 37-49.

Evans, H. C. (1982). Entomogenous fungi in tropical forest ecosystems: an appraisal. Ecol. entomol. 7: 47-60.

Evans, H. C. & Samson, R. A. (1977). *Sporodiniella umbellata*, an entomogenous fungus of the Mucorales from cocoa farms in Ecuador. Can. J. Bot. 55: 2981-2984.

Evans, H. C. & Samson, R. A. (1982). *Cordyceps* species and their anamorphs pathogenic on ants (Formicidae) in tropical forest ecosystems. I. The *Cephalotes* (Myrmicinae) complex. Trans. Br. mycol. Soc. 79: 431-453.

Evans, H. C. & Samson, R. A. (1984). *Cordyceps* species and their anamorphs pathogenic on ants (Formicidae) in tropical forest ecosystems II. The *Camponotus* (Formicinae) complex. Trans. Br. mycol. Soc. 81: 127-150.

Evlakhova, A. A. & Chekourina, T. A. (1962). L'activité de défense de la cuticule de la punaise des cereales *(Eurygaster integriceps Put.)* contre les microorganismes végétaux. Entomophaga, Memoire Hors Serie 2, Colloquium Int. Path. Insectes, Paris, pp. 137-141.

Fargues, J. (1981). Spécificité des Hyphomycètes entomopathogénes et résistance interspècifique des larves d'insectes. Theses Doct. d'Etat., Univ. Paris 6, 2 vol.

Fargues, J. (1984). Adhesion of the fungal spore to the insect cuticle in relation of pathogenicity. In: Infection processes of fungi (Eds. D. W. Roberts & J. R. Aist). Publs. Rockefeller Fdn, pp. 90-110.

Fargues, J. & Remaudière, G. (1977). Considerations on the specificity of entomopathogenic fungi. Mycopathologia 62: 31-37.

Fargues, J., Robert, P. H. & Reisinger, O. (1979). Formulation des productions de masse de l'Hyphomycète entomopathogène *Beauveria* en vue des applications phytosanitaires. Annls. zool. ecol. amim. 11: 247-257.

Fawcett, H. S. (1944). Fungus and bacterial diseases of insects as factors in biological control. Bot. Rev. 10: 327-348.

Federici, B. A. (1981). Mosquito control by the fungi *Culicinomyces*, *Lagenidium* and *Coelomomyces*. In: Microbial control of pests and plant diseases 1970-1980 (Ed. H. D. Burges), Academic Press, London & New York, pp. 555-572.

Ferron, P. (1977). Influence of relative humidity on the development of fungal infection caused by *Beauveria bassiana* (Fungi imperfecti, Moniliales) in imagines of *Acanthoscelides obtectus* (Col. Bruchidae). Entomophaga 22: 393-396.

Ferron, P. (1978). Biological control of insect pests by entomogenous fungi. A. Rev. Ent. 23: 409-442.

Ferron, P. (1981). Pest control by the fungi *Beauveria* and *Metarhizium*. In: Microbial control of pests and plant diseases 1970-1980, (Ed. H. D. Burges), Academic Press, London & New York, pp. 465-482.

Filshie, B. K. (1970). The fine structure and deposition of the larval cuticle of the sheep blowfly *(Lucilia cuprina)*. Tissue Cell 2: 479-489.

Filshie, B. K. (1982). Fine structure of the cuticle of insects and other arthropods. In: Insect ultrastructure, Vol 1. (Eds. R. C. King & H. Akai), Plenum Press, New York & London, pp. 281-312.

Fisher, F. E. (1950). Two new species of *Hirsutella* Patouillard. Mycologia 42: 290-297.

Fuxa, J. R. (1984). Dispersion and spread of the entomopathogenic fungus *Nomuraea rileyi* (Moniliales, Moniliaceae) in a soybean field. Environ. Entomol. 13: 252-258.

Garraway, M. O. & Evans, R. C. (1984). Fungal nutrition and physiology. John Wiley & Sons, New York,

Giard, A. (1893). L'*Isaria densa* (Link) Fries, champignon parasite du hanneton commun *(Melolontha vulgaris* L.). Bull. Scient. France et Belgique 24: 1-112.

Goettel, M. S., Sigler, S. & Carmichael, J. W. (1984). Studies on the mosquito pathogenic hyphomycete *Culicinomyces clavisporus*. Mycologia 76: 614-625.

Govleke, C. G. (1977). Biological reclamation of solid wastes. Rodale press, Emmaus, Pennsylvania.

Goral, V. M. (1979). Effect of cultivation conditions on the entomopathogenic properties of muscardine fungi. In Proc. First joint US/USSR conference on the production, selection and standardization of entomopathogenic fungi of the US/USSR joint working group on the production of substance by microbiological means (Ed. C. M. Ignoffo), pp. 217-228.

Gottwald, T. R. & Tedders, W. L. (1984). Colonization, transmission and longevity of *Beauveria bassiana* and *Metarhizium anisopliae* (Deuteromycotina: Hyphomycetes) on pecan weevil larvae (Coleoptera: Curculionidae) in the soil. Environ. Entomol. 13: 557-560.

Götz, P. & Boman, H. G. (1985). Insect immunity. In: Comprehensive insect physiology, biochemistry and pharmacology, Vol. 3. (Eds. G. A. Kerkut & L. I. Gilbert), Pergamon Press, Oxford, pp. 453-485.

Gray, R. C. (1858). Notices of insects that are known to form the bases of fungoid parasites. Privately printed, London, 22 pp.

Grobler, J. H., MacLeod, D. M. & Delyzer, A. J. (1962). The fungus *Empusa aphidis* Hoffman parasitic on the woolly pine needle aphid, *Schizolachnus pini-radiatae* (Davidson). Can. ent. 94: 46-49.

Grula, E. A., Burton, R. L., Smith, R., Maplo, T. L., Cheung, P. Y. K., Pekrul, S., Champlin, F. R., Grula, M., & Abegaz, B. (1979). Biochemical basis for entomopathogenicity of *Beauveria bassiana*. In: Proc. First joint US/USSR conference on the production, selection and standardization of entomopathogenic fungi of the US/USSR joint working group on the production of substance by microbiological means (Ed. C. M. Ignoffo), pp. 192-216.

Grula, E. A., Woods, S. P. & Russell, H. (1984). Studies utilizing *Beauveria bassiana* as an entomopathogen. In: Infection processes of fungi (Eds. D. W. Roberts & J. R. Aist), Publs. Rockefeller Fdn, pp. 147-152.

Hall, I. M. & Dunn, P. H. (1957). Entomophthorous fungi parasitic on the spotted alfalfa aphid. Hilgardia 27: 159-181.

Hall, I. M. & Dunn, P. H. (1959). The effect of certain insecticides and fungicides on fungi pathogenic to the spotted alfalfa aphid. J. Econ. entomol. 52: 28-29.

Häll, L. & Söderhäll, K. (1982). Purification and properties of a protease inhibitor from crayfish hemolymph. J. Invert. Path. 39: 29-37.

Häll, L. & Söderhäll, K. (1983). Isolation and properties of a protease inhibitor in crayfish *(Astacus astacus)* cuticle. Comp. Biochem. Physiol. 76B: 699-702.

Hall, R. A. (1981). The fungus *Verticillium lecanii* as a microbial insecticide against aphids and scales. In: Microbial control of pests and plant diseases 1970-1980 (Ed. H. D. Burges), Academic Press, London & New York, pp. 483-498.

Hall, R. A. (1982). Deuteromycetes: virulence and bioassay design. Proc. Third Int. Colloquium Invert. Path., Brighton, pp. 191-196.

Hall, R. A. (1984). Epizootic potential for aphids of different isolates of the fungus *Verticillium lecanii*. Entomophaga 29: 311-321.

Hall, R. A. (1985). Whitefly control by fungi. In: Biological pest control (Eds. N. W. Hussey & N. Scopes), Blandford Press, Poole, Dorset, pp. 116-118,

Hall, R. A. & Espinosa Becerril, A. (1981). The coconut mite, *Eriophyes guerreronis* with special references to the problem in Mexico. Proc. Br. Crop Protect. Conf. Pest. Disease, Brighton, pp. 113-120.

Hall, R. A. & Latgé, J. P. (1980). Etude de quelques facteurs stimulant la formation in vitro de blastospores de *Verticillium lecanii* (Zimm.) Viegas. C. r. heb. Seanc. Acad. Sci. Paris 291 D: 75-78.

Hancock, J. G. (1976). Multiple forms of endo-pectatelyase formed in culture and in infected squash hypocotyls by *Hyphomyces solanii* f.sp. *cucurbitae*. Phytopathology 66: 40-45.

Harmstorf, B. & Götz, P. (1987). Investigations of the prophenoloxidase activating system in haemolymph of *Chironomus* (Diptera). Dev. Comp. Immunol. 10, 630.

Harper, A. M. (1958). Notes on behaviour of *Pemphigus betae* Doane (Homoptera: Aphididae) infected with *Entomophthora aphidis* Hoffm. Can. entomol. 90: 439-440.

Harper, J. D., Herbert, D. A. & Moore, R. E. (1984). Trapping patterns of *Entomophthora gammae* (Weiser) (Entomophthorales: Entomophthoraceae) conidia in a soybean field infested with soybean looper, *Pseudoplusia includens* (Walker)(Lepidoptera: Noctuidae). Environ. Entomol. 13:1186-1190.

Harris, M. R. (1948). A Phycomycete parastic on aphids. Phytopathology 38:118-122.

Hawker, L. E. (1966). Environmental influences on reproduction. In: The fungi; an advanced treatise (Eds. G. C. Ainsworth & A. S. Sussman), Academic Press, New York & London, pp. 435-469.

Heale, J. B. (1982). Genetic studies on fungi attacking insects. Proc. Third Int. Colloquium Invert. Path., Brighton, pp. 25-27.

Heale, J. B. (1988). The potential impact of fungal genetics and molecular biology on biological control with particular references to entomopathogen. In: Fungi in biological control systems (Ed. M. N. Burge), Manchester Univ. Press (in press).

Heitefuss, R. (1982). General review of active defense mechanisms in plants against pathogens. In: Active defense mechanisms in plants (Ed. R. K. S. Wood), Plenum Press, New York & London, pp. 1-18.

Hergenhahn, H. G. & Söderhäll, K. (1985). a_2-macroglobulin-like activity in plasma of the crayfish *Pacifastacus leniusculus*. Comp. Biochem. Physiol. 81: 838-845.

Hergenhahn, H. G., Aspan, A. & Söderhäll, K. (1986). Purification and properties of an inhibitor to prophenoloxidase activation from plasma of the crayfish *Pacifastacus leniusculus*. In: Fundamental and applied aspects of invertebrate pathology (Eds. R. A. Samson, J. M. Vlak & D. Peters). Foundat. Fourth Int. Colloq. Invert. Path., Wageningen, Netherlands, p. 465.

Humber, R. A. (1976). The systematics of the genus *Strongwellsea* (Zygomycetes: Entomophthorales). Mycologia 68:1042-1060.

Humber, R. A. (1981). *Erynia* (Zygomycetes: Entomophthorales): Validations and new species. Mycotaxon 13: 471-480.

Humber, R. A. (1984). Foundations for an evolutionary classification of the Entomophthorales (Zygomycetes). In: Fungus-insect relationships (Eds. Q. Wheeler & M. Blackwell), Columbia University Press, New York, pp. 166-183.

Humber, R. A. & Ramoska, W. A. (1986). Variations in entomophthoralean life cycles: practical implications. In: Fundamental and applied aspects of invertebrate pathology (Eds. R. A. Samson, J. M. Vlak & D. Peters). Foundat. Fourth Int. Colloq. Invert. Path. Wageningen, Netherlands, pp. 190-193.

Humphreys, A. M., Inch, J. M. M., Gillespie, A. T. & Trinci, A. P. J. (1986). Blastospore production in *Paecilomyces* spp. In: Fundamental and applied aspects of invertebrate pathology (Eds. R. A. Samson, J. M. Vlak & D. Peters). Foundat. Fourth Int. Colloq. Invert. Path. Wageningen, Netherlands, p. 239.

Hussey, N. W. & Tinsley, T. W. (1981). Impressions of insect pathology in the people's republic of China. In: Microbial control of pests and plant diseases 1970-1980 (Ed. H. D. Burges), Academic Press, London & New York, pp. 785-795

Hutchinson, J. A. (1962). Studies on a new *Entomophthora* attacking calyptrate flies. Mycologia 54: 258-271.

Huxham, I. M., Lackie, A. M. & McCorkindale, N. J. (1986). An in vitro assay to investigate activation and suppression by a pathogenic fungus of prophenoloxidase by insect haemocytes. In: Fundamental and applied aspects of invertebrate pathology (Eds. R. A. Samson, J. M. Vlak & D. Peters). Foundat. Fourth Int. Colloq. Invert. Path. Wageningen, Netherlands, p. 463.

Hywel-Jones, N. L. (1986). The formation of tetradiate conidia of *Erynia conica*. In: Fundamental and applied aspects of invertebrate pathology (Eds. R. A. Samson, J. M. Vlak & D. Peters). Foundat. Fourth Int. Colloq. Invert. Path., Wageningen, Netherlands, p. 227.

Hywel-Jones, N. L. & Webster, J. (1986). Mode of infection of *Simulium* by *Erynia conica*. Trans. Br. mycol. Soc. 87: 381-387.

Ignoffo, C. M. (1981). The fungus *Nomuraea rileyi* as a microbial insecticide. In: Microbial control of pests and plant diseases 1970-1980 (Ed. H. D. Burges), Academic Press, London & New York, pp. 513-538.

Ignoffo, C. M. (1982). Environmental persistence of *Nomuraea rileyi*. Proc. Third Int. Colloquium Invert. Path., Brighton, pp. 331-335.

Ignoffo, C. M., Garcia, C. & Droha, M. J. (1982). Susceptibility of larvae of *Trichoplusia ni* and *Anticarsia gemmatalis* to intrahemocoelic injections of conidia and blastospores of *Nomuraea rileyi*. J. Invert. Path. 39: 198-202.

Jackson, C. W. & Heale, J. B. (1983). Protoplast fusion to overcome vegetative incompatibility in *Verticillium lecanii* parasexual genetics. In: Protoplasts 1983, Poster Proc., 6th Int. Protoplast Symp. Basel (Ed. I. Potryleus), Birkhauser, Basel, p 318-319.

Jackson, C. W., Heale, J. B. & Hall, R. A. (1985). Traits associated with virulence to the aphid *Macrosiphoniella samborni* in eighteen isolates of *Verticillium lecanii*. Ann. Appl. Biol. 106: 39-48.

Jaques, R. P. & Morris, O. N. (1981). Compatibility of pathogens with other methods of pest control and with different crops. In: Microbial control of pests and plant diseases 1970-1980 (Ed. H. D. Burges), Academic Press, London & New York, pp. 695-715.

Jaronski, S. T. (1982). Oomycetes in mosquito control. Proc. Third Int. Colloquium Invert. Path., Brighton, pp. 420-424.

Johansson, M. W. & Söderhäll, K. (1985). Exocytosis of the prophenoloxidase activating system from crayfish haemocytes. J. Comp. Physiol. B. 156: 175-181.

Johnson, J. A., Hall, I. M. & Arakawa, K. Y. (1984). Epizootiology of *Erynia phytonomi* (Zygomycetes: Entomophthorales) and *Beauveria bassiana* (Deuteromycotina: Moniliales) parasitizing the Egyptian alfalfa weevil (Coleoptera: Curculionidae) in southern California. Environ. entomol. 13: 95-99.

Johnston, J. R. (1915). The entomogenous fungi of Porto Rico. Bulletin of the Board of Commissioners of Agriculture, Porto Rico, No. 10: 1-33.

Katerere, Y. (1983). The fungus *Entomophthora planchoniana* Cornu (non Thaxter) on the pine needle aphid, *Eulachnus rileyi* (Williams) Zimbabwe. commonw. For. Rev. 62: 271-273.

Keller, S. (1986a). Quantitative ecological evaluation of the May beetle pathogen, *Beauveria brongniartii*, and its practical application. In: Fundamental and applied aspects of invertebrate pathology (Eds. R. A. Samson, J. M. Vlak & D. Peters). Foundat. Fourth Int. Colloq. Invert. Path., Wageningen, Netherlands, p. 178-181.

Keller, S. (1986b). Control of May beetle grubs (*Melolontha melolontha* L.) with the fungus *Beauveria brongniartii* (Sacc.) Petch. In: Fundamental and applied aspects of invertebrate pathology (Eds. R. A. Samson, J. M. Vlak & D. Peters). Foundat. Fourth Int. Colloq. Invert. Path., Wageningen, Netherlands, pp. 525-528.

Keller, S. & Suter, H. (1980). Epizootiologische Untersuchungen über das *Entomophthora*-Auftreten bei feldbaulich wichtigen Blattlauarten. Acta Oecol. Oecol. Appl. 1: 63-81.

Kerwin, J. L. (1982). Chemical control of the germination of asexual spores of *Entomophthora culicis*, a fungus parastic on dipterans. J. gen. Microbiol. 128: 2179-2186.

Kerwin, J. L. (1984). Fatty acid regulation of the germination of *Erynia variabilis* conidia on adults and puparia of the lesser housefly, *Fannia canicularis*. Can. J. Microbiol. 30: 158-161.

Kerwin, J. L., Simmons, C. A. & Washino, R. K. (1986). Oosporogenesis by *Lagenidium giganteum*. J. Invert. Path. 47: 258-270.

Kerwin, J. L., & Washino, R. K. (1986a). Ground and aerial application of the sexual and asexual stages of *Lagenidium giganteum* (Oomycetes: Lagenidiales) for mosquito control. J. Amer. Mosq. Control Assoc. 2: 182-189.

Kerwin, J. L. & Washino, R. K. (1986b). Cuticular regulation of host recognition and spore germination by entomopathogenic fungi. In: Fundamental and applied aspects of invertebrate pathology (Eds. R. A. Samson, J. M. Vlak & D. Peters) Foundat. Fourth Int. Colloq. Invert. Path. Wageningen, Netherlands, pp. 423-425.

Kerwin, J. L. & Washino, R. K. (1986c). Regulation of oosporogenesis by *Lagenidium giganteum*: promotion of sexual reproduction by unsaturated fatty acids and sterol availability. Can. J. Microbiol. 32: 294-300.

Kerwin, J. L. & Washino, R. K. (1983). Sterol induction of sexual reproduction in *Lagenidium giganteum*. Exp. Mycol. 7: 109-115.

Killick, H. T. (1986). UV protectans in viral control of insects in relation to spray droplet size. In: Fundamental and applied aspects of invertebrate pathology (Eds. R. A. Samson, J. M. Vlak & D. Peters). Foundat. Fourth Int. Colloq. Invert. Path. Wageningen, Netherlands, pp. 620-623.

King, D. S. & Humber, R. A. (1981). Identification of the Entomophthorales. In: Microbial Control of Pests and Plant Diseases 1970-1980 (Ed. H. D. Burges), Academic Press, London & New York, pp. 107-127.

Kish, L. P. & Allen, G. E. (1978). The biology and ecology of *Nomuraea rileyi* and a program for predicting its incidence on *Anticarsia gemmatalis* in soybean. IFAS, Univ. Florida, 48 pp.

Kobayasi, Y. (1941). The genus *Cordyceps* and its allies. Science Reports of the Tokyo Bunrika Daigaku Sect. B. 84: 53-260.

Koidsumi, K. (1957). Antifungal action of cuticular lipids in insects. J. Insect Physiol. 1: 40-51.

Komano, H. D., Mizuno, D. & Natori, S. (1981). A possible mechanism of induction of insect lectin. J. Biol. Chem. 256: 7087-7089.

Kononova, E. V. (1979). Selection of commercial strains of the fungus, *Beauveria bassiana*. In: Proc. First joint US/USSR conference on the production, selection and standardization of entomopathogenic fungi of the US/USSR joint working group on the production of substances by microbiological means (Ed. C. M. Ignoffo), pp. 173-191.

Krassilstschik, I. M. (1888). La production industrielle des parasites végétaux pour la destruction des insectes nuisibles. Bull. Scient. Fr. Belg. 19: 461-472.

Kristiansen, B. & Chamberlain, H. E. (1983). Fermenter design. In: The filamentous fungi. Vol. 4 Fungal technology (Eds. J. E. Smith, D. R. Berry & B. Kristiansen), Edward Arnold Publ. London, pp. 1-19.

Lackie, A. M. (1980). Invertebrate immunity. Parasitology 80: 393-412.

Lackie, A. M. (1986). Immune mechanisms in invertebrate vectors. Oxford University Press, 300 pp.

Lackie, A. M. & Vasta, G. R. (1986). Biochemical characterization and biological role of a serum lectin from the cockroach *Periplaneta americana*. Dev. Comp. Immunol. 10, 631.

Lambiase, J. T. & Yendol, W. G. (1977). The fine structure of *Entomophthora apiculata* and its penetration of *Trichoplusia ni*. Can. J. Microbiol. 23: 452-464.

Latgé, J. P. (1975). Croissance et sporulation de 6 espèces d'Entomophthorales 1. Infuence de la nutrition carbonée. Entomophaga 20: 201-207.

Latgé, J. P. (1980). Sporulation de *Entomophthora obscura* Hall & Dunn en culture liquide. Can. J. Microbiol. 26: 1038-1048.

Latgé, J. P. (1981). Comparison des exigences nutritionelles des Entomophthorales. Ann. Microbiol. (Inst. Pasteur) 132B: 299-306.

Latgé, J. P. (1982). Production of Entomophthorales. Proc. Third Int. Colloquium Invert. Path. Brighton, pp. 164-169.

Latgé, J. P. (1983). *Conidiobolus obscurus* et les Entomophthorales pathogènes de pucerons. Thèse Doct Sci. Univ. Paris XI, 2 vol.

Latgé, J. P. & Moletta, R. (1983). Cinetique de la croissance mycelienne de *Conidiobolus obscurus*. Ann. Microbiol. (Inst. Pasteur) 134A: 267-279.

Latgé, J. P. & Papierok, B. (1982). The potential use of Chytridiomycete and Zygomycete fungi in vector control programmes. Proc. Third Int. Colloquium Invert. Path. Brighton, pp. 425-428.

Latgé, J. P. & Papierok, B. (1988). Fungal diseases and other pathogens. In: Aphids, their biology, natural Enemies and Control (Eds. P. Harrewijn & A. K. Minks), Elsevier, Amsterdam (in press).

Latgé, J. P. & Perry, D. F. (1980). Perfectionnements apportés aux procédés de préparation de spores durables d'Entomophthorales pathogènes d'insectes, préparation des spores ainsi obtenues et compositions phytosanitaires contenant les dites préparations. French Patent 80 24 769 (11.21.1980)

Latgé, J. P., & Sanglier, J. J. (1985). Optimisation de la croissance et de la sporulation de *Conidiobolus obscurus*. Can. J. Bot. 63: 68-85.

Latgé, J. P., Remaudière, G., Soper, R. S., Madore, C. D. & Diaquin, M. (1977a). Growth and sporulation of *Entomophthora virulenta* on semi-defined media. J. Invert. Path. 31: 225-233.

Latgé, J. P., Soper, R. S. & Madore, C. D. (1977b). Media suitable for industrial production of *Entomophthora virulenta* zygospores. Biotech. Bioengin. 19: 1269-1284.

Latgé, J. P., Perry, D., Papierok, B., Coremans-Pelseneer, J., Remaudière, G. & Reisinger, O. (1978a). Germination des azygospores de *Entomophthora obscura* Hall & Dunn, role du sol. C. R. Acad. Sci., Paris 287D: 943-946.

Latgé, J. P., Remaudière, G. & Diaquin, M. (1978b). Un nouveau milieu pour la croissance des champignons Entomophthorales pathogenès d'Aphides. Ann. Microbiol. (Inst. Pasteur) 129B: 463-476.

Latgé, J. P., Perry, D. Reisinger, O., Papierok, B. & Remaudière, G. (1979). Induction de la formation des spores de resistance d'*Entomophthora obscura* Hall & Dunn. C. R. Acad. Sci., Paris 288 D: 599-601.

Latgé, J. P., Papierok, B. & Sampedro, L. (1982). Agressivité de *Conidiobolus obscurus* vis à vis du puceron du pois I. Comportement des conidies sur la cuticule avant la pénétration du tube germinatif dans l'insecte. Entomophaga 27: 323-330.

Latgé, J. P., Silvie, P., Papierok, B., Remaudière, G., Dedryver, C. & Rabasse J. M. (1983). Advantages and disadvantages of *Conidiobolus obscurus* in the biological control of aphids. In: Aphid antagonists (Ed. R. Cavalloro), Balkema, Rotterdam, pp. 20-32.

Latgé, J. P., Fournet, B., Cole, G. T., Dubourdieu, D. & Tong, N. H. (1984a). Composition chimique et ultrastructurale des parois des corps hyphaux et des azygospores de *Conidiobolus obscurus*. Can. J. Microbiol. 30: 1507-1521.

Latgé, J. P., Monsigny, M., Prévost, M. C., Roche, A. C., Kieda, C. & Fournet, B. (1984b). Carbohydrate-binding proteins in the entomogenous fungus *Conidiobolus obscurus*. Biol. Cell 51: 52.

Latgé, J. P., Sampedro, L. & Hall, R. (1984c). Agressivité de *Conidiobolus obscurus* vis à vis du puceron du pois III. Activités enzymatiques exocellulaires. Entomophaga 29: 185-201.

Latgé, J. P., Beauvais, A. & Vey, A. (1986a). Wall synthesis in the Entomophthorales and its role in the immune reaction of infected insects. Dev. comp. Immunol., 10, 639.

Latgé, J. P, Cole, G. T., Horisberger, M. & Prévost, M. C. (1986b). Ultrastructure and chemical composition of the ballistospore wall of *Conidiobolus obscurus*. Exp. Mycol. 10: 99-113.

Latgé, J. P., Eilenberg, J., Beauvais, A. & Prévost, M. C. (1987a). *Entomophthora muscae* protoplasts grown in vitro. Protoplasma (in press).

Latgé, J. P., Sampedro, L., Brey, P. & Diaquin, M. (1987b). Aggressiveness of *Conidiobolus obscurus* against the pea aphid. Influence of cuticular compounds on spore germination. J. gen. Microbiol. 133 1987-1997.

Latgé, J. P., Cabrera Cabrera, R. I. & Prévost, M. C. (1988a). Microcycle conidiation in *Hirsutella thompsonii*. Can. J. Microbiol. (in press).

Latgé, J. P., Perry, D. F., Prevost, M. C. & R. A. Samson (1988b). Ultrastructural studies of ballistospores of *Conidiobolus*, *Erynia* and related Entomophthorales. Can. J. Bot. (submitted)

Laulan, A., Lestage, J., Bouc, A. M. & Chateaureynaud-Duprat, C. (1986). Evidence of antibody function and complement function in invertebrate protection mechanisms: *Lumbricus terrestris*. Dev. Comp. Immunol 10, 642.

Leonard, C., Ratcliffe, N. A. & Rowley A. F. (1985). The role of prophenoloxidase activation in non-self recognition and phagocytosis by insect blood cells. J. Insect Physiol. 31: 789-799.

Le Ru, B. (1986). The role of *Neozygites fumosa* in regulation of cassava meatybug populations. In: Fundamental and applied aspects of invertebrate pathology (Eds. R. A. Samson, J. M. Vlak & D. Peters), Foundat. Fourth Int. Colloq. Invert. Path., Wageningen, Netherlands, pp. 163-166.

Lewis, A. J. (1949). Natural control of the destructive sweetclover weevil *Sitona cylindricollis* Fabr. by an entomogenous fungus parasite. Phytopathology 39: 501.

Lichtwardt, R. W. (1986). The Trichomycetes. Fungal associates of arthopods. Springer Verlag, New York/Berlin.

Lightner, D. V. (1981). Fungal disease of marine crustacea. In: Pathogenesis of Vertebrate microbial diseases (Ed. E. W. Davidson), Allanheld, Osmun Publ. Ottawa, New Jersey, pp. 451-484.

Lisansky, S. G. & Hall, R. A. (1983). Fungal control of insects. In: The Filamentous Fungi. Vol 4. Fungal Technology. (Eds. J. E. Smith, D. R. Berry & B. Kristiansen), Edward Arnold Publ., London, pp. 337-345.

Livolant, F. Giraud, M. M. & Bouligand, Y. (1978). A goniometric effect observed in sections of twisted fibrous material. Biol. Cell. 31: 159-168.

Locke, M. (1984). Structure of insect cuticle. In: Infection processes of fungi (Eds. D. W. Roberts & J. R. Aist), Publs. Rockefeller Fdn, pp. 38-53.

Loos-Frank, B. & Zimmermann, G. (1976). Ueber eine dem Dicrocoelium-Befall analoge Verhaltensaenderung bei Ameisen der Gattung *Formica* durch einen Pilx der Gattung *Entomophthora*. Zeitschrift für Parasitenkunde 49: 281-289.

Los, L. M. & Allen, W. A. (1983). Incidence of *Zoophthora phytonomi* (Zygomycetes: Entomophthorales) in *Hypera postica* (Coleptera: Curculionidae) larvae in Virginia. Environ. Entomol. 12: 1318-1321.

MacLeod, D. M. (1956). Notes on the genus *Empusa* Cohn. Can. J. Bot. 34: 16-26.

MacLeod, D. M.. (1960). Nutritional studies on the genus *Hirsutella* II. Acid hydrolysed casein and casamino acids combination as sources of nitrogen. J. Invert. Path. 2: 139-146.

MacLeod, D. M. (1963). Entomophthorales infections. In: Insect pathology, an advanced treatise. Vol. 2. (Ed. E. A. Steinhaus), Academic Press, London & New York, pp. 189-231.

MacLeod, D. M., Tyrrell, D. & Welton, M. A. (1980). Isolation and growth of the grasshopper pathogen *Entomophthora grylli*. J. Invert. Path. 36: 85-89.

Maddox, I. S. & Richert, S. H. (1977). Use of response surface methodology for the rapid optimization of microbiological media. J. Appl. Microbiol. 43: 197-204.

Madelin, M. F. (1966). Fungal parasites of insects. A. Rev. Ent. 11: 423-448.

Magoon, J. & Messing-Al-Aidroos, K. (1984). Determination of ploidy of sectors formed by mitotic recombination in the entomopathogenic fungus *Metarhizium anisopliae*. Trans. Br. mycol. Soc. 82: 95-98.

Mains, E. B. (1958). North American entomogenous species of *Cordyceps*. Mycologia 50: 169-222.

Major, R. H. (1944). Agostino Bassi and the parasitic theory of disease. Bulletin of Historical Medicine 16: 97-107.

Marikovsky, P. I. (1962). On some features of behaviour of the ant *Formica rufa* L. infected with fungous disease. Insectes Sociaux 9: 173-179.

Marques, E. J., Villas Boas, A. M. & Pereira, C. E. F. (1981). Orientacoes tecnicas para a producao do fungo entomogeno *Metarhizium anisopliae* em laboratorios setoriais. Boletim Tecnico Planalsucar, Piracicaba 3: 5-23.

Massee, G. (1895). A revision of the genus *Cordyceps*. Ann. Bot. 9: 1-44.

Massee, G. (1901). South African locust fungus. Kew Bulletin 1901: 94-99.

Matcham, S. E., Jordan, B. R. & Wood, D. A. (1984). Methods for assessment of fungal growth on solid substrates. In: Microbial methods for environmental biotechnology (Eds. J. M. Grainger & J. M. Lynch), Academic Press, London & New York, pp. 5-18.

Matewele, P., Hall, R. A. & Burges, H. D. (1986). Sporulation of *Culicinomyces clavisporus* on various nitrogen and carbon sources in submerged culture. In: Fundamental and applied aspects of invertebrate pathology (Eds. R. A. Samson, J. M. Vlak & D. Peters). Foundation of the Fourth International Colloquium of Invertebrate Pathology, Wageningen, Netherlands. p. 216

Mathieson, J. (1949). *Cordyceps aphodii*, a new species, on pasture cockchafer grubs. Trans. Br. mycol. Soc. 32: 113-136.

Matta, A. (1982). Mechanisms in non-host resistance. In: Active defence mechanisms in plants (Ed. R. K. S. Wood), Plenum Press, New York & London, pp. 119-142.

McAlpine, D. (1900). The systematic position of the locust fungus imported from the Cape. Agricultural Gazette of N. S. W. Australia, Miscellaneous Publication no. 364: 1-3.

Mc Cabe, D. & Soper, R. S. (1985). Preparation of an entomopathogenic fungal insect control agent. US patent 4530 834 (July 23, 1985).

McCauley, V. J. E., Zaccharuck, R. Y. & Tinline, R. D. (1968). Histopathology of green muscardine in larvae of four species of Elateridae (Coleoptera). J. Insect Path. 12: 444-459.

McCoy, C. W. (1981). Pest control by the fungus *Hirsutella thompsonii*. In: Microbial control of pests and plant diseases 1970-1980 (Ed. H. D. Burges), Academic Press, London & New York, pp. 449-512.

McCoy, C. W. (1982). Hyphomycetes: field use and effectiveness. Proc. Third Int. Colloquium Invert. Path., Brighton, pp. 197-201.

McCoy, C. W. (1986). Factors governing the efficacy of *Hirsutella thompsonii* in the field. In: Fundamental and applied aspects of invertebrate pathology (Eds. R. A. Samson, J. M. Vlak & D. Peters). Foundat. Fourth Int. Colloquium Invert. Path., Wageningen, Netherlands, pp. 171-174.

McInnis, T. M. (1971). A physiological and biochemical investigation of the aquatic phycomycete *Lagenidium* sp., a facultative parasite of certain mosquito larvae. Ph. D. thesis, University of N. Carolina.

McGuire, M. R., Maddox, J. V., Morris, M. J. & Armbrust, E. J. (1986). The occurrence of *Erynia radicans* in an Illinois *Empoasca fabae* field population. In: Fundamental and applied aspects of invertebrate pathology (Eds. R. A. Samson, J. M. Vlak & D. Peters). Foundat. Fourth Int. Colloquium Invert. Path., Wageningen, Netherlands, p. 232.

Michel, B. (1981). Recherches expérimentales sur la pénétration des champignons pathogènes chez les insectes. Thèse 3e cycle. Univ. Sci. Tech. Languedoc (Montpellier).

Millstein, J. J., Brown, G. C. & Nordin, G. L. (1982). Microclimatic humidity influence on conidial discharge in *Erynia* sp. (Entomophthorales: Entomophthoraceae), an entomopathogenic fungus of the alfalfa weevil (Coleoptera: Curculionidae). Environ. Entomol. 11: 1166-1169.

Milner, R. J. (1982). On the occurrence of pea aphids, *Acyrthosiphon pisum* resistant to isolates of the fungal pathogen *Erynia neoaphidis*. Ent. Exp. Appl. 32: 23-27.

Minter, D. M., Brady, B. L. & Hall, R. A. (1983). Five hyphomycetes isolated from eriophyid mites. Trans. Br. mycol. Soc. 81: 455-471.

Monsigny, M. (1984). The role of carbohydrates in cell recognition, endogenous lectins. Biol. Cell 51: 294.

Moo-Young, M., Moreira, A. R. & Tengerdy, R. P. (1983). Principles of solid-substrate fermentation. In: The filamentous fungi. Vol 4. Fungal technology (Eds. J. E. Smith, D. R. Berry & B. Kristiansen), Edward Arnold Publ. London, pp. 117-144.

Morrill, A. W. & Back, E. A. (1912). Natural control of whiteflies in Florida. Bulletin of the USDA Bureau of Entomology 102: 1-78.

Morrow, B. J., Boucias, D. G. & Latgé, J. P. (1986). Analysis of in vitro growth of the dimorphic fungus *Nomuraea rileyi*: emphasis on cell wall composition. In: Fundamental and applied aspects of invertebrate pathology (Eds. R. A.

Samson, J. M. Vlak & D. Peters). Foundat. Fourth Int. Colloquium Invert. Path. Wageningen, Netherlands, p. 256.

Moss, S. T. & Descals, E. (1986). A previously undescribed stage in the lifecycle of Harpellales (Trichomycetes). Mycologia 78: 213-222.

Most, B. H. & Quinlan, R. J. (1986). Formulation of biological pesticides. In: Fundamental and applied aspects of invertebrate pathology (Eds. R. A. Samson, J. M. Vlak & D. Peters). Foundat. Fourth Int. Colloquium Invert. Path. Wageningen, Netherlands, pp. 624-627.

Natori, S. (1986). The role of lectins in immune reactions in insects. In: Fundamental and applied aspects of invertebrate pathology (Eds. R. A. Samson, J. M. Vlak & D. Peters). Foundat. Fourth Int. Colloquium Invert. Path., Wageningen, Netherlands, pp. 411-414.

Nicholson, R. L. (1984). Adhesion of fungi to the plant cuticle. In: Infection processes of fungi (Eds. D. W. Roberts & J. R. Aist), Publs. Rockefeller Fdn, pp. 74-89.

Nicholson, R. L. & Moraes, W. B. C. (1980). Survival of *Colletotrichum graminicola:* importance of the spore matrix. Phytopathology 70: 255-261.

Nolan, R. A. (1985). Protoplasts from Entomophthorales. In: Fungal protoplasts application in biochemistry and genetics (Eds. J. F. Peberdy & Ferenczy, L.), Marcel Dekker publ., New York, pp. 73-112.

Nordin, G. L., Brown, G. C. & Millstein, J. A. (1983). Epizootic phenology of *Erynia* disease of the alfalfa weevil, *Hyperica postica* (Gyllenhal)(Coleoptera: Curculionidae), in Central Kentucky. Environ. Entomol. 12: 1350-1355.

Nyhlen, L. & Unestam, T. (1975). Ultrastructure of the penetration of the crayfish integument by the fungal parasite *Aphanomyces astaci* oomycete. J. Invert. Path. 26: 353-366.

Ouchi, S. (1983). Recognition and specificity of plant disease. In: Infection process of fungi (Eds. D. W. Roberts & J. R. Aist). Publ. Rockefeller Fdn., pp. 175-184.

Otvos, I. S., Macleod, D. M. & Tyrell, D. (1973). Two species of *Entomophthora* pathogenic to the eastern hemlock looper (Lepidoptera: Geometridae) in Newfoundland. Can. Ent. 105: 1435-1441.

Pace, G. W. (1980). Rheology of mycelial fermentation broths. In: The filamentous fungi. Vo.. 4. Fungal biotechnology (Eds. J. E. Smith, D. R. Berry & B. Kristiansen), Academic Press, London & New York, pp. 95-110.

Pace, G. W. & Righelato, R. C. (1980). Production of extracellular microbial polysaccharides. Adv. Biochem. Eng. 15: 41-70.

Pant, C. (1986). Ecology and control of vectors: use of biological agents as a component of integrated vector control programmes. In: Fundamental and applied aspects of invertebrate pathology (Eds. R. A. Samson, J. M. Vlak & D. Peters). Foundat. Fourth Int. Colloquium Invert. Path., Wageningen, Netherlands, pp. 501-509.

Papierok, B. (1982). Entomophthorales: virulence and bioassay design. Proc. Third Int. Colloquium Invert. Path., Brighton, pp. 176-181.

Paris, S. (1980). Etude physiologique, biochimique et génétique des caractères de *B. brongniartii* (Sacc.) Petch liés à la pathogénicité de ce champignon pour le hanneton commun *Melolontha melolontha*. Thèse Doct Sc. Université Paris XI.

Paris, S. & Ferron, P. (1979). Study of the virulence of some mutants of *Beauveria brongniartii* (= *B. tenella*). J. Invert. Path. 34: 71-77.

Parkin, J. (1906). Fungi parasitic upon scale insects (Coccidae and Aleyrodidae): a general account with special reference to Ceylon forms. Annls Roy. Bot. Gard., Peradeniya 3: 11-82.

Pekrul, S. & Grula, E. A. (1979). Mode of infection of the corn earworm *(Heliothis zea)* by Beauveria bassiana as revealed by scanning electron microscopy. J. Invert. Path. 34: 238-247.

Pendland, J. C. (1982). Resistant structures in the entomogenous hyphomycete *Nomuraea rileyi:* an ultrastructural study. Can. J. Bot. 60: 1569-1576.

Pendland, J. C. & Boucias, D. G. (1984a). Ultrastructure of conidial germination in the entomopathogenic fungus *Nomuraea rileyi*. J. Invert. Path. 43: 432-434.

Pendland, J. C. & Boucias, D. G. (1984b). Use of labeled lectins to investigate cell wall surface of the entomogenous hyphomycete *Nomuraea rileyi*. Mycopathologia 87: 141-148.

Pendland, J. C. & Boucias, D. G. (1985) Hemagglutinin activity in the hemolymph of *Anticarsia gemmatalis* larvae infected with the fungus *Nomuraea rileyi*. Dev. Comp. Immunol. 9: 21-30.

Pendland, J. C. & Boucias, D. G. (1986a). Lectin binding characteristics of several entomogenous hyphomycetes: possible relationship to insect hemagglutinins. Mycologia 78: 818-824.

Pendland, J. C. & Boucias, D. G. (1986b). Characteristics of a galactose-bomding hemagglutinin from hemolymph of *Spodoptera exigua* larvae. Dev. Comp. Immunol. 10, 477-487.

Perry, D. F. & Latgé, J. P. (1982). Dormancy and germination of *Conidiobolus obscurus* azygospores. Trans. Br. mycol. Soc. 78: 221-225.

Perry, D. F. & Régnière, J. (1986). The role of fungal pathogens in spruce budworm population dynamics: frequency and temporal relationships. In: Fundamental and applied aspects of invertebrate pathology (Eds. R. A. Samson, J. M. Vlak & D. Peters). Foundat. Fourth Int. Colloquium Invert. Path., Wageningen, Netherlands, pp. 167-170.

Perry, D. F. & Whitfield, G. H. (1984). The interrelationships between microbial entomopathogens and insect hosts: a system study approach with particular reference to the Entomophthorales and the eastern spruce budworm. In: Invertebrate-microbial interactions (Eds. J. M. Anderson, A. D. M. Rayner & D. W. H. Walton), Cambridge Univ. Press, pp. 307-331.

Persson, M. & Söderhäll, K. (1983). *Pacifastacus leniusculus* and its resistance to the parasitic fungus *Aphanomyces astaci* Sikora. In: Freshwater crayfish v. (Ed. C. Goldman), AVI Publishing Company, Westport, Conn., pp. 292-298.

Persson, M., Hall, L. & Söderhäll, K. (1984). Comparison of peptidase activities in some fungi pathogenic to arthropods. J. Invert. Path. 44: 342-348.

Petch, T. (1925). Entomogenous fungi and their use in controlling insects pests. Bulletin of the Department of Agriculture, Ceylon no 71, Government Printer, Colombo, 40 pp.

Petch, T. (1935). Notes on entomogenous fungi. Trans. Br. mycol. Soc. 19: 161-194.

Petch, T, (1937). Notes on entomogenous fungi. Trans. Br. mycol. Soc. 21: 34-67.

Pierre, J. S. & Dedryver, C. A. (1984). Un modèle de regression multiple appliqué à la prévision des pullulations d'un puceron des céréales, *Sitobion avenae* F. sur blé d'hiver. Acta Oecologica oecol. Applic. 5: 153-172.

Pillai, J. S. (1982). The biology and pathology of the imperfect fungi with vector control potential. Proc. Third Int. Colloquium Invert. Path., Brighton, pp. 404-408.

Pirt, S. J. (1975). Principles of microbe and cell cultivation. Blackwell Sc. Publication, Oxford, 274 pp.

Poinar, G. O. & Thomas, G. M: (1982). An entomophthoralean fungus from Dominican amber. Mycologia 74: 332-334.

Poinar, G. O. & Thomas, G. M. (1984). A fossil entomogenous fungus from Dominican amber. Experientia 40: 578-579.

Poprawski, T. J., Maniania, N. K. & Robert, P. H. (1986). Entomotoxicity of destruxin to nymphs of the leafhopper, *Empoasca vitis*, (Homoptera: Cicadellidae). In: Fundamental and applied aspects of invertebrate pathology (Eds. R. A. Samson, J. M. Vlak & D. Peters), Foundat. Fourth Int. Colloquium Invert. Path., Wageningen, Netherlands, pp. 259.

Powell, M. J. (1976). Ultrastructural changes with cell surface of *Coelomomyces punctatus* infecting mosquito larvae. Can. J. Bot. 54: 1419-1437.

Rabasse, J. M. & Dedryver, C. A. (1982). Facteurs de limitations des populations d'*Aphis fabae* dans l'ouest de la France. IV. Nouvelles données sur le déroulement des épizooties à Entomophthoraceae sur féverole de Printemps. Entomophaga 27: 39-53.

Ratault, C. & Vey, A. (1977). Production d'estérase et de N-acétyl-glucosaminidase dans le tégument du coleoptère *Oryctes rhinoceros* par le champignon *Metarhizium anisopliae*. Entomophaga 22: 289-294.

Ratcliffe, N. A. (1985). Invertebrate immunity, a primer for the non-specialist. Immunol. Letters 10: 253-270.

Ratcliffe, N. A. & Rowley, A. F. (1979). Role of hemocytes in defence against biological agents. In: Insect Hemocytes (Ed. A. P. Gupta), Cambridge Univ. Press, pp. 331-414.

Ratcliffe, N. A., Leonard, C. & Rowley, A. F. (1984). Prophenoloxidase activation: non-self recognition and cell cooperation in insect immunity. Science 226: 557-559.

Remaudière, G. & Keller, S. (1980). Revision systématique des genres d'Entomophtoraceae à potentialité entomopathogène. Mycotaxon 11: 323-338.

Remaudière, G., Latgé, J. P. & Michel, M. F. (1981). Ecologie comparée des Entomophthoracées pathogènes de pucerons en France littorale et continentale. Entomophaga 26: 157-178.

Renwrantz, L. (1983). Involvement of agglutinins (lectins) in invertebrate defence reactions: the immuno-biological importance of carbohydrate-specific binding molecules. Dev. Comp. Immunol. 7: 603-608.

Reuss, M. Debus, D. & Zoll, G. (1982). Rheological properties of fermentation fluids. Chem. Eng. 381: 233-236.

Riba, G. & Glandard, A. (1980). Mise au point d'un milieu nutritif pour la culture profonde du champignon entomopathogène *Nomuraea rileyi*. Entomophaga 25: 317-322.

Riba, G., Glandard, A., Ravelojaon, A. M. & Ferron, P. (1980). Isolement de recombines mitotiques stables de type 'intermédiaire' chez *Metarhizium anisopliae* (Metchnikoff) par hybridation de biotypes sauvages. C. R. Acad. Sci., Paris 291 D: 657-660.

Righelato, R. C. (1975). Growth kinetics of mycelial fungi. In: The filamentous fungi. Vol 1. Industrial mycology (Eds. J. E. Smith & D. R. Berry), Edward Arnold Publ., London, pp. 79-103.

Righelato, R. C. (1978). The kinetics of mycelial growth. In: Fungal walls and hyphal growth (Eds. J. H. Burnett & A. P. J. Trinci), Cambridge Univ. Press, pp. 385-401.

Roberts, D. W. (1981). Toxins of entomopathogenic fungi. In: Microbial control of pests and plant diseases 1970-1980 (Ed. H. D. Burges), Academic Press, London & New York, pp. 441-464.

Roberts, D. W. & Humber, R. A. (1984). Entomopathogenic fungi. In: Infection processes of fungi (Eds. D. W. Roberts and J. R. Aist), Publs. Rockefeller Fdn, pp. 1-12.

Roberts, D. W. & Sweeney, A. W. (1982). Production of fungi imperfecti with vector control potential. Proc. Third Int. Colloquium Invert. Path. Brighton, pp. 409-413.

Roberts, D. W. & Wraight, S. P. (1986). Current status on the use of insect pathogens as biocontrol agents in agriculture: fungi. In: Fundamental and applied aspects of invertebrate pathology (Eds. R. A. Samson, J. M. Vlak & D. Peters), Foundat. Fourth Int. Colloquium Invert. Path. Wageningen, Netherlands, pp. 510-513.

Rolfs, P. H. (1897). A fungus disease of the San José scale *(Sphaerostilbe coccophila* Tul.). Bulletin of the Florida Agricultural Experiment Station no 41: 513-542.

Rolfs, P. H. & Fawcett, H. S. (1908). Fungus diseases of scale insects and whitefly. Bulletin of the Florida Agricultural Experiment Station no. 94: 1-17.

Rorer, J. B. (1913). The use of the green muscardine in the control of some sugar cane pests. Phytopathology 3: 88-92.

Saito, T. & Aoki, J. (1983). Toxicity of free fatty acids on the larval surface of two lepidopterous insects towards *Beauveria bassiana* (Bals.) Vuill. and *Paecilomyces fumosoroseus* (Wize) Brown et Smith (Deuteromycetes: Moniliales) Appl. Ent. Zool. 18: 225-233.

Sampedro, L., Uziel, A. & Latgé, J. P. (1984). Agressivité de *Conidiobolus obscurus* vis a vis du puceron du pois II. Mode de germination in vitro des conidies primaires de souches d'aggressivités différentes. Mycopathologia 86: 3-19.

Samsinakova, A., Bajan, C., Kalakova, S., Kmitowa, K. & Wojciechowska, M. (1977). The effect of some entomophagous fungi on the Colorado potato beetle and their enzyme activity. Bull. Acad. pol. Sci. 25: 521-526.

Samsinakova, A., Misikova, S. & Leopold, J. (1971). Action of enzymatic systems of *Beauveria bassiana* on the cuticle of the greater wax moth larvae *(Galleria mellonella)*. J. Invert. Path. 18: 322-330.

Samson, R. A. (1982). Laboratory culture and maintenance of entomopathogenic fungi. Proc. Third Int. Colloquium Invert. Path., Brighton, pp. 182-187.

Samson, R. A., McCoy, C. W. & O'Donnell, K. L. (1980). Taxonomy of the acarine parasite *Hirsutella thompsonii*. Mycologia 72: 359-377.

Samson, R. A. & Evans, H. C. (1973). Notes on entomogenous fungi from Ghana. I. The genera *Gibellula* and *Pseudogibellula*. Acta bot. Neerl. 22: 522-528.

Samson, R. A., Evans, H. C. & van de Klashorst, G. (1981). Notes on entomogenous fungi from Ghana V. The genera *Stilbella* and *Polycephalomyces*. Proc. Kon. Nederl. Akad. Wetensch. Ser. C 84: 289-301.

Samson, R. A., Evans, H. C. & Hoekstra, E. S. (1982). Notes on entomogenous fungi from Ghana VI. The genus *Cordyceps*. Proc. Kon. Nederl. Akad. Wetensch. Ser. C 85: 589-605.

Samson, R. A. & Evans, H. C. (1982). Two new *Beauveria* spp. from South America. J. Invert. Path. 39: 93-97.

Samson, R. A., Rombach, M. C. & Seifert, K. A. (1984). *Hirsutella guignardii* and *Stilbella kervillei*, two troglobiotic entomogenous Hyphomycetes. Persoonia 12: 123-134.

Samson, R. A. & Rombach, M. C. (1985). Biology of the fungi *Verticillium* and *Aschersonia*. In: Biological pest control, (Eds. N. W. Hussey & N. Scopes). Blandford Press, Poole, Dorset, pp. 34-42.

Samuels, K. D. Z., Heale, J. B. & Llewellyn, M. J. (1985). Strain selection and improvement in *Metarhizium anisopliae* for enhanced pathogenicity towards *Nilaparvata lugens*. Abstr. XVIII SIP meeting, Sault Ste Marie, p. 32.

Schabel, H. G. (1978). Percutaneous infection of *Hylobius pales* by *Metarhizium anisopliae*. J. Invert. Path. 31: 180-187.

Seymour, R. L. (1984). *Leptolegnia chapmanii*, an oomycete pathogen of mosquito larvae. Mycologia 76: 670-674.

Sinclair, C. G. & Mavituna, F. (1983). Mass and energy transfer. In: The filamentous fungi. Vol 4. Fungal technology (Eds. J. E. Smith, D. R. Berry & B. Kristiansen). Edward Arnold Publ. London. pp. 20-76.

Smith, J. E. (1978). Asexual sporulation in filamentous fungi. In: The filamentous fungi. Vol 3. Developmental mycology (Eds. J. E. Smith & D. R. Berry), Edward Arnold Publ., London, pp. 214-239.

Smith, J. E., Anderson, J. G., Deans, S. G. & Berry, D. R. (1981). Biochemistry of microcycle conidiation. In: Biology of conidial fungi. (Eds. G. T. Cole & B. Kendrick). Academic Press, New York & London, pp. 329-356.

Smith, R. J., Pekrul, S. & Grula, E. A. (1981). Requirement for sequential enzymatic activities for penetration of the integument of the corn earworm (*Heliothis zea*). J. Invert. Path. 38: 335-344.

Smith, R. J. & Grula, E. A. (1982). Toxic components on the larval surface of the corn earworm (*Heliothis zea*) and their effects on germination of *Beauveria bassiana*. J. Invert. Path. 39: 15-22.

Smith, V. J. and Söderhäll, K. (1983a). ß-1,3-glucan activation of crustacean haemocytes in vitro and in vivo. Biol. Bull. 164: 299-314.

Smith, V. J. & Söderhäll, K. (1983b). Induction of degranulation and lysis of haemocytes in the freshwater crayfish *Astacus* astacus by components of the prophenoloxidase activating system in vitro. Cell Tiss. Res. 233: 295-303.

Snow, F. H. (1896). Contagious diseases of the chinch bug. Ann. Rept. dir. Kansas Univ. Expt. Sta. 5: 7-55.

Söderhäll, K. (1981). Fungal cell wall ß 1-3-glucans induce clotting and phenoloxidase attachment to foreign surfaces of crayfish hemocyte lysate. Dev. Comp. Immunol. 5: 565-573.

Söderhäll, K. (1982). Prophenoloxidase activating system and melanization: a recognition mechanism of arthropods? A review. Dev. Comp. Immunol. 6: 601-611.

Söderhäll, K. (1983). ß-1,3-glucan enhancement of protease activity in crayfish hemocyte lysate. Comp. Biochem. Physiol. ß 74: 221-224.

Söderhäll, K. & Ajaxon, R. (1982). Effect of quinones and melanin on mycelial growth of *Aphanomyces* spp and extra cellular protease of *A. astaci*, a parasite on crayfish. J. Invert. Path. 39: 105-109.

Söderhäll, K. & Smith, V. J. (1983). Separation of the haemocyte populations of *Carcinus maenes* and other marine decapods, and prophenoloxidase distribution. Dev. Comp. Immunol. 7: 229-239.

Söderhäll, K. & Smith, V. J. (1986). The prophenoloxidase activity system: the biochemistry of its activation and role in arthropod cellular Immunity, with special references to crustaceans. In: Invertebrate Immunity (Ed. M. Brehelin), Springer Verlag, Berlin, pp. 208-223.

Söderhäll, K., Hall, L., Unestam, T. & Nyhlen, L. (1979). Attachment of phenoloxidase to fungal cell walls in arthropod immunity. J. Invert. Path. 34: 285-294.

Söderhäll, K., Vey, A. & Ramstedt, M. (1984). Hemocyte lysate enhancement of fungal spore encapsulation by crayfish hemocytes. Dev. Comp. Immunol. 8: 23-29.

Soper, R. S. (1974). The genus *Massospora*, entomopathogenic for cicadas. Part I. Taxonomy of the genus. Mycotaxon 1: 13-40.

Soper, R. S. (1982). Commercial mycoinsecticides. Proc. Third Int. Colloquium Invert. Path. Brighton, pp. 98-102.

Soper, R. S., Delyzer, A. J. & Smith, L. F. R. (1976). The genus *Massospora*, entomopathogenic for cicadas. Part II. Biology of *Massospora levispora* and its host *Okanagana rimosa* with notes on *Massospora cicadina* on the periodical cicadas. Annals entomol. Soc. America 69: 89-95.

Soper, R. S. & MacLeod, D. M. (1981). Descriptive epizootiology of an aphid mycosis. USDA Technical Bulletin No. 1632, 17 pp.

Soper, R. S. & Ward, M. G. (1981). Production, formulation and application of fungi for insect control. In: Biological control in crop production. (Ed. G. C. Papavizas), Allanheld, Osmun Publ., Ottawa, New Jersey, pp. 161-180.

South, F. W. (1910). The control of scale insects in the British West Indies by means of fungoid parasites. West Indian Bulletin 11: 1-30.

Speare, A. T. (1912). Fungi parasitic upon insects injurious to sugarcane. Report of the experiment station of the Hawaiian Sugar Planters Association Bulletin 12: 1-62.

Speare, A. T. & Colley, R. H. (1912). The artificial use of the brown-tail fungus in Massachusetts. Wright and Potter, Boston, 29 pp.

Speare, A. T. & Yothers, W. W. (1924). Is there an entomogenous fungus attacking the citrus rust mite in Florida? Science 60: 41-42.

Steinhaus, E. A. (1946). Insect microbiology. Comstock Publishing Co., Ithaca, New York, 763 pp.

Steinhaus, E. A. (1949). Principles of Insect Pathology. McGraw-Hill Book Co., New York, 757 pp.

Steinhaus, E. A. (1956). Microbial control, the emergence of an idea. Hilgardia 26: 107-157.

Steinhaus, E. A. (1975). Disease in a minor chord. Ohio State University Press, Columbus, 488 pp.

St. Leger, R. J., Charnley, A. K. & Cooper, R. M. (1986a). Cuticle-degrading enzymes of entomopathogenic fungi: synthesis in culture on cuticle. J. Invert. Path. 48: 85-95.

St. Leger, R. J., Cooper, R. M. & Charnley, A. K. (1986b). Cuticle degrading enzymes of entomopathogenic fungi: cuticle degradation in vitro by enzymes from entomopathogens. J. Invert. Path. 47: 167-177.

St. Leger, R. J., Charnley, A. K. & Cooper, R. M. (1986c). Mechanisms of interaction with insect cuticle. J. Invert. Path. 47: 295-301.

St. Leger, R. J., Charnley, A. K. & Cooper, R. M. (1986d). Proteases as pathogenicity determinants of entomopathogenic fungi. In: Fundamental and applied aspects of invertebrate pathology (Eds. R. A. Samson, J. M. Vlak & D. Peters). Foundat. Fourth Int. Colloquium Invert. Path., Wageningen, Netherlands, pp. 428-431.

Strand, M. A., Bailey, C. H. & Laird, M. (1977). Pathogens of Simuliidae (black flies). WHO Bulletin 55: 213-237.

Swan, D. I. (1974). A review of the work on predators, parasites and pathogens for the control of *Oryctes rhinoceros* (L.) in the Pacific area. Commonwealth Institute of Biological Control, Miscellaneous Publication no. 7, 64 pp.

Sweeney, A. W. (1981). An undescribed species of *Smittium* (Trichomycetes) pathogenic to mosquito larvae in Australia. Trans. Br. mycol. Soc. 77: 55-60.

Sweeney, A. W. (1982). Field evaluation of fungal pathogens of mosquito larvae with particular reference to *Culicinomyces*. Proc. Third Int. Colloquium Invert. Path., Brighton, pp. 414-418.

Sweeney, A. W. (1986). Laboratory and field observations on persistence and recycling potential of *Culicnomyces clavisporus*. In: Fundamental and applied aspects of invertebrate pathology (Eds. R. A. Samson, J. M. Vlak & D. Peters). Foundat. Fourth Int. Colloquium Invert. Path., Wageningen, Netherlands, pp. 175-177.

Sweeney, A. W., Inman, A. O., Bland, C. E. & Wright, R. G. (1984). The fine structure of *Culicinomyces clavisporus* invading mosquito larvae. J. Invert. Path. 42: 224-243.

Tarrant, C. A. & Soper, R. (1986). Evidence for the vertical transmission of *Coelomycidium simulii* (Myceteae: Chytridiomycetes). In: Fundamental and applied aspects of invertebrate pathology (Eds. R. A. Samson, J. M. Vlak & D. Peters). Foundat. Fourth Int. Colloquium Invert. Path., Wageningen, Netherlands, p. 212.

Teh, J. S. (1974). Toxicity of short-chain fatty acids and alcohols towards *Cladosporium resinae*. Appl. Microbiol. 28: 840-844.

Thaxter, R. (1888). The Entomophthoreae of the United States. Memoirs of the Boston Society of Natural History 4: 133-201.

Thaxter, R. (1896). Contribution toward a monograph of the Laboulbeniaceae, Part. I. American Academy of Arts and Science 12: 187-429.

Thaxter, R. (1908). Contribution toward a monograph of the Laboulbeniaceae, Part. II. American Academy of Arts and Science 13: 219-469.

Thaxter, R. (1924). Contribution toward a monograph of the Laboulbeniaceae, Part. III. American Academy of Arts and Science 14: 431-580.

Thaxter, R. (1931). Contribution toward a monograph of the Laboulbeniaceae, Part. IV. American Academy of Arts and Science 16: 1-435.

Touzé, A. & Esquerré-Tugayé, M. T. (1982). Defence mechanisms of plants against varietal non-specific pathogens. In: Active defence mechanisms in plants (Ed. R. K. S. Wood). Plenum Press, New York, pp. 103-117.

Travland, L. B. (1979a). Initiation of infection of mosquito larvae *(Culiseta inornata)* by *Coelomomyces psorophorae*. J. Invert. Path. 33: 95-105.

Travland, L. B. (1979b). Structure of the motile cells of *Coelomomyces psorophorae* and function of the zygote in encystment on a host. Can. J. Bot. 57: 1021-1035.

Trinci, A. P. J. (1971). Infuence of the width of the peripheral growth zone on the radial growth rate of fungal colonies on solid media. J. gen. Microbiol. 67: 325-344.

Trinci, A. P. J. (1978). The duplication cycle and vegetative development in moulds. In: The filamentous fungi. Vol. 3. Developmental mycology (Eds. J. E. Smith & D. R. Berry). Edward Arnold Publ., London, pp. 132-163.

Tyrrell, D. & MacLeod, D. M. (1972). Spontaneous formation of protoplasts by a species of *Entomophthora*. J. Invert. Path. 19: 354-360.

Ullyett, G. C. & Schonken, D. B. (1940). A fungus disease of *Plutella maculipennis* Curt. in South Africa, with notes on the use of entomogenous fungi in insect control. Science Bulletin of the South African Department of Agriculture and Forestry No. 21, 24 pp.

Umphlett, C. J. (1973). A note to identify a certain isolate of *Lagenidium* which kills mosquito larvae. Mycologia 65: 970-972.

Unestam, T. (1965). Studies on the crayfish plague fungus *Aphanomyces astaci* I. Some factors affecting growth in vitro. Physiol. Plant. 18: 483-505.

Unestam, T. & Ajaxon, R. (1975). Phenol oxidation in soft cuticle and blood of crayfish compared with that in other arthropods and activation of the phenoloxidases by fungal and other cell walls. J. Invert. Path. 27: 287-295.

Uziel, A., Schwartz, A. & Kenneth R. G. (1981). Positive phototropism and hydromorphogenesis in germination structures of *Conidiobolus* species. Israel J. Bot. 30: 75-80.

Uziel, A. & Kenneth, R. (1986). Survival capacity of capilli conidia over primary conidia at low humidity in *Erynia* (Subgen. *Zoophthora*) and *Neozygites fresenii* (Zygomycetes: Entomophthorales). In: Fundamental and applied aspects of invertebrate pathology (Eds. R. A. Samson, J. M. Vlak & D. Peters), Foundat. Fourth Int. Colloquium Invert. Path., Wageningen, Netherlands, p. 231.

Van den Bosch, R. (1971). Biological control of insects. A. Rev. Ecol. System. 2: 45-66.

Van den Ende, H. (1978). Sexual morphogenesis in the Phycomycetes. In: The filamentous fungi. Vol 3. Developmental mycology (Eds. J. E. Smith & D. R. Berry), Edward Arnold Publ., London, pp. 257-274.

Van Winkelhoff, A. J. & McCoy, C. W. (1984). Conidiation of *Hirsutella thompsonii* var. *synnematosa* in submerged culture. J. Invert. Path. 43: 59-68.

Vey, A. (1971a). Recherches sur la réaction hémocytaire anticryptogamique de type granulome chez les insectes. Thèse Doct es Sciences Université Paul Sabatier, Toulouse, 250 pp.

Vey, A. (1971b). Etude des reactions cellulaires anticryptogamiques chez *Galleria mellonella* L.: structure et ultrastructure des granulomes à *Aspergillus niger* v. Tiegh. Annls zool. ecol. anim. 3: 17-30.

Vey, A. (1977). Etude in vitro des réactions hémocytaires d'invertébrès et de leur sensibilité à certaines influences biochimiques. Ann. Parasitol. 52: 75-77.

Vey, A. & Fargues, J. (1977). Histological and ultrastructural studies of *Beauveria bassiana* infection in *Leptinotarsa decemlineata* larvae during ecdysis. J. Invert. Path. 31: 207-215.

Vey, A. & Götz, P. (1975). Humoral encapsulation in Diptera (Insecta): comparative studies in vitro. Parasitology 70: 77-86.

Vey, A. & Vago, C. (1971). Reactions anticryptogamiques de type granulome chez les insectes. Ann. Inst. Pasteur 121: 527-532.

Vey, A., Quiot, J. M. & Vago, C. (1973). Mise en évidence et étude de l'action d'une mycotoxine, la beauvericine, sur des cellules d'insects cultivées in vitro. C. R. Acad. Sci., Paris 276 D: 2489-2492.

Vey, A. Bouletreau, M., Quiot, J. M. & Vago C. (1975). Etude in vitro en microcinématographie vis à vis d'agents bacteriens et cryptogamiques. Entomophaga 20: 337-351.

Viegas, A. P. (1939). Um amigo do fazendeiro *Verticillium lecanii* (Zimm.) N. Comb. o causador do halo branco do *Coccus viridis* Green. Revista do Instituto do Cafe, Sao Paulo 14: 754-772.

Warner, S. A. & Domnas, A. J. (1981). Evidence for a cycloartenol- based sterol synthetic pathway in *Lagenidium* spp. Exp. Mycol. 5: 184-188.

Weiser, J. & Pillai, J. S. (1981). *Tolypocladium cylindrosporum* (Deuteromycetes, Moniliaceae) a new pathogen of mosquito larvae. Entomophaga 26: 357-361.

Weiser, J., Matha, V., Tryachov, N. D. & Gelbic, I. (1985). *Entomophthora grylli* destruction of locust *Oxya-hyla intricata* in Vietnam. International Rice Research Newsletter 10:16-17.

WHO (1984). Report of the seventh meeting of the scientific working group on biological control of vectors, Geneva, 31 pp.

Wilding, N. (1975). *Entomophthora* infecting pea aphids. Trans. Royal entomol. .Soc. London 127: 171-183.

Wilding, N. (1981). Pest control by Entomophthorales. In: Microbial control of pests and plant diseases, 1970-1980 (Ed. H. D. Burges), Academic Press, London, New York. pp. 539-554.

Wilding, N. (1982). Entomophthorales: field use and effectiveness. Proc. Third Int. Colloquium Invert. Path., Brighton, pp. 170-175.

Wilding, N., Latteur, G. & Dedryver, C. A. (1986a). Evaluation of Entomophthorales for aphid control: laboratory and field data. In: Fundamental and applied aspects of invertebrate pathology (Eds. R. A. Samson, J. M. Vlak & D. Peters). Foundat. Fourth Int. Colloquium Invert. Path. Wageningen, Netherlands, pp. 159-162.

Wilding, N., Mardell, S. K. & Brobyn, P. J. (1986b). Introducing *Erynia neoaphidis* into a field population of *Aphis fabae*: form of the inoculum and effect of irrigation. Ann. Appl. Biol. 108: 373-385.

Willets, H. J. (1978). Sclerotium formation. In: The filamentous fungi. Vol 3. Developmental mycology (Eds. J. E. Smith & D. R. Berry). Edward Arnold Publ., London, pp. 197-213.

Willis, J. H. (1959). Australian species of the fungal genus *Cordyceps* (Fr.) Link. Muelleria 1: 67-89.

Wittler, R., Baumgartl, H., Schugerl, K. & Lubbers, D. W. (1984). Oxygen transfer in *Penicillium chrysogenum* pellets. Proc. Third Congres on Biotechnology 1: 513-520.

Woods, S. P. & Grula, E. A. (1984). Utilizable surface nutrients on *Heliothis zea* available for growth of *Beauveria bassiana*. J. Invert. Path. 43: 259-269.

Wraight, S. P., Galaini-Wraight, S., Carruthers, R. I. & Roberts, D. W. (1986). Field transmission of *Erynia radicans* to Empoasca leafhopper in alfalfa following application of a dry mycelial preparation. In: Fundamental and applied aspects of invertebrate pathology (Eds. R. A. Samson, J. M. Vlak & D. Peters), Foundat. Fourth Int. Colloquium Invert. Path. Wageningen, Netherlands, p. 233.

Yeaton, R. (1981). Invertebrate lectins II. Diversity of specificity, biological synthesis and function in recognition. Dev. Comp. Immunol. 5: 535-545.

Yoshida, H., Ochiai, M. & Ashida, M (1986). β-1,3-glucan receptor and peptidoglycan receptor are present as separate entities within insect prophenoloxidase activating system. Biochem. Biophys. Res. Commun., 143, 1177-1184.

Zaccharuck, R. Y. (1970a). Fine structure of the fungus *Metarrhizium anisopliae* infecting three species of larval Elateridae (Coleoptera) II. Conidial germ tubes and appressoria. J. Invert. Path. 15: 81-91.

Zaccharuck, R. Y. (1970b). Fine structure of the fungus *Metarrhizium anisopliae* infecting three species of larval Elateridae (Coleoptera) III. Penetration of the host integument. J. Invert. Path. 15: 372-396.

Zaccharuck, R. Y. (1981). Fungal diseases of terrestrial insects. In: Pathogenesis of invertebrate microbial disease (Ed. E. D. W. Davidson), Allanhed, Osmun Publ. pp. 367-402.

Zebold, S. L., Whisler, H. C., Shemanchuk, J. & Travland, L. B. (1979). Host specificty and penetration in the mosquito pathogen *Coelomomyces psorophorae*. Can. J. Bot. 57: 2766-2770.

Zimmermann, G. (1981). Gewachshausversuche zur Bekampfung der gejunchten Dickmanrublers, *Otiorhynchus sulcatus* (F.) mit dem Pilz *Metarhizium anisopliae*. Nachrichtenbl. Deutsch. Pflanzenschutzd. 33: 103-108.

Zimmermann, G. & Simons, W. R. (1986). Experiences with biological control of the black vine weevil, *Otiorhynchus sulcatus* (F). In: Fundamental and applied aspects of invertebrate pathology (Eds. R. A. Samson, J. M. Vlak & D. Peters). Foundat. Fourth Int. Colloquium Invert. Path. Wageningen, Netherlands, pp. 529-533.

Index

Abacarus hystrix, 148
Abarus, 96
Acanthoscelides obtectus, 132
Acantocyclops vernalis, 19
Achillea millefolia, 145
Acremonium, 5, 10, 16
Acridium purpuriferum, 168
Acrostalagmus, 14
Acyrthosiphon pisum, 45, 133
Aedes, 140, 141, 170
    A. aegypti, 22, 125
Aegerita, 5, 7
    A. webberi, 167
Aeneolamia, 166, 169
    A. albo-fasciata, 149
    A. saccharina, 165
Aethus species, 112
agricultural ecosystems, 145
Agrotis segetum, 48
Agrotis sputator, 43
Akanthomyces, 5, 9, 10, 15, 96
    A. aculeatus, 18, 75
    A. gracilis, 18, 76
    A. pistillariiformis, 18, 77
allergic responses, 153
Amoebidiales, 26
Amoebidium parasiticum, 17, 26
anamorphs, 7
Anisoplia austriaca, 146
Anopheles, 140, 170
    A. quadrimaculatus, 19
Anticarsia gemmatalis, 130, 132, 134, 135, 146
Aphanomyces, 128
    A. astaci, 17, 23, 135
Aphides, 169
Aphis fabae, 50, 147, 169

Aphodius howitti, 148
Aphodius tasmaniae, 166
aquatic Entomophthorales, 141
aquatic habitats, 140
arable crops, 146
Aschersonia, 5, 11, 15, 143, 145, 151, 167, 168
    A. aleyrodis, 6, 18, 78, 145, 167
    A. cubensis, 18, 79
    A. goldiana, 145, 167
    A. turbinata, 18, 80
Ascomycota, 9
Ascosphaera, 5, 9, 14
    A. aggregata, 18, 52
    A. apis, 18, 53
Asellaria aselli, 17, 25
Asellus aquaticus, 25
Aspergillus, 5, 11, 15, 138
    A. niger, 136
    A. parasiticus, 11, 18, 81
Aspidiotus perniciosus, 167
Astacus astacus, 134
Athalia rosae, 42
Atractium, 11
Atricordyceps, 5, 9
attachment of the spore, 128, 130

Bacillus thuringiensis, 171
Beauveria, 5, 11, 16, 137, 141, 142, 145, 146, 166, 171
    B. amorpha, 84
    B. bassiana, 2, 6, 18, 82, 83, 130, 131, 132, 133, 134, 138, 142, 143, 147, 150, 155, 159, 161, 163, 166, 167
    B. brongniartii, 18, 83, 134, 148, 149, 153, 166

    B. velata, 18, 84
biological control, 165
blastospores, 154
Blissus leucopterus, 166
Boudierella, 8
boverin, 167
Brachycaudus helichrysi, 145

Callibaphus species, 62, 76
Calonectria, 5, 7, 12, 15
    C. pruinosa, 18, 54
Camnula pellucida, 149
Camponotus, 59, 66, 99, 142
carbon source, 161
Cephalosporium, 10
    C. lecanii, 167
Cephalotes atratus, 142
Choristoneura fumiferana, 143
Chytridiomycetes, 140
Chytridiomycota, 7
Chytridiomycota-Blastocladiales, 7
Chytridiomycota-Chytridiales, 7
Cicadidae, 110
Cladosporium, 5
Clathroconium, 11
Cleonus punctiventris, 146
Coccus species, 126
Coccus viridis, 167
Coelomomyces, 5, 7, 17, 128, 130, 137, 140, 155, 170
    C. dodgei, 17, 19
    C. opifexi, 20
    C. psorophorae, 20, 130, 131
    C. punctatus, 20
Coelomycidium, 5, 7
    C. simulii, 17, 21, 140
Cofana species, 102

Cofana spectra, 148
Conidiobolus, 5, 6, 8, 14
    C. apiculatus, 17, 28, 149
    C. coronatus, 17, 29, 149, 169
    C. major, 17, 30, 169
    C. obscurus, 17, 31, 129, 130, 132, 133, 135, 147, 155, 159, 161, 163, 164
    C. thromboides, 160, 161
Corallomyces, 10
Cordycepioideus, 5, 9
    C. bisporus, 18, 55
    C. octosporus, 18, 55
Cordyceps, 2, 4, 5, 5, 9, 10, 11, 12, 13, 14, 15, 141, 142, 144, 148, 150, 151, 154, 171, 172
    C. aphodii, 148, 151
    C. australis, 18, 56, 142
    C. calocerioides, 18, 57
    C. dipterigena, 98
    C. entomorhiza, 123
    C. gunnii, 18, 58, 144
    C. hawkesii, 144
    C. lloydii, 18, 59, 99
    C. lloydii var. binata, 59
    C. martialis, 18, 60
    C. militaris, 18, 61, 133, 136, 160, 171
    C. nutans, 18, 62
    C. polyartha, 18, 63
    C. robertsii, 2, 144
    C. sinensis, 2
    C. sobolifera, 18, 64
    C. sphecocephala, 2
    C. taylori, 144
    C. tuberculata, 18, 65
    C. unilateralis, 18, 66, 67, 142, 143

Couchia, 5
Culex, 138, 140, 170
  C. pipiens, 134
Culicinomyces, 5, 11, 16, 130
  C. clavisporus, 6, 18, 85, 141, 170
Cydia pomonella, 167

defence system, 137
Delacroixia, 8
Delia antiqua, 147
Dendrolimus punctatus, 167
Desmidiospora, 12
development of fungus inside host, 135
Dialeurodes species, 168
Discofusarium, 11

Eccrinales, 17
Empoasca fabae, 169
Empusa, 8
  E. caroliniana, 17
encapsulation, 136
Encarsia formosa, 81
endocuticle, 128
Engyodontium, 5, 11, 16
  E. aranearum, 18, 86
Enterobryus species, 17, 27
Entomophaga, 5, 6, 8, 14
  E. aulicae, 6, 17, 32, 137, 143, 169
  E. caroliniana, 17, 33
  E. grylli, 17, 34, 144, 148, 149, 169
  E. kansana, 144
  E. tenthredinis, 17, 35
Entomophthora, 3, 5, 8, 14, 149, 169
  E. culicis, 17, 39
  E. muscae, 6, 17, 36, 37, 129, 137, 147, 148
  E. planchoniana, 17, 38, 143
  E. tenthredinis, 17
Entomophthorales, 168
epicuticle, 128
epizootics, 145
Eriophyes guerreronis, 145
Erynia, 5, 6, 8, 14, 149
  E. aquatica, 17, 39
  E. aulicae, 143
  E. blunckii, 17, 42
  E. canadensis, 143
  E. castrans, 17, 40, 148
  E. conica, 17, 41, 141
  E. dipterigena, 17, 42
  E. elateridiphaga, 17, 43
  E. formicae, 144
  E. gammae, 18, 43, 147

E. megasperma, 144
E. myrmecophaga, 144
E. neoaphidis, 18, 45, 131, 133, 145, 147, 149, 169
E. nouryi, 148
E. phytonnomi, 147
E. planchoniana, 147
E. plecopteri, 18, 47, 141
E. radicans, 18, 46, 143, 145, 146, 149, 169
E. rhizospora, 18, 47, 141
E. variabilis, 132
E. virescens, 18, 48
Eryniopsis, 8
Eulachnus rileyi, 143
exocuticle, 128

Fannia canicularis, 132
Filobasidiella depauperata, 5
forest habitats, 141
Formica, 144
formulation, 154, 163
fossil, 1
Funicularis, 5, 11
Fusarium, 5, 7, 10, 11, 15, 138
  F. coccophilum, 18, 87, 151, 167

Galleria mellonella, 133, 136
genetic manipulation, 153
germination, 132, 133
germination of the spore, 131
Gibellula, 5, 7, 10, 11, 12, 15, 16, 143, 155
  G. alata, 18, 88
  G. leiopus, 18, 89
  G. pulchra, 18, 90
α-glucanase, 134
$\beta(1, 3)$ glucan receptor, 136
$\beta(1, 3)$ glucopyransoyl, 136
Granulomanus, 7, 10, 12, 16, 18, 72, 91
growth kinetics, 155

haemagglutinins, 130
haemocyte, 128
Harposporium, 9
Heliothis virescens, 168
Heliothis zea, 134, 138
Hepialis species, 114
Hirsutella, 5, 6, 7, 9, 15, 16, 141, 142, 143, 144, 145, 149, 150, 155, 168
  H. aphidis, 144
  H. citriformis, 18, 92
  H. entomophila, 18, 93

H. formicarum, 67
H. gigantea, 143
H. jonesii, 18, 94
H. sausserei, 18, 95
H. thompsonii, 18, 96, 129, 145, 153, 159, 160, 161, 163, 164, 168
H. versicolor, 18, 97
Hylemya flies, 148
Hymenostilbe, 5, 9, 12, 16, 62, 101, 141, 150
  H. dipterigena, 18, 98
  H. formicarum, 6, 18, 99
  H. muscaria, 18, 100
  H. species, 18
Hypera brunneipennis, 147
Hypera postica, 147
hyphal bodies, 154
hyphal segments, 154
Hypocrella, 5, 10, 11, 15, 142, 143, 151
  H. amomi, 18, 68

Idiocerus species, 97
industrial fermentation, 161
industrial nutrients, 161
Inscctiola, 10
Isaria, 5, 12
Ischnaspsis species, 70
Isotomorus palustris, 25

keys to genera, 14

Lagenidium, 5, 8
  L. giganteum, 17, 22, 130, 138, 140, 159, 160, 161, 170
Lambdina fiscellaria, 143
Lechnidium, 11
Legeriomyces species, 17, 24
Lepidosaphes, 71
  L. species, 122
Leptinotarsa decemlineata, 146, 167
Leptolegnia, 5, 8
  L. chapmanii, 140
liquid media, 161

Macrosiphoniella sanborni, 131
Mahanarva posticata, 166
Malacasoma disstria, 144
mass production, 161, 163
Massospora, 5, 8, 144, 149
  M. cicadina, 17, 40
  M. levispora, 144
Megachile rotundata, 52
melanin, 137

Melanopus, 144
Melanospora parasitica, 5
Melolontha larvae, 166
Melolontha melolontha, 134, 146, 148
Meristacrum, 5, 9
Metarhizium, 5, 12, 15, 137, 141, 142, 146, 148, 171
  M. album, 18, 102
  M. anisopliae, 2, 6, 103, 130, 132, 134, 143, 146, 153, 155, 163, 165, 166, 170
  M. anisopliae var. anisopliae, 18, 104
  M. anisopliae var. majus, 18, 105
  M. flavoviride, 18, 106
Microcera, 11
microcycle, 153
Mucor, 138, 169
  M. racemosus, 168
mucus, 129
muscardine silkworms, 2
mycotal, 152
Mygalidae, 57, 107
Myriangium, 5, 10, 14
  M. duriaei, 18, 69, 145
Myriophagus, 5, 7
Myzus ascalonicus, 145
Myzus ornatus, 145
Myzus persicae, 169

N-acetylglucosamine, 130
natural control, 145, 171
natural epizootics, 143
Nectria, 5, 7, 10, 10, 11, 14
  N. auranticola, 145
  N. flammea, 18, 70
Neleus species, 58
Neozygites, 5, 6, 9, 14
  N. adjarica, 18, 49
  N. fresenii, 18, 50, 131, 147, 149
  N. fumosa, 18, 49, 147
  N. turbinatum, 9
Nephotettix species, 94
Nilaparvata lugens, 170
Nilaparvata species, 92, 106
nitrogen source, 161
Nomuraea, 5, 9, 12, 15, 141, 146
  N. atypicola, 12, 18, 107
  N. rileyi, 18, 108, 129, 130, 131, 137, 146, 153, 154, 163, 168
non-forest habitats, 144
nutritional requirements, 154
Nygmia phaeorrhoea, 169

Okanagama rimosa, 144
Oomycota, 8
Oomycota-Lagenidiales, 8
Oomycota-Saprolegniales, 8
Ophiocordyceps, 9
Orchesellaria mauguioi, 17, 25
Oryctes, 103, 105
Ostrinia furnacalis, 166
Ostrinia nubialis, 167
Otiorhynchus sulcatus, 146, 166
Oxyahyla, 148

Pacifastacus leniusculus, 23, 135
Paecilomyces, 5, 9, 10, 12, 15, 64, 141, 145, 146, 150
   P. amoeneroseus, 18, 109
   P. cicadae, 18, 110
   P. farinosus, 18, 111, 143
   P. fumosa-roseus, 131
   P. lilacinus, 18, 112
   P. tenuipes, 18, 113
Paltothyreus tarsatus, 56, 73, 76, 117, 119, 121, 124, 142
Paraisaria, 5, 9, 13, 16
   P. dubia, 18, 114
pastures, 148
Pemphigus betae, 148
penetration of the host integument, 133
Penicillium, 5
   P. chrysogenum, 158
Pericystis, 9
Periplaneta americana, 137
Peziotrichum, 7, 10
Phenacoccus manihoti, 147
Phorodon humuli, 145
Phyllocoptruta oleivora, 168
Pieris brassica, 131
Pleurodesmospora, 5, 13, 15
   P. coccorum, 18, 115
Plutella, 146
   P. maculipennis, 146
Podonectria, 5, 10, 13, 14
   P. coccicola, 18, 71, 145
Polistes, 95
Polycephalomyces, 5, 13, 16
   P. ramosus, 18, 116
Polyrhachis species, 66
production on solid media, 163
production strategy, 163
prophenoloxidase, 135
Propylaea, 28
Prosapia, 169
   P. simulans, 149

protoplast formation, 137
Pseudogibellula, 5, 9, 10, 13, 16
   P. formicarum, 18, 117
Pseudomicrocera, 11
Pseudoplusia includens, 147
Psila rosae, 148

quinones, 137

recovery, 163
rheology, 158

Sarcophaga, 135
scale insects, 142
Scapanes australis, 105
Schistocerca gregaria, 137
Schizolachnus pini-radiatae, 143
Sciaridae, 28
sclerotia, 154
Septobasidiales, 149
Septobasidium, 5
Simulidae, 141
Sitobion avenae, 147
Sorosporella, 5, 13
Spicaria, 12
Spodoptera exigua, 108
Sporodiniella, 5, 9, 145
   S. umbellata, 18, 51
Sporothrix, 5, 9, 13, 16
   S. insectorum, 18, 119
   S. isarioides, 18, 118
sporulation, 159
Sterigmatocystis, 11
Stilbella, 5, 9, 10, 13, 15, 142, 144, 145
   S. buquetii var. buquetii, 18, 120
   S. buquetii var. formicarum, 18, 121
Stilbum, 13
storage, 164
Strongwellsea, 8, 17, 40, 148
submerged culture, 157
subthalli, 154
surface culture, 156
synanamorphs, 7
Syngliocladium, 13
Synnematium, 12

Tabanomyces, 9
Taeniella carcini, 17, 27
Tarichium gammae, 18
Tenvirostitermes species, 55
Tetracrium, 5, 10, 13, 16, 71
   T. coccicolum, 18, 122

Tetranacrium, 5, 10, 13
Tetranychus urticae, 49
Thaxterosporium, 9
Thyrophygus species, 27
Tilachlidiopsis, 13
   T. nigra, 18, 123
Tilachlidium, 5, 10, 14
   T. liberianum, 18, 124
Tipula paludosa, 33
Tipulidae, 141
Tolypocladium, 5, 11, 14, 16
   T. cylindrosporum, 18, 125, 141, 170
Torrubia, 9
Torrubiella, 5, 7, 10, 12, 13, 14, 15, 16, 143, 150, 151
   T. arachnophila, 18, 72
   T. carnata, 18, 73
   T. rubra, 18, 74
tree crops, 145
Trichomycetes, 5, 17, 26
Trichosterigma, 12
Triplosporium, 9
Troglobiomyces, 12
troglobiotic, 144
tropical forests, 141

Umbonia species, 51

vector fungi, 170
vertalec, 152
Verticillium, 5, 9, 10, 14, 16, 141, 143, 151
   V. lecanii, 18, 126, 127, 129, 131, 134, 143, 150, 151, 152, 153, 154, 155, 159, 164, 167

Wigeana, 166

Zoophthora, 8
Zulia, 166
Zygomycota-Entomophthorales, 8

3 0490 0334825 +

RANDALL LIBRARY-UNCW